T0337841

Fundamentals of Network Planning and Optimisation 2G/3G/4G: Evolution to 5G

All content is personal and does not represent company views

Fundamentals of Network Planning and Optimisation 2G/3G/4G: Evolution to 5G

Second Edition

Ajay R. Mishra

Ericsson
India

This edition first published 2018
© 2018 John Wiley & Sons Ltd

Edition History
John Wiley & Sons Ltd (1e, 2004)

The right of Ajay R. Mishra to be identified as the author of this work has been asserted in accordance with law.

Registered Offices
John Wiley & Sons, Inc., 111 River Street, Hoboken, NJ 07030, USA
John Wiley & Sons Ltd, The Atrium, Southern Gate, Chichester, West Sussex, PO19 8SQ, UK

Editorial Office
The Atrium, Southern Gate, Chichester, West Sussex, PO19 8SQ, UK

For details of our global editorial offices, customer services, and more information about Wiley products visit us at www.wiley.com.

Wiley also publishes its books in a variety of electronic formats and by print-on-demand. Some content that appears in standard print versions of this book may not be available in other formats.

Library of Congress Cataloging-in-Publication Data

Names: Mishra, Ajay R., author.
Title: Fundamentals of network planning and optimisation 2G/3G/4G : evolution to 5G / Ajay R. Mishra, Ericsson, IN.
Other titles: Fundamentals of cellular network planning and optimisation
Description: Second edition. | Hoboken, NJ, USA : Wiley, 2018. | Revised edition of: Fundamentals of cellular network planning and optimisation | Includes bibliographical references and index. |
Identifiers: LCCN 2018023713 (print) | LCCN 2018026686 (ebook) | ISBN 9781119331704 (Adobe PDF) | ISBN 9781119331766 (ePub) | ISBN 9781119331711 (hardcover)
Subjects: LCSH: Cell phone systems–Design and construction. | Wireless communication systems–Design and construction. | Computer networks–Design and construction.
Classification: LCC TK6570.M6 (ebook) | LCC TK6570.M6 M54 2018 (print) | DDC 621.3845/6–dc23
LC record available at https://lccn.loc.gov/2018023713

Cover Design: Wiley
Cover Image: ©StationaryTraveller/iStockphoto

Set in 10/12pt Warnock by SPi Global, Pondicherry, India

Printed in the UK

Dedicated
to
The Lotus Feet of my Guru

Contents

Foreword by Aruna Sundararajan

The most intriguing aspect of 5G is that it promises to solve problems of the future, as compared with other evolutionary technologies in telecommunication, which have all tried to solve existing problems. It is this aspect of 5G that makes it so much more difficult to develop policies and frame points of view on how the 5G ecosystem will evolve. Therefore, it might not be wrong to say that if we are to adopt 5G and reap its promised benefits, it would require a fundamentally different way of thinking and policy making.

Policy making in developing countries has always been criticised for being reactive. Most of the problems that we try to solve today could have been solved at a much lower cost and therefore provided much better returns had we tried to solve them earlier. 5G provides a big opportunity in that sense. It may not just be a big leap in terms of technological breakthrough, but also provide an opportunity to bring a paradigm change in our policy making approach, from being reactive to proactive, from being rigid to being flexible.

One other aspect of telecommunications is that it is a cross cutting technology that has a profound impact on almost every aspect of our lives. Any evolutionary change in this technology is bound to have far reaching repercussions. 5G, with its promise of enabling commercialisation of use cases such as self-driving cars and machine-to-machine communication, has policy impacts far beyond the telecommunications sector.

While there are unanswered questions when it comes to providing a point of view on how the 5G ecosystem will evolve, the fact that it will be here, and that it will have a deep impact is almost beyond doubt. Therefore, it is in our interest to be one of the key stakeholders in shaping the way this technology evolves.

The Government of India is taking a number of proactive steps to be an active player in this evolution. I would urge all stakeholders, including the industry players and standard setting bodies to help create a baseline roadmap that would synergise everybody's efforts in this direction.

The success of any technology is dependent on the efficiency of the designed networks. End-to-end perspective becomes even more important in the fast evolution of the technology scenario. This is where this book plays an important role – giving insights on designing and planning networks across various technologies from 2G to 5G and in all three domains – Radio, Transmission, and Core.

I congratulate the author, who has worked in the Telecom Industry for more than two decades, on making a valuable effort towards demystifying various aspects of network technologies – from 2G to 5G evolution, which I am sure will be extremely helpful to network design engineers and telecom professionals alike. I look forward to more such contributions as the technology evolves and matures over a period.

Aruna Sundararajan
Secretary, Department of Telecommunication
Government of India
New Delhi, India

Foreword by Rainer Deutschmann

The mobile telecommunications industry is a key enabler for economic growth and innovation. It contributes 4.5% to the global GDP, comprises today about 8 bn SIM connections and is expected to add 20 bn IoT connections by 2025. However, industry revenues are stagnating, even declining in some markets, despite vastly growing mobile traffic with above 40% CAGR in the next few years. This, along with commercial 5G launches, drives network CAPEX intensities above 15% of revenue. Competitive pressure intensifies between established players, and a new breed of disruptive Digital Telco emerges with dramatically lower production cost. As a reference example, Reliance Jio in India captured 100 million customers and gained mobile broadband market leadership in less than six months post market launch.

In this market environment, operators have no choice but to become Digital Telcos. McKinsey estimates that a full digital transformation can double the cash flow margin within five years. Many operators had experimented with off-core OTT business models, but focus is largely back to the roots – network and service customer experience, security and trust, interoperability, and efficiency. Customers rightfully expect a consistent outdoor and indoor coverage with high bandwidth, low latency and an enticing voice and video experience. Network efficiencies are gained from full migration to IP and virtualisation, automation, employing open source, network sharing, refarming and ultimately retirement of legacy networks.

The Digital Telco is the ideal foundation of 5G use cases – enhanced mobile broadband, IoT, and mission-critical applications. These use cases will require network performance to increase 10-fold over current levels across all network parameters, as measured by latency, throughput, reliability, scale, and power management. Digital Telco networks will be an integral part of pervasive artificial intelligence (AI) – AI in every single end-point, be it the smartphone, speaker, camera, drone, or car. In fact, digital networks themselves become part of the AI cloud, as safe harbour of customer data and trusted source of insights and identity. We will see a convergence of the cloud and the edge, with seamless interworking of training complex AI models in the cloud and real-time operation on the edge. As densification progresses and with the proliferation of small cells, the underlying fibre infrastructure will further gain importance, as will smart spectrum management, carrier aggregation, and spatial diversity (MIMO).

The author has spent more than 20 years in the Telecom Industry, both on vendor and operator sides. This new edition of his book covers end-to-end network planning and optimisation from 2G to 5G. He emphasises the planning process itself and develops all necessary foundations and concepts for planners. The book gives new planners a flying start, experienced planners a broader and deeper view, and even beyond engineers anyone with a genuine interest in one of the most important assets of a Digital Telco – the network – a solid basis to help shape the future of their company.

Rainer Deutschmann
Group Chief Operating Officer
Dialog Axiata PLC,
Colombo, Sri Lanka

Preface

This edition arrives a good fourteen years after the first edition. Around 2002/2003, we were deciphering 4G/LTE and now in 2018, we are in the middle of defining 5G standards. It seems that the mobile world has moved two steps forward – 3G to 4G and then 4G to 5G – and what huge steps they are. By the time this book reaches the market, some serious trials of 5G could already be taking place.

There are more than five billion unique mobile subscribers across the world. This is a staggering number – but there is still a lot to be done. 5G is knocking at the door with latencies of less than 1 millisecond and data rates of more than 10 Gbps. As we move into this scenario with many countries still having GSM systems in place, the challenges for the communication service providers (operators) are to provide quality networks and services to their subscribers at low cost.

This scenario of the legacy of GSM networks combined with WCDMA and LTE while working towards 5G networks – along with the Internet of Things (IoT), cloud technology, artificial intelligence, virtualisation, etc., in the network ecosystem – makes things exciting and challenging for network planning and optimisation engineers. 5G technology that will not be backward compatible will be coexisting with the 2G, 3G and 4G technologies. Thus, interoperability issues and issues related to each element becoming more and more intelligent will need to be catered for.

There are already some excellent books on the market on various technologies but, like before, it was felt that there was a gap not covering planning engineers who are designing the networks. This book tries to fill that gap – understanding the utilisation of various aspects of technologies from network planning aspects. The gap between network technology and network design is fulfilled.

This book covers radio, transmission and core at the same time. Apart from the planning and optimisation aspects for GSM, GPRS, EDGE, WCDMA, LTE for all radio, transmission and core, this book also delves into 5G technology. This will help the planning and optimisation engineers to understand not only the technological evolution but also the changes and modifications taking place during planning and optimisation processes.

This book has been divided into four parts. Each of these parts is dealing with 2G, 3G, 4G and 5G networks. There are ten chapters in this book. Parts I, II, and III deal with 2G (GSM, GPRS and EDGE), 3G (WCDMA) and 4G (LTE) networks, respectively. These three parts deal with the fundamentals of radio, transmission, and core network planning and optimisation. Parts I and II have three chapters, each dealing exclusively with planning and optimisation of these three sections of the network. Part III contains two chapters focusing on LTE radio and core network planning and optimisation.

Part I contains three chapters that focus on 2G (GSM, GPRS and EDGE) network planning and optimisation.

Chapter 1 is an overview of mobile networks. It contains a brief history of mobile networks and their evolution. Some concepts from information theory are explained briefly that are relevant to network planning engineers. At the end of the chapter, some of the technologies that support mobile networks are also mentioned, such as optical technology.

Chapter 2 focuses on 2G (GSM, GPRS and EDGE) radio network planning and optimisation. The basics of radio network planning are best understood using this technology. It was possible to reduce the length of this chapter by putting some of the concepts in the WCDMA or LTE radio planning chapters; I felt it best to put the basics in this chapter.

Chapters 3 and 4 follow a similar approach for transmission and core network planning. These chapters cover scope, pre-planning, detailed planning and optimisation for transmission and core networks, respectively.

Part II focuses on 3G (WCDMA) network planning and optimisation.

Chapters 4, 5 and 6 focus on radio, transmission and core network planning and optimisation. The areas covered go beyond 3G into 3.5G or HSPA as well.

Part III contains three chapters and focuses on the 4G (LTE) network planning and optimisation.

Chapters 8 and 9 deal with radio and core network planning and optimisation, respectively.

Part IV gives an overview of 5G networks.

We are in the midst of 5G recommendations which are expected to be complete by 2020 and by then there will also be some global 5G network deployments. Chapter 10 is devoted to giving a brief overview of what is in store for planning engineers.

The book contains a few appendices covering some very exciting technology areas including IoT, massive MIMO, artificial intelligence, block chain, spectrum management and a 3GPP standards overview. I hope that planning engineers will appreciate an insight into technical areas that impact their daily work using current technologies. Erlang B tables are provided to help planning engineers in their day-to-day work.

Finally, there is a list of recommended reading in the Bibliography.

Acknowledgements

It is an absolute pleasure for me to thank all who have inspired me to take on this project and who have supported me all the way to its completion.

My first big thank you is due to my colleague from Ericsson, Nishant Batra, who encouraged me to write this second edition of the book.

My gratitude also goes to Chandrasekaran Vasudevan and Andrew Thuan from Ericsson who supported me in taking on this challenging assignment.

My humblest gratitude goes to Aruna Sundararajan and Rainer Deutschmann for donating time from their extremely busy schedules to write visionary forewords for the book.

Many thanks are also due to senior experts from across the globe who contributed in writing appendices on some of the most advanced concepts and technologies – to Jeevan Talegaonkar, Swapnaja Deshpande, Priyanka Ray, Guninder Preet Singh, Ramy Ahmed Fathy, Asit Kadayan, and Pieter Geldenhuys.

My thanks also go to my colleagues, Koushik Basu, Anna Bestard, Vikas Khera, Sudarshan Sen Gupta, Shikha Singh, and Marianne Bermundo, who devoted time to review the project.

Another thank you is due to the Wiley editorial team of Peter Mitchell, Sandra Grayson, and Anita Yadav, who wanted me to write this second edition a good 12 years after the publication of the first edition, and who supported me throughout the process.

My thanks are also due to Reliance JIO's Pankaj Pawar and Sunil Dutt – whose support I can always count on.

My gratitude goes to my mentors from university, the Indian Space Research Organisation, and industry whose guidance was invaluable to me – in particular to Uday Gilankar, G.P. Srivastava, K.K. Sood, and Anil Jain.

I would like to thank my parents, Mrs Sarojini Devi Mishra and Mr Bhumitra Mishra, whose blessings always inspire me. And finally, I would like to thank my wife, Shilpy, and my children, Krishna and Om, for all their understanding during the writing of this book.

List of Abbreviations

3GPP	Third Generation Partnership Project
5GPPP	Fifth Generation Public Private Partnership
8-PSK	Octagonal Phase Shift Keying
AAA	Authentication, Authorisation and Accounting
AAL	ATM Adaptation Layer
AAS	Active Antenna System
ABR	Available Bit Rate
AC	Authentication Centre
AC	Admission Control
ADM	Application Development Maintenance
AF	Antenna Filter
A-GPS	Assisted GPS
AH	Authentication Header
AI	Artificial Intelligence
AIR	Authentication Information Request
AKA	Authentication and Key Agreement
AM	Application Manager
AM	Acknowledged Mode
AMC	Adaptive Modulation and Coding
AMR	Adaptive Multi-Rate
AMR	Adaptive Mean Rate
ANR	Automatic Neighbour Relation
ANSI	American National Standard Institute
ARIB	Alliance of Radio Industries and Business
ARP	Allocation and Retention Priority
ARP	Allocation/Retention Protocol
ARQ	Automatic Repeat Request
ARQ	Accessibility, Retainability, Quality
AS	Access Stratum
AS	Application Server
ASIC	Application-Specific Integrated Circuit
ATM	Asynchronous Transport Mode
AUC	Authentication Centre
AXC	ATM Cross Connect
BB	Base Band

BCCH	Broadcast Control Channel
BCH	Broadcast Channel
BER	Bit Error Rate
BG	Border Gateway
BGCF	Breakout Gateway Control Function
BLE	Bluetooth Low Energy
BLER	Block Error Rate
BMC	Broadcast-Multicast Control
BPR	Business Process Reengineering
BS	Base Station
BSC	Base Station Controller
BSIC	Base Station Identity Code
BSR	Buffer Status Report
BSS	Base Station Subsystem
BTS	Base Transceiver Station
C/I	Channel to Interference Ratio
CA	Carrier Aggregation
CAC	Call Admission Control
CAC	Connection Admission Control
CBR	Constant Bit Rate
CC	Chase Combing
CCCH	Common Control Channel
CCH	Control Channel
CCM	Common Channel Management
CCO	Coverage and Capacity Optimisation
CDF	Cumulative Distribution Function
CDV	Cell Delay Variation
CDVT	Cell Delay Variation Tolerance
CER	Cell Error Ratio
CG	Charging Gateway
CL	Convergence Layer
CLP	Cell Loss Priority
CLR	Cell Loss Ratio
CMR	Cell Mis-insertion Ratio
CN	Core Network
C-NBAP	Common NBAP
CPICH	Common Pilot Channel
CP-OFDM	Cyclic Prefix-based OFDM
CPS	Connection Processing Solutions
CQI	Channel Quality Information
CS	Circuit Switched
CS-1, CS-2	Coding Schemes
CSCF	Call Session Control Function
CSFB	Circuit Switch Fall Back
CSI	Channel State Information
CSR	Call Success Rate
CTD	Cell Transfer Delay

CU	Control Unit
DCCH	Dedicated Control Channel
DCH	Data Control Channel
DCM	Dedicated Channel Management
DCN	Data Communication Network
DCR	Dropped Call Rate
DeNB	Donor eNB
DFT	Discrete Fourier Transform
DFTs-OFDM	Discrete Fourier Transform-Spread-OFDM
DL-SCH	Downlink Shared Channel
DLT	Distributed Ledger Technology
DMS	Dealer Management System
D-NBAP	Dedicated NBAP
DNS	Domain Name System
DOCSIS	Data Over Cable Services Interface Specifications
DoT	Department of Telecommunications
DPCS	Destination Point Codes
DPoS	Delegated Proof of Stake
DRNC	Drifting RNC
DRX	Discontinuous Repetition Cycle
DSCH	Downlink Shared Channel
DSP	Digital Signal Processor
DSSS	Direct Sequence Spread Spectrum
DS-WCDMA_FDD	Direct Sequence WCDMA Frequency Division Duplex
DS-WCDMA_TDD	Direct Sequence WCDMA Time Division Duplex
DTCH	Dedicated Traffic Channel
DTMF	Dual Tone Multi-Frequency
D-TxAA	Double Transmit Antenna Array
E2E	End-to-End
E-AGCH	Enhanced Absolute Grant Channel
E-CID	Enhanced Cell ID
ECSD	Enhanced Circuit Switched Data
EDAP	EGPRS Dynamic Pool
E-DCH	Enhanced Dedicated Channel
EDGE	Enhanced Data Rates in GSM Environment
E-DPCCH	Enhanced Dedicated Physical Control Channel
EEO	End-to-End Orchestration
E-HICH	E-DCH HARQ Indicator Channel
eICIC	Enhanced Inter-Cell Interference Coordination
EIR	Equipment Identity Register
EIRP	Effective Isotropic Radiative Power
e-MBB	Enhanced Mobile Broadband
EMS	Element Management System
EPC	Evolved Packet Core
EPS	Evolved Packet System
E-RGCH	Enhanced Relative Grant Channel
ERP	Enterprise Resource Planning

ES	Error Seconds
ESP	Encapsulation Security Payload
E-TFCI	E-DCH Transport Format Combination Indicator
ETSI	European Telecommunication Standard Institute
E-UTRAN	Evolved UMTS Terrestrial Radio Access Network
FAACH	Fast Associated Control Channel
FACH	Forward Access Channel
FBMC	Filter Bank Multi-Carrier
FCCH	Frequency Correction Channel
FDM	Frequency Division Multiplexing
FEC	Forward Error Correction
FER	Frame Error Rate
FFR	Fractional Frequency Re-use
FH	Frequency Hopping
FM	Fade Margin
FPLMTS	Future Public Land Mobile Telecommunications System
FQDN	Fully Qualified Domain Name
FR	Full Rate
FSS	Frequency Selective Scheduling
FW	Firewalls
GbE	Gigabit Ethernet
GBR	Guaranteed Bit Rate
GEPON	Gigabit Ethernet Passive Optical Network
GFC	Generic Flow Control
GGSN	Gateway GPRS Support Node
GMSC	Gateway Mobile Switching Centre
GMSK	Gaussian Minimum Phase Shift Keying
GPRS	General Packet Radio Services
GPU	Graphic Processing Unit
GSM	Global System for Mobile Communication
GSW	Group Switch
GUMMEI	Global MME Identity
GUTI	Global Unique Temporary Identity
HAPS	High Altitude Platform Stations
HARQ	Hybrid Automatic Repeat Request
HEC	Header Error Control
HeNB	Home eNB
HII	High Interference Indicator
HLR	Home Location Register
HO	Handover
HR	Half Rate
HSDPA	High Speed Downlink Packet Access
HS-DPCCH	High Speed Dedicated Physical Control Channel
HS-DSCH	High Speed Downlink Shared Channel
HSPA	High Speed Packet Access
HS-PDSCH	High Speed Physical Downlink Shared Channel
HSS	Home Subscriber Server

HS-SCCH	High Speed Shared Control Channel
HSUPA	High Speed Uplink Packet Access
ICI	Inter-Cell Interference
ICIC	Inter-Cell Interference Coordination
I-CSCF	Interrogating Call Session Control Function
IFU	Interface Unit
IM	Interference Margin
IMEI	International Mobile Equipment Identity
IMR	IP Multimedia Register
IMS	IP Multimedia Subsystem
IMSI	International Mobile Subscriber Identity
IMS-MGW	IP Multimedia Subsystem-Media Gateway Function
IN	Intelligent Network
INAP	IN Application Protocol
IoT	Internet of Things
IP	Internet Protocol
IR	Incremental Redundancy
ISI	Inter-Symbol Interference
ITU	International Telecommunication Union
IVR	Interactive Voice Response
IWF	Inter-Working Function
KPI	Key Performance Indicator
LAA	Licensed-Assisted Access
LAN	Local Area Network
LCB	Loop Control Bit
LCM	Life Cycle Management
LIG	Legal Interception Gateway
LLC	Logical Link Control
LNA	Low Noise Amplifier
LOS	Line of Sight
LPWA	Low Power Wide Area
LPWAN	Low Power Wide Area Network
LTE	Long Term Evolution
M2M	Machine-to-Machine
MANET	Mobile Ad-hoc Network
MANO	Management, Automation and Orchestration
MBR	Maximum Bit Rate
MBS	Maximum Burst Size
MCB	Master Control Bit
MC-CDMA	Multi-Carrier Code Division Duplex
MCCH	Multicast Control Channel
MCH	Multicast Channel
Mcps	Million Chips per Second
MCR	Minimum Cell Rate
MCS-1 to MCS-9	Modulation and Coding Schemes
MCWCDMA	Multi-Carrier WCDMA
MGCF	Media Gateway Control Function

MGW	Media Gateway
MIB	Master Information Block
MIMO	Multiple-Input Multiple-Output
MISO	Multiple-Input Single-Output
ML	Machine Learning
MM	Mobility Management
MME	Mobility Management Entity
m-MTC	Massive Machine Type Communication
MPLS	Multiprotocol Label Switching
MRF	Media Resource Function
MRFC	Multimedia Resource Function Controller
MRFP	Multimedia Resource Function Processor
MS	Mobile Station
MSC	Mobile Switching Centre
MSISDN	Mobile Subscriber ISDN number
MSK	Minimum Phase Shift Keying
MT	Mobile Termination
MTC	Machine-Type Communication
M-TMSI	MME-Temporary Mobile Subscriber Identity
MTTR	Mean Time to Repair
MU-MIMO	Multi User MIMO
MUX	Multiplexing Unit
NAICS	Network-Assisted Interference Cancellation and Suppression
NAPTR	Name Authority Printer
NAS	Non-Access Stratum
NAS	Non-Access Security
NB	Narrowband
NBAP	Node B Application Part
NE	Network Element
NFAP	National Frequency Allocation Plan
NFV	Network Function Virtualisation
NFVI	Network Function Virtualisation Infrastructure
NFVO	NFV Orchestrator
NGC	Next Generation Core
NGN	Next Generation Network
NGPON2	Next Generation PON2
NLOS	No Line of Sight
NMS	Network Management System
NMT	Nordic Mobile Telephone
Non-GBR	Non-Guaranteed Bit Rate
NR	New Radio
NRT	Non-Real-Time Data
NRT-VBR	Non-Real-Time Variable Bit Rate
NSS	Network Switching Subsystem
NSS	Network Subsystem
OEM	Original Equipment Manufacturer
OFDM	Orthogonal Frequency Division Multiplexing

OI	Overload Indicator
OSI	Open System Interconnection
OSS/BSS	Operations Support Systems and Business Support Systems
OSVF	Orthogonal Variable Spreading Factor
OTDOA	Observed Time Difference of Arrival
OVSF	Orthogonal Variable Spreading Factor
P2P	Peer-to-Peer
PA	Power Amplifier
PAN	Private Area Network
PAPR	Peak to Average Power Ratio
PAPU	Packet Processing Unit
PBCH	Physical Broadcast Channel
PBFT	Practical Byzantine Fault Tolerance
PCCH	Paging Control Channel
PCFICH	Physical Control Format Indicator Channel
PCH	Paging Channel
PCI	Physical Cell Identity
PCR	Peak Cell Rate
PCRF	Policy and Charging Rules Function
P-CSCF	Proxy Call Session Control Function
PCU	Packet Control Unit
PDCCH	Physical Downlink Control Channel
PDCH	Packet Data Channel
PDCP	Packet Data Convergence Protocol
PDF	Policy Decision Function
PDN	Packet Data Network
PDP	Packet Data Protocol
P-FFR	Partial Frequency Re-use
P-GW	Packet Data Network Gateway
PHICH	Physical Hybrid ARQ Indicator Channel
PICH	Paging Indication Channel
PIM	Physical Infrastructure Manager
PM	Physical Medium
PMRTS	Public Mobile Radio Trunking Service
PNF	Physical Network Function
PoA	Proof of Authority
PoB	Proof of Burn
PoC	Proof of Capacity
PoET	Proof of Elapsed Time
POH	Path Overhead
PON	Passive Optical Network
PoS	Proof of Stake
PoW	Proof of Work
PRACH	Physical Random-Access Channel
PRB	Physical Resource Block
PRC	Primary Reference Clock
PS	Packet Switched

PT	Payload Type
PUCCH	Physical Uplink Control Channel
PUSCH	Physical Uplink Shared Channel
PVC	Permanent Virtual Connection
QCI	Quality Class Identifier
QoS	Quality of Service
QPSK	Quadrature Phase Shift Keying
RA	Random Access
RACH	Random Access Channel
RAN	Radio Access Network
RAT	Radio Access Technologies
RF	Radio Frequency
RFID	Radio Frequency Identification
RIT	Radio Interface Technology
RLC	Radio Link Control
RLM	Radio Link Management
RN	Relay Node
RNC	Radio Network Controller
RNSAP	Radio Network Subsystem Application Part
RNTP	Relative Narrowband Transmit Power
RRC	Radio Resource Control
RRM	Radio Resource Management
RS	Reference Signal
RSRP	Reference Signal Received Power
RSRQ	Reference Signal Received Quality
RSSI	Received Signal Strength Indicator
RT-VBR	Real-Time Variable Bit Rate
S and M	Summing and Multiplexing
SA	Service Assurance
SAACH	Slow Associated Control Channels
SAAL	Signalling ATM Adaptation Layer
SAE	System Architecture Evolution
SAR	Segmentation and Reassembly
SC-FDMA	Single Carrier Frequency Division Multiple Access
SCFT	Single Cell Functionality Test
SCH	Synchronisation Channel
SCR	Sustainable Cell Rate
S-CSCF	Serving Call Session Control Function
SDMA	Space Division Multiple Access
SDN	Software Defined Network
SDU	Service Data Unit
SECBR	Severely Errored Cell Block Ratio
SES	Severely Error Seconds
S-FFR	Soft Frequency Reuse
SFN	Single Frequency Network
SGF	Signalling Gateway Function
SGSN	Serving GPRS Support Node

S-GW	Serving Gateway
SHO	Soft Handover
SI	Self-Interference
SIM	Subscriber identity module
SIMO	Single-Input Multiple-Output
SINR	Signal to Interference and Noise Ratio
SIP	Session Initiation Protocol
SIR	Signal to Interference Ratio
SISO	Single Input and Single Output
SL	Signalling Link
SLA	Service Level Agreement
SLF	Subscription Locator Function
SLS	Signalling Link Set
SMSC	Short Message Service Centre
SNDCP	Sub-Network Dependent Convergence Protocol
SNR	Signal to Noise Ratio
SO	Subscribers Originating
SOH	Section Overhead
SON	Self-Organising Network
SP	Signal Processor
SPC	Signalling Point Code
SR	Scheduling Request
SRNC	Serving Radio Network Controller
SRS	Sounding Reference Signal
SR-VCC	Single Radio Voice Call Continuity
SSU	Synchronisation Supply Unit
ST	Subscribers Terminating
SU-MIMO	Single-User MIMO
SVC	Switched Virtual Connection
SVD	Singular Value Decomposition
TA	Terminal Adapter
TA	Tracking Area
TAs	Timing Advances
TAU	Tracking Area Update
TC	Transmission Convergence
TCH	Traffic Channel
TCSM	Transcoder Sub-Multiplexer
TDMA	Time Division Multiple Access
TE	Terminal Equipment
TETRA	Terrestrial Trunked Radio
TM	Transparent Mode
TRA	Telecom Regulatory Authority
TRX	Transceiver
TRXM	Transceiver Management
TRXSIG	Transceiver Signalling
TS	Timeslot
TSP	Telecom Service Provider

TTI	Transmit Time Interval
UBR	Unspecified Bit Rate
UE	User Equipment
UF	Universal Filtered
ULR	Update Location Request
UL-SCH	Uplink Shared Channel
UM	Unacknowledged Mode
UMTS	Universal Mobile Telecommunications System
UNB	Ultra-Narrow Band
UPA	Ubiquitous Personal Assistant
UPF	Usage Parameter Function
u-RLLC	Ultra-Reliable and Low Latency Communication
UTDOA	Uplink Time Difference of Arrival
UWB	Ultra Wideband
VAF	Voice Activity Factor
VANET	Vehicle Ad-hoc Networks
VAS	Value Added Services
VC	Virtual Channel
VIM	Virtualised Infrastructure Manager
VLR	Visitor Location Register
VMD	Variable Messaging Display
VMS	Voice Mail System
VNF	Virtualised Network Function
VNFM	VNF Manager
VoIP	Voice Over IP
VoLGA	Voice Over LTE via Generic Access
VoLTE	Voice Over LTE
VP	Virtual Path
VPI	Virtual Path Identifier
VPN	Virtual Private Network
VRRP	Virtual Router Redundancy Protocol
VTS	Vehicular Transportation System
V-V	Vehicle-to-Vehicle
WAN	Wide Area Network
WDM	Wavelength Division Multiplexing
WMSC	Wideband CDMA Mobile Switching Centre
WPC	Wireless Planning and Coordination
WRC	World Radiocommunication Conference

1

Overview of Mobile Networks

1.1 Introduction

Mobile networks are differentiated from each other with the word 'generation', such as 'first generation', 'second generation', etc. This is quite correct because there is a big 'generation gap' between the technologies.

The first-generation mobile systems were the analogue (or semi-analogue) systems, which came in the early 1980s, also called NMTs (Nordic Mobile Telephones). They offered mainly speech and related services and were highly incompatible with each other. Thus, their main limitations were the limited amount of services offered and their incompatible nature.

An increase in the necessity for a system that catered to mobile communication needs and also offered increased compatibility with other systems, resulted in the birth of the second-generation mobile systems. International bodies played a key role in evolving a system that would provide better services and was more transparent and compatible with networks globally. But, unfortunately, the second-generation network standards could not fulfil the dream of having just one set of standards for networks globally. The standards in Europe differed from the standards in Japan and that of the Americas and so on. Of all the standards, the GSM went all the way in fulfilling the technical and the commercial expectations.

But, again, none of the standards in the second generation was able to fulfil the globalisation dream of the standardisation bodies. This would be fulfilled by the third-generation mobile systems. Also, it is expected that the third-generation systems will be predominantly data traffic oriented as compared with the second-generation networks that were carrying predominantly voice traffic.

The major standardisation bodies that play an important role in defining the specifications for the mobile technology are:

- ITU (International Telecommunication Union): The ITU with headquarters in Geneva, Switzerland is an international organisation within the United Nations, where governments and the private sector coordinate global telecom networks and services. The ITU-T is one of the three sectors of ITU, which produces the quality standards covering all the fields of telecommunications.
- ETSI (European Telecommunication Standard Institute): This body was primarily responsible for the development of the specifications for the GSM. Due to the technical and commercial success of the GSM, this body will also play an important role in

Fundamentals of Network Planning and Optimisation 2G/3G/4G: Evolution to 5G,
Second Edition. Ajay R. Mishra.
© 2018 John Wiley & Sons Ltd. Published 2018 by John Wiley & Sons Ltd.

the development of the third-generation mobile systems. ETSI mainly develops the telecommunication standards throughout Europe and beyond.

- ARIB (Alliance of Radio Industries and Business): This body is predominant in the Australasian region and is playing an important role in the development of the third-generation mobile systems. ARIB basically serves as a standards developing organisation for radio technology.
- ANSI (American National Standard Institute): ANSI currently provides a forum for over 270 ANSI-accredited standards developers representing approximately 200 distinct organisations in the private and public sectors. This body has been responsible for the standards development for the American networks.
- 3GPP (Third Generation Partnership Project): This body was created to maintain the complete control of the Specification design and process for the third-generation networks. The result of the 3GPP work is a complete set of specifications that will maintain the global nature of the 3G networks.
- 5GPPP (Fifth Generation Public Private Partnership): This is driven by the EU (European Commission) and ICT Industry under the EU's Horizon 2020 initiative. More than 30 members are part of this consortium including Industry bodies, SMEs, research bodies, etc.

1.2 Mobile Network Evolution

Mobile network evolution has been categorised into 'generations' as shown in Figure 1.1. A brief overview on each generation is given below.

1.2.1 First-generation System (Analogue System)

The first-generation mobile system started in the 1980s was based on analogue transmission techniques. At that time, there was no worldwide (not even Europe-wide) coordination for the development of the technical standards for the system. Nordic counties deployed NMTs, while UK and Ireland went for a Total Access Communication System or TACS, and so on. Roaming was not possible and efficient use of the frequency spectrum was not there.

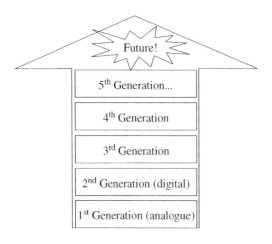

Figure 1.1 Evolution of mobile networks.

1.2.2 Second-generation System (Digital System)

By the mid-1980s, the European commission started a series of activities to liberalise the communications sector, including mobile communication. This resulted in the creation of ETSI, which inherited all the standardisation activities in Europe. This saw the birth of the first specifications and the network based on the digital technology; it was called the Global System for Mobile Communication or GSM. Since the first networks at the beginning of 1991, GSM has gradually evolved to meet the requirements of data traffic and many more services than the original networks that were capable of handling mainly voice traffic.

- GSM (Global System for Mobile communication): The main elements of this system are the mobile, BTS (Base Transceiver Station) and BSC (Base Station Controllers) in the BSS (Base Station Subsystem) and the MSC (Mobile Switching Centre), VLR (Visitor Location Register), HLR (Home Location Register), AC (Authentication Centre), EIR (Equipment Identity Register) in the NSS (Network Switching Subsystem). This network can provide all the basic services such as speech and the data services up to 9.6 kbps, e.g. fax, etc. This GSM network also has an extension to the fixed telephony networks.
- GSM and VAS (Value Added Services): The next advancement in the GSM system was the addition of two platforms, the 'Voice Mail System' (VMS) and the 'Short Message Service Centre' (SMSC). The SMS proved to be incredibly commercially successful so much so that in some networks, the SMS traffic constitutes a major part of the total traffic of the network. Along with the VAS, IN (INtelligent services.) also made its mark in the GSM system, with its advantage of giving the operators the chance to create a whole range of new services. Fraud management and 'pre-paid' services are the result of the IN service.
- GSM and GPRS (General Packet Radio Services): As the requirement for sending data on the air-interface increased, new elements such as SGSN (Serving GPRS Support Node) and GGSN (Gateway GPRS Support Node) were added to the existing GSM system. These elements made it possible to send the packet data on the air-interface. This part of the network handling the packet data is also called the packet core network. In addition to the SGSN and GGSN, it also contains the IP routers, firewall servers and the DNS (Domain Name Server). This enables wireless access to the Internet and the bit rate reaching to 150 kbps in optimum conditions.
- GSM and EDGE (Enhanced Data rates in GSM Environment): Now that both the voice and data traffic were moving on the system, the need to increase the data rate was felt. This was done by using more sophisticated coding methods over the Internet and thus increasing the data rate up to 384 kbps.

1.2.3 Third-generation Networks (WCDMA in UMTS)

In EDGE, very high movement of the data was possible, but still, the packet transfer on the air-interface behaves like a circuit switch call. Thus, part of this packet connection efficiency is lost in the circuit switch environment. Moreover, the standards for developing the networks till the second generation were different for different parts of the world. Hence, it was decided to have a network that provides services independent of the technology platform and whose network design standards are the same globally.

Thus, 3G was born. In Europe, it was called UMTS (Universal Terrestrial Mobile System), which is ETSI driven. IMT-2000 is the ITU-T name for the third-generation system while cdma2000 is the name of the American 3G variant. WCDMA is the air-interface technology for the UMTS. The main components include BS (Base Station) or Node B, RNC (Radio Network Controller) apart from WMSC (Wideband CDMA Mobile Switching Centre) and SGSN/GGSN. This platform offers many services that are based on the Internet, along with video phoning, imaging, etc.

1.2.4   Fourth-generation Networks (LTE)

Further advancements to mobile networks technology led to LTE or Long Term Evolution – a technology referred to as 4G. First proposed by NTT DoCoMo, the main aim of LTE was to increase the speed and capacity while reducing latency of the mobile networks. The mobile networks become simpler in architecture while moving towards an ALL-IP system. The air-interface used in LTE is OFDM (Orthogonal Frequency Division Multiplexing). Key elements include base stations called eNodeB, and Core elements include MME, P-GW and S-GW.

1.2.5   Fifth-generation Networks

5G or fifth-generation networks will probably be the networks for which technology evolution will occur. The new elements in 5G Radio would be NR or New Radio and a network that will be a truly convergent network that would include 4G LTE, Non-3GPP access technologies, Wi-Fi and core including NFV, SDN, Internet of Things (IoT), Cloud. 5G networks are expected to launch in 2019/2020, connecting billions of devices, a multifold increase in data, a latency as low as 1 ms and data rates as high as 100 Mbps.

1.3   Information Theory

1.3.1   Multiple Access Techniques

The basic concept of multiple access is to permit the transmitting station to transmit to the receiving station without any interference. Sending the carriers separated by frequency, time, and code can achieve this.

1.3.1.1   Frequency Division Multiple Access (FDMA)

This is the most traditional technique in radio communications, relying on the separation of the frequencies between the carriers. All that is required is that all the transmitters should be transmitting on different frequencies and that their modulation does not cause the carrier bandwidths to overlap. There should be as many as possible users that should be utilising the frequencies. This multiple access method is used in the first-generation or the analogue cellular networks. The advantage of the FDMA system is that transmission can be without coordination or synchronisation. The constraint in the FDMA system is the availability of the frequency.

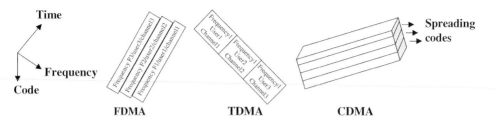

Figure 1.2 Generic multiple access methods.

1.3.1.2 Time Division Multiple Access (TDMA)

As the mobile communication moved on from first to second generation, the FDMA was not considered an effective way for frequency utilisation. Thus, to utilise already scarce frequency resources, TDMA was used. Thus, as shown in the Figure 1.2, many users can use the same frequency as each frequency can be divided into small slots of time called time slots, which are generated continuously. Thus, many users can log on to the same frequency.

1.3.1.3 Code Division Multiple Access (CDMA)

By utilising the spread spectrum technique, CDMA combines modulation and multiple access to achieve a certain degree of information efficiency and protection. Initially developed for the military applications, it gradually developed into a system that promised better bandwidth and service quality in an environment of spectral congestion and interference. In this technology, every user is assigned a separate code(s) depending upon the transaction. One user may have several codes in certain conditions. Thus, separation is not based on frequency or time, but on the basis of codes. These codes are nothing but very long sequences of bits having a higher bit rate than the original information. The major advantage of using the CDMA is that there is no plan for frequency re-use, the number of channels is greater, there is optimum utilisation of bandwidth, and the confidentiality of the information is well protected.

1.3.1.4 Orthogonal Frequency Division Multiple Access

FDM or frequency division multiplexing is a technique in which many signals are combined on single communication channels wherein each of these subchannels has a different frequency within the main channel. Due to this multiplexing, a high bandwidth channel that is complex in nature is produced. A demultiplexer at the receiver is used to separate the channels. The subchannel signal transmission is done at maximum speed. Now orthogonality is applied to FDM. Orthogonality signifies that the subcarriers are at 90° with respect to each other. This is possible by assigning the subcarrier frequencies to these channels. The application of orthogonality not only results in reduction of cross-talk between these carriers but also negates the presence of a guard band, while at the same time, higher spectrum efficiency is utilised.

1.3.2 Modulations

1.3.2.1 Gaussian Minimum Phase Shift Keying (GMSK)

GMSK is the modulation method for signals in GSM. This is a special kind of modulation method derived from minimum phase shift keying (MSK). This falls under the frequency modulation scheme. The main disadvantage of MSK is that it has a relatively wide spectrum of operation, but as in GSM, the frequency is scarce, hence GMSK was chosen to be the modulation method as it utilises the limited frequency resources better. GMSK modulation works with two frequencies and is able to shift easily between the two. The major advantage of GMSK is that it does not contain any amplitude modulation portion and the required bandwidth of the transmission frequency is 200 kHz, which is an acceptable bandwidth by GSM standards. This is the modulation scheme used in GSM and GPRS networks.

1.3.2.2 Octagonal Phase Shift Keying (8-PSK)

The reason behind the enhancement of the data in the 2.5-generation networks such as the GPRS/EGPRS is the introduction of octagonal phase shift keying or 8-PSK. In this scheme, the modulated signal is able to carry three bits per modulated symbol over the radio path as compared with one bit in the GMSK modulated path. But, this increase in data throughput is at the cost of the decrease in the sensitivity of the radio signal. Thereby the highest data rates being provided within a limited coverage. This is the modulation scheme used in the EGPRS/EDGE networks.

1.3.2.3 Quadrature Phase Shift Keying (QPSK)

To demodulate the output of the frequency modulation, phase shift keying or PSK has been used as a preferred modulation scheme. In PSK, the phase of the transmitted waveform is changed instead of its frequency. In PSK, the number of phase changes is two while a step forward is the assumption that the number of phase changes is more than two, i.e. four, which is the case with QPSK. This enables the carrier to carry four bits instead of two, effectively doubling the capacity of the carrier. For this reason, QPSK is the modulation scheme chosen in WCDMA.

1.3.2.4 Quadrature Amplitude Modulation (QAM) and Discrete Fourier Transformation (DFT) in OFDM

QAM and DFT are modulation techniques used in OFDM. In the QAM method, the input data stream is encoded using QAM symbols. A QAM symbol block is inputted in the parallel to serial converter. This result is an in-phase signal. These signals are then transmitted over the radio channel and demultiplexed at the receiver. These QAM signals are filtered and recovered at the receiver using the parallel to serial converter and QAM demodulator. This system requires a narrower bandwidth, i.e. better spectral efficiency than most of the other systems. However, this is not the most spectrally efficient system as there is frequency spillage due to adjacent frequency sub-bands (in subchannels). This leads to the need for some amount of guard band and increased spacing between the sub-bands – leading to lower spectral efficiency. In the DFT modulation, though the fundamental QAM is used, the only difference is the addition of FFT. An inverse FFT is applied to the output of the QAM encoder resulting in complex time domain samples.

1.3.3 OSI Reference Model

The basic idea behind the development of the Open System Interconnection (OSI) Reference Model by the ITU was to separate the various parts that form a

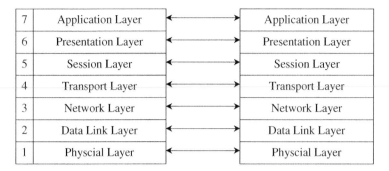

7	Application Layer		Application Layer
6	Presentation Layer		Presentation Layer
5	Session Layer		Session Layer
4	Transport Layer		Transport Layer
3	Network Layer		Network Layer
2	Data Link Layer		Data Link Layer
1	Physcial Layer		Physcial Layer

Figure 1.3 OSI reference model.

communication system. This was possible by layering and modularisation of the functions that were performed by various layers (parts of the communication system). Although initially developed for communication between the computers, this model is being extensively used in the telecommunication field, especially in mobile communication.

1.3.3.1 Basic Function of the OSI Reference Model

Each layer shown in Figure 1.3 communicates with the layer above or below it. No two layers that lie above each other are dependent. The lower layer does not worry about the content of the information that it is receiving from the layer above. Thus, communication between adjacent layers is direct, while with the other layers it is indirect. Each node has the same reference model. When communicating with the other nodes, each layer can communicate with its counterpart in that node, e.g. physical layer with physical layer, transport layer with transport layer. This means that all the messages are exchanged at the same level/layer between two network elements and this is known as a peer-to-peer protocol. All the data exchange in a mobile network belongs to a peer-to-peer protocol.

1.3.3.2 Seven Layers of OSI Reference Model

1.3.3.2.1 Layer 1: Physical Layer

The physical layer is called so because of its 'physical' nature, i.e. it can be copper wire, an optical fibre cable, radio transmission or even a satellite connection. This layer is responsible for the actual transmission of data. This layer transmits the information that it receives from layer 2 without any changes except for the information needed to synchronise with the physical layer of the next node where the information is to be sent.

1.3.3.2.2 Layer 2: Data Link Layer

The function of this layer is to pack the data. The data packaging is based on a high-level data link control protocol. This layer combines the data into packets or frames and sends it to layer 1 or the physical layer for transmission. Layer 2 does the error detection and correction and forms an important part in the protocol testing as the information from layer 3 (data packet format) is sent to layer 2 to be framed into packets that can be transferred over layer 1.

1.3.3.2.3   Layer 3: Network Layer

This layer is responsible for giving all the information related to the path that a data packet has to take and the final destination it has to reach. Thus, this layer gives the routing information for the data packets.

1.3.3.2.4   Layer 4: Transport Layer

This layer is a boundary between the physical elements and logical elements in a network and provides a communication service to the higher layers. This layer checks the consistency of the message by performing end-to-end data control. This layer can perform error detection (but no error correction); it can cater for the reduced flow rate to enable retransmission of data. Thus, layer 4 provides flow control, error detection and multiplexing of the several transport connections on one network connection.

1.3.3.2.5   Layer 5: Session Layer

This layer enables synchronisation between two applications. Both nodes use layer 5, for coordination of the communication between them. This means that it does the application identification but not the management of the application.

1.3.3.2.6   Layer 6: Presentation Layer

This layer basically defines and prepares the data before it is sent to the application layer. This layer presents the data to both sides of the network in the same way. This layer is capable of identifying the type of the data and changes the length by compression or decompression depending upon the need, before sending it to the application layer.

1.3.3.2.7   Layer 7: Application Layer

The application layer itself does not contain any application but acts as an interface between the communication process (layers 1–6) and the application itself. Layer 1 is medium dependent while layer 7 is application dependent.

1.4   Second-generation Mobile Network

Of all the second-generation mobile systems, the Global System for Mobile communication or GSM is the most widely used. In this section we will briefly go through the important constituents of this system. The GSM system is divided into three major parts as shown in Figure 1.4: Base Station Subsystem or BSS, Network Subsystem or NSS and Network Management System or NMS.

1.4.1   Base Station Subsystem (BSS)

The BSS consists of the Base Transceiver Station or BTS, Base Station Controller or BSC and Transcoder Sub-Multiplexer (TCSM). The TCSM is sometimes physically located at the MSC, hence the BSC also has three standardised interfaces to the fixed network, namely A_{bis}, A and X.25.

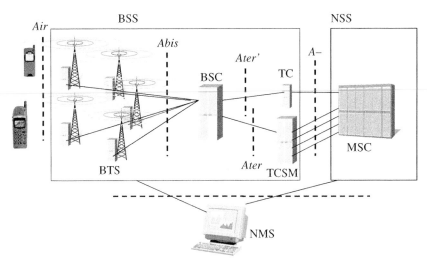

Figure 1.4 GSM architecture.

1.4.1.1 Base Transceiver Station (BTS)

This manages the interface between the network and the mobile station. Hence, it performs the important function of acting as a hub for the whole of the network infrastructure. Mobile terminals are linked to the BTS through the air-interface. Transmission and reception at the BTS with the mobile is done via omnidirectional or directional antennas (usually having 120° sectors). The major functions of the base station are transmission of signals in the desired format, coding and decoding of the signals, countering the effects of multi-path transmission by using equalisation algorithms, encryption of the data streams, measurements of quality and received signal power, and operation and management of the base station equipment itself, etc.

1.4.1.2 Base Station Controller (BSC)

This controls the radio subsystem, especially the base stations. The major functions of the base station include the management functions of the radio resources and handover. It is also responsible for the control of the power transmitted. It manages the O&M and its signalling, security configurations and alarms.

1.4.2 Network Subsystem (NSS)

The NSS acts as an interface between the GSM network and the public networks PSTN/ISDN. The main components of the NSS are MSC, HLR, VLR, AUC, and EIR.

1.4.2.1 Mobile Switching Centre (MSC)

MSC or Switch as it is generally called, is the single most important element of the NSS as it is responsible for the switching functions that are necessary for the interconnection between the mobile users and that of mobile and the fixed network users. For this purpose, MSC makes use of the three major components of the NSS, i.e. HLR, VLR, and AUC.

1.4.2.2   Home Location Register (HLR)

HLR contains the information related to each mobile subscriber. Each subscriber mobile has some information that contains data such as the kind of subscription, services that the user can use, the subscriber's current location, and the mobile equipment status. The database in the HLR remains intact and unchanged until the termination of the subscription.

1.4.2.3   Visitor Location Register (VLR)

VLR comes into action once the subscriber enters the coverage region. Unlike the HLR, VLR is dynamic in nature and interacts with the HLR when recoding the data of a particular mobile subscriber. When the subscriber moves to another region, the database of the subscriber is also shifted to the VLR of the new region.

1.4.2.4   Authentication Centre (AUC)

The AUC or AC is responsible for policing actions in network. This has all the data that is required to protect the network against false subscribers and to protect the calls of regular subscribers. There are two major keys in the GSM standards, one is the encryption of the mobile users and other is the authentication of the mobile users. The encryption keys are held both in the mobile equipment and the AUC and the information is protected against unauthorised access.

1.4.2.5   Equipment Identity Register (EIR)

Each item of mobile equipment has its own personal identification, which is denoted by a number called the International Mobile Equipment Identity (IMEI). The number is installed during the manufacture of the equipment itself stating conformation to the GSM standards. Thus, whenever a call is made, the network would check the identity number and if this number is not found on the approved list of the authorised equipment, access is denied. The EIR contains this list of the authorised numbers and allows the IMEI to be verified.

1.4.3   Network Management System (NMS)

The main task of the NMS is to ensure that the running of the network is smooth. For this purpose, it has four major tasks to perform: network monitoring, network development, network measurements, and fault management. Once the network is up and running, the NMS takes over the responsibility of monitoring the performance of the network. If it sees some faults, it would generate the relevant alarms. Some of the faults may be corrected through the NMS itself (mostly software oriented) while for others, site visits would be required. NMS is also responsible for the collection of data and the analysis of its performance, thereby leading to accurate decisions related to the optimisation of the network. The capacity and the configuration of the NMS are dependent upon the size (both in terms of capacity and geographical area) and the technological needs of the network.

1.4.4   Interfaces and Signalling in GSM

As can be seen from Figure 1.4, there are some interfaces and signalling involved in the GSM system. Here, we will briefly discuss interfaces and some signalling.

1.4.4.1 Interfaces

1.4.4.1.1 Air Interface

The air interface is the central interface and most important interface in every mobile system. The importance of this interface arises from the fact that this is the only interface that the mobile subscriber is exposed to and that the quality of this interface is a critical factor for the success of the mobile network. The quality of this air interface primarily depends upon the efficient usage of the frequency spectrum that is assigned to it.

In the FDMA system, one specific frequency is allocated to one user engaged in a call. When there are numerous calls, the network tends to get overloaded leading to failure of the system. In a full rate (FR) system, 8 time slots (TSs) are mapped on every frequency while in the half rate (HR) system, 16 TSs are mapped on every frequency. In the TDMA system, only impulse-like signals are sent periodically, unlike the FDMA system where signals are assigned permanently. Thus, by combining the advantages of both techniques, TDMA allows seven other channels to be served on the same frequency (FR). GSM uses the GMSK modulation technique.

For the FR system:

Every impulse on frequency is called a burst. Every burst corresponds to a TS.

1 TDMA frame = 8 bursts
Uplink frequency band = 890–915 MHz
Downlink frequency band = 935–960 MHz

Frequency division multiplexing divides each of the frequency ranges into 124 channels of 200 kHz width. Since this is required for both transmission and reception, 124 duplex communication channels are produced.

Now, channel bit rate = D per unit time.

As one TDMA frame has 8 bursts, hence each burst has a bit rate $d = D/8$.

This means that each TS is of 577 µs duration. Thus, one frame has a duration of 4.615 ms.

There are many kinds of frames in the air-interface such as Multiframe, Superframe, and Hyperframes as shown in Table 1.1. Each frame has a duration of 577 µs and carries a communication channel in which a message element called a packet is transmitted periodically.

The physical channel of the TDMA frames carry logical channels that transport user data and signalling information. Thus, there are traffic channels and signalling channels in the TDMA frame. The traffic channels (TCH) either transmit data or voice signal and can be HR or FR. HR provides a bit rate for coded speech of 6.5 kbps and FR has double

Table 1.1 GSM frame hierarchy.

Multiframe$_{(51)}$ = 51 TDMA frames
Multiframe$_{(26)}$ = 26 TDMA frames
Superframe$_{(51)}$ = 51 multiframes
Superframe$_{(26)}$ = 26 multiframes
Hyperframe = 2048 superframes

capacity, i.e. 13 kbps. The signalling channels may contain information such as broadcast channels (e.g. BCCH, SCH, FCH); common control channels (e.g. AGCH, PCH, RACH); dedicated channels (e.g. DCCH, SDCCH); and associated channels (e.g. FACCH, SACCH, SDCCH).

1.4.4.1.2 A_{bis} Interface

The A_{bis} interface is the interface between the BTS and the BSC. It is a PCM interface, i.e. defined by the 2 Mbps PCM link. Thus, it has a transmission rate of 2.048 Mbps, having 32 channels of 64 kbps each. As the traffic channel is 13 kbps on the air-interface, while A_{bis} is 64 kbps, hence multiplexing and transcoding do conversion from 64 kbps on the A_{bis} interface to 13 kbps on the air-interface. Four channels are multiplexed into one PCM channel while traffic channels are transcoded up to 64 kbps. TCSMs are usually located at the MSC to save the transmission costs but can also be located at the BSC site. LAPD signalling along with the TRX management (TRXM), common channel management (CCM), radio link management (RLM), and dedicated channel management (DCM) form a part of this interface.

1.4.4.1.3 A Interface

The A interface is present between the TCSM and MSC or physically between the MSC and BSC (generally, TCSMs are physically located at the MSC). This interface consists of one or more PCM links each having a capacity of 2048 Mbps. There are two parts of the A interface: one from the BTS to the TRAU where the transmitted payload is compressed; and the other between the TRAU and MSC where all the data is uncompressed. The TRAU is typically located between the MSC and BSC and should be taken into consideration when dealing with this interface. SS7 signalling is present on the A interface.

1.4.4.2 Signalling

1.4.4.2.1 LAPD_m

This stands for 'modified Link Access Protocol for D-Channel'. This is a modified and optimised version of the LAPD signalling for the GSM air-interface. The frame structure consists of 23 bytes and is present in three formats A-, B- and A_{bis}. Both the A- and B-formats are used for both the uplink and downlink, while the A_{bis} is only used for the downlink.

1.4.4.2.2 SS7

This provides the basis of all the signalling traffic on all of the NSS interfaces. SS7 or Signalling system No. 7 is a signalling standard developed by the ITU. This provides the protocols by which the network elements in the mobile (and telephone) networks can exchange information. It is used between the BSC and MSC. It is capable of managing the signalling information in complex networks. It is used for call set up and call management, features such as roaming, authentication, call forwarding, etc. It is not allocated permanently and is required only for the call set up and call release function. It is also known as Common Channel Signalling No. 7 or CCS#7.

1.4.4.2.3 X.25

This links the BSC to the O&M centre. It is an ITU-developed signalling protocol, allowing communication between remote devices. This is a packet switched data network protocol that allows both data and control information flow between the host and the network. By utilising connection-oriented services, it makes sure that the packets are transferred in order.

1.5   Third-generation Mobile Networks

Third-generation mobile networks are designed for multimedia communication, thereby enhancing image and video quality, and increasing data rates within public and private networks. In the standardisation forums, WCDMA technology emerged as the most widely adopted third-generation air-interface. The specification was created by the 3GPP and the name WCDMA is widely used for both the FDD and TDD operations. 3G networks consist of two major parts: the Radio Access Network (RAN); and the Core Network (CN), as shown in Figure 1.5. RAN consists of both the radio and transmission parts.

1.5.1   Radio Access Network (RAN)

The main network elements in this part of the network are the base station (BS) and the radio network controller (RNC). The major functions include management of the radio resources and telecommunication management.

1.5.1.1   Base Station (BS)

The BS in 3G is also known as Node B. The BS is an important entity as an interface between the network and the WCDMA air-interface. As in second-generation networks, transmission and reception of the signals from the BS is done through omnidirectional or directional antennas. The main functions of the BS include channel coding, interleaving, rate adaptation, spreading, etc., along with the processing of the air-interface.

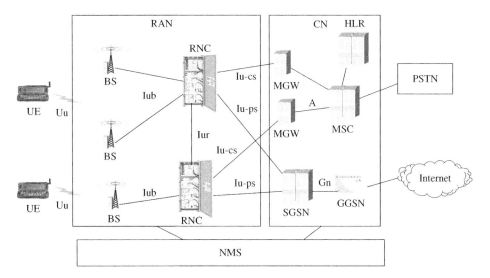

Figure 1.5 Third-generation system (WCDMA).

1.5.1.2   Radio Network Controller (RNC)

This acts as an interface between the BS and the CN. RNC is responsible for control of the radio resources. And, unlike in GSM, the RNC in conjunction with the BS would be able to handle all the radio resource functions without the involvement of the CN. The major functions of the RNC involve load and congestion control of the cells, admission control and code allocation, routing of the data between the I_{ub} and I_{ur} interfaces, etc.

1.5.2   Core Network (CN)

The CN in 3G networks, consist of two domains: a circuit switched (CS) domain; and a packet switched (PS) domain. The CS part handles the real time traffic and the PS part handles the non-real time traffic. Both these domains are connected to the other networks, e.g. CS to the PSTN and PS to the public IP network. The protocol design of the UE and UTRAN is based on the new WCDMA technology but the CN definitions have been adopted from the GSM specifications. Major elements of the CN are WMSC/VLR, HLR, MGW (Media Gateway) on the CS side and SGSN (Serving GPRS Support Node) and GGSN (Gateway GPRS Support Node) on the PS side.

1.5.2.1   WCDMA Mobile Switching Centre (WMSC) and VLR (Visitor Location Register)

The switch and database are responsible for call control activities. WMSC is used for the CS transactions and the VLR function holds information on the subscriber visiting the region, which includes the mobile's location within the region.

1.5.2.2   Gateway Mobile Switching Centre (GMSC)

This is the interface between the mobile network and the external CS networks. This establishes the call connections that are coming in and going out of the network. It also finds the correct WMSC/VLR for the call path connection.

1.5.2.3   Home Location Register (HLR)

This is the database that contains all the information related to the mobile user and the kind of services subscribed to. A new database entry is done when a new user is added to the system and this entry is kept until the user is subscribed to the network. It also stores the UE location in the system.

1.5.2.4   Serving GPRS Support Node (SGSN)

The SGSN maintains an interface between the RAN and the PS domain of the network. This is mainly responsible for mobility management issues like the registration and update of the UE, paging related activities, and security issues for the PS network.

1.5.2.5   Gateway GPRS Support Node (GGSN)

This acts as an interface between the 3G network and the external PS networks. Its functions are similar to the GMSC in the CS domain of CN, but for the PS domain.

1.5.2.6   Network Management System in 3G Networks

As the network technology evolved from 2G to 3G, so did the network management systems. The NMS in 3G systems will be capable of managing packet switched data also, as against voice and circuit switched data in 2G systems. The management systems in 3G would be more efficient, i.e. more work would be possible from the NMS rather than visiting the sites. These systems would be able to optimise and improve the system quality in a more efficient way. The management systems are also expected to handle both the multi-technology, i.e. 2G to 3G, and multi-vendor environments.

1.5.3 Interfaces and Signalling in 3G Networks

As seen from Figure 1.5, there are some interfaces and signalling involved in 3G systems. Here, we will briefly discuss interfaces and some signalling.

1.5.3.1 Interfaces
1.5.3.1.1 Uu/WCDMA Air-Interface
Uu or the WCDMA air-interface is the most important interface in 3G networks. The Uu interface works on WCDMA principles wherein all the users are assigned one code, which varies with the transaction. As shown in Figure 1.2, each user uses a separate spreading code. Unlike with GSM, here every user uses the same frequency band. There are three variants of the CDMA, which are direct sequence WCDMA frequency division duplex (DS-WCDMA-FDD), direct sequence WCDMA time division duplex (DS-WCDMA-TDD), and multi-carrier WCDMA (MC-WCDMA). In the initial phase of the WCDMA network rollouts, the FDD variant would be used. Initially, 'wideband' was introduced because the Euro-Japanese CDMA version used a wider bandwidth than the American version of CDMA.

1.5.3.1.2 Frequency Bands in WCDMA-FDD
The air-interface transmission directions are separated at different frequencies and a duplex distance of 190 MHz (Figure 1.6). The uplink frequency band is 1920–1980 MHz and the downlink is from 2110 to 2170 MHz. There are several bandwidths defined for

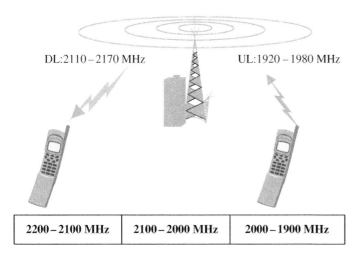

| 2200 – 2100 MHz | 2100 – 2000 MHz | 2000 – 1900 MHz |

Figure 1.6 Frequency bands for WCDMA.

the WCDMA system, e.g. 5, 10 and 20 MHz. However, 5 MHz is the one that is currently being used in network development. Though the bandwidth is 5 MHz, the effective bandwidth would be 3.84 MHz, as the guard band would take in 0.6 MHz from either side (i.e. 1.2 MHz in total makes up the guard band).

1.5.3.1.3 I_{ub} Interface

This is the interface that connects the BS to the RNC. This is standardised as an open interface unlike in the GSM where the interface between the BTS and BSC was not an open one. This consists of common signalling links and traffic termination points. Each of these traffic termination points is controlled by dedicated signalling links. One traffic termination point is capable of controlling more than one cell. I_{ub} interface signalling consists of a Node B Application Part (NBAP), which has two important constituents, i.e. a Common NBAP (C-NBAP) and Dedicated NBAP (D-NBAP), which are used for common and dedicated signalling links, respectively.

1.5.3.1.4 I_{ur} Interface

This is a unique interface in WCDMA networks. I_{ur} is an interface between the RNC and RNC. There was no such interface in the GSM network, i.e. between the BSC and BSC. This interface was designed for supporting the inter-RNC soft-handover functionality. This interface also supports the mobility between the RNC, both the common and dedicated channel traffic and resource management (e.g. transfer of the cell measurements between the cells). The signalling protocol used for this interface is the Radio Network Subsystem Application Part (RNSAP).

1.5.3.1.5 I_u Interface

This interface connects the RAN to the CN. It is an open interface and handles the switching, routing and control functions for both the CS and PS traffic. Thus, the interface that connects the RAN to the CS part of the CN is called the $I_{u\text{-}cs}$ interface while the one that connects the RAN to the PS part of the CN is called the $I_{u\text{-}ps}$ interface.

1.5.3.2 Signalling

The signalling in the third-generation's WCDMA networks is in three planes: the transport plane; the control plane; and the user plane. This is because the 3G network is understood in three layers that contain data flows. These three layers are transport, control and user plane layers as shown in Figure 1.7.

The transport plane is a means to provide connection between a UE and network (i.e. air-interface). It contains three layers: physical; data; and network. The physical layer is the WCDMA layer (TDD/FDD), while the data layer is responsible for setting-up/maintaining/deleting the radio link, protection, error corrections, etc. The data layer also controls the physical layer. The network layer basically contains functions that are required for the transport control plane. The control plane contains signalling related to the services that are handled by the network. The I_{ub} interface is maintained by NBAP; in the I_u interface it is RNAPA while in I_{ur} it is RNSAP. For signalling between the application and the destination on the physical layer, user plane signalling is utilised, e.g. on the U_u interface it is DPDCH.

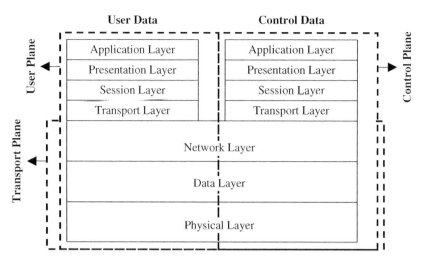

Figure 1.7 OSI model in 3G.

1.6 Fourth-generation Mobile Networks

The fourth-generation or 4G networks are the next generation networks beyond 3G, promising a data rate of 100 Mbps. This fully IP-based integrated system technology provides a comprehensive IP solution where users can experience voice, data, and multimedia – anytime and anywhere. Long Term Evolution or LTE is the most famous 4G technology and is based on 3GPP Release 8 (December 2008). So, what is the fundamental motivation behind the development of 4G technology? The motivation is the need for higher rates beyond that of HSPA, high QoS and low latency. To achieve this, the LTE system is a whole packet switch system that has low complexity (even in architecture) that provides higher QoS and data rates. The network has Access Network and the Evolved Packet System (EPS). EPS is a pure IP-based system with Internet Protocol that will carry both real time services and data. The air-interface technology used in LTE is. The architecture in LTE as shown in Figure 1.8 is quite simplified with Access Network consisting of BSs called eNodeB and a CN consisting of EPS.

1.6.1 Access Network

The Access Network consists of Evolved UMTS Terrestrial Radio Access Network (E-UTRAN). E-UTRAN performs functions such as: connection mobility control, radio resource management, dynamic resource controller, radio bearer control, eNodeB measurement configuration and provisioning. There are two interfaces to E-UTRAN – one towards the other BS called X2 and the other towards EPC called S1.

1.6.1.1 eNodeB

Evolved NodeB or eNodeB is the enhanced base station based on 3GPP standards, providing the RRM and LTE air-interface for the Access Network. Each eNodeB that controls the UEs in one or more cells is called the serving eNodeB. There are two

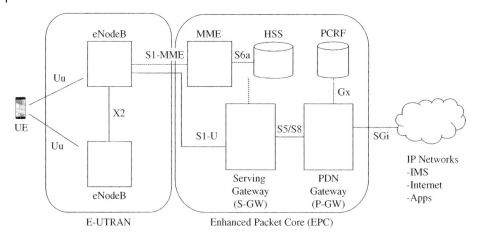

Figure 1.8 4G/LTE network.

fundamental functions performed by eNodeB – performing the low-level operations of UEs connected to it (e.g. handover) and sending the radio transmissions to all UEs connected to it.

1.6.2   Evolved Packet Core Network

The evolved packet core has a few network elements. The key ones discussed include the Mobility Management Entity (MME), Serving Gateway (S-GW), Packet data network gateway (P-GW) and Home Subscriber Server (HSS).

1.6.2.1   Mobility Management Entity (MME)

The MME is the entity that interfaces with the access network. Key functions include load sharing, MME pooling, paging, bearer activations/deactivations, congestion management, etc. On the EPC side, it is responsible for attaching the UE to the S-GW. Both the signalling and control plane functioning are functions handled by the MME. The interaction between the LTE and 2G/3G network also happens through the MME.

1.6.2.2   Serving Gateway (S-GW)

The S-GW connects the data between eNodeB and the P-GW. It also supports the handovers between eNodeBs and also between LTE and 2G/3G technologies.

1.6.2.3   Packet Data Network Gateway (P-GW)

The P-GW's main function is connecting the UE to the external packet data networks using the SGi interface. This is similar to GGSN in the EPGRS networks. Multiple Packet Data Networks (PDNs) can be accesses by the one UE using connectivity to multiple PDNs.

1.6.2.4   Home Subscriber Server (HSS)

The HSS is similar to ones used in GSM and UMTS networks. It is a central database that has information – both user related and subscription related – about the network operator's subscribers.

1.6.3   Interfaces and Signalling

Though there are few interfaces in the LTE network, we will discuss the two most visible interfaces: X2 and S1.

1.6.3.1   X2

This is the interface between two eNodeBs. There are two components of the X2 interface: X2-UP and X2-CP, for user plane and control plane, respectively. X2-UP tunnels the packets between eNodeBs using the tunnelling protocol, GTP-U (over UDP or IP). In the case of X2-CP, the transport layer is defined between the two eNodeBs using the SCTP interface over the IP. Some of the functions performed by X2-CP include general management, error handling, intra-LTE mobility support, handover cancellation, etc.

1.6.3.2   S1

This is the interface between E-UTRAN and MME. As usual, there are two components: user plane and control plane, i.e. S1-UP and S1-CP. S1-UP is responsible for transferring the data packets (non-guaranteed delivery) between the eNodeB and S-GW. As in the case of X1, the transport network player is built on IP transport and GTP-U. S1-CP is responsible for controlling many UEs through the SCTP or Stream Control Transmission Protocol. It is also responsible for the handover signalling procedure, paging procedure, etc.

1.7   Fifth-generation Networks

Also called 5G networks, these are designed and are currently in the process of being standardised. Till 4G, we were talking about the technology, but in 5G we speak about a networked society. Almost in all of the technologies – from 1G to 4G, the lead was in Europe, but for the first time, we will see 5G expected to be launched in Asia (e.g. in Korea and Japan). Some of the key attributes of the 5G network will be:

- Data rates: 1–10 Gbps
- Latency: 1–10 ms
- Reliability: 99.99%
- Operating frequency range: apart from regular frequency in the sub-GHz range, it will also operate in the range of 6–100 GHz

As mentioned before, in a few years from now, there will be a fully networked society – 5G along with IoT and Cloud are key elements that will make it happen. To reach this stage, there are a few challenges associated with the 5G technology: apart from catering to data rates and latency, the other key challenges are to be able to cater to 1000 times higher data, a 100 times higher number of connected devices, 10 times less energy consumption and, of course, higher battery life up to 10 years. The devices in the 5G networks will not just be UEs but also sensors and other devices that will support making a totally networked society.

LTE will have advancements in technology leading to LTE-evolved and later on to full 5G networks. Fully evolved 5G networks are expected to be up and running by 2020. As shown in Figure 1.9, New Radio technology called NR (as defined in 3GPP) will be the

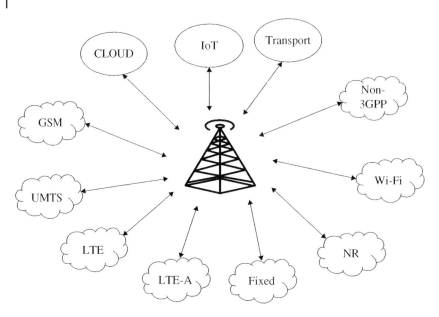

Figure 1.9 5G networks.

hallmark of the 5G radio network. These networks will be compatible with the legacy mobile and PSTN network apart from Wi-Fi. Also, M2M (and later IoT), Cloud, applications, transport networks will be an integral part of the ecosystem.

1.8 Supporting Technologies

As we move into advanced technology environments, it is good to know about some of the technologies both in wired and wireless domains that would help network planning engineers in their understanding of networks.

1.8.1 ADSL2+

Digital Subscriber Line or DSL is a common technology used for bringing broadband to homes. Various types of DSL include SDSL, ADSL, and ADSL2+. SDSL and ADSL are symmetric and asymmetric digital subscriber lines, respectively – the major difference being that the bandwidth can be allocated to the user in ADSL. In the former, both the upload and download speed is the same while in latter the speeds are different (or asymmetric); the download speed is higher than upload in ADSL. The major advantage that ADSL has is the ability to have the DSL and a telephone unit on the same two wires and they can be used simultaneously. This is because ADSL does not take up the entire bandwidth. SDSL uses the whole bandwidth and leaves no room for a telephone unit. The bandwidth for the telephone is allocated to the upload speed and explains the much higher bandwidth despite using the same two wires. ADSL is standardised technology. If the number of downstream channels is doubled, it becomes ADSL2+, leading to data rates as high as 24 Mbps downstream and up to 1.4 Mbps upstream (depending on the distance from the digital subscriber line access multiplexer or DSLAM to the customer premises). ITU G.992.5 defines the standard for ADSL2+.

1.8.2 VDSL2

Very-high-bit-rate DSL is a DSL technology that provides faster data transmission than ADSL. The data transmission rates go up to 52 Mbps downstream and 16 Mbps upstream in twisted cable and up to 85 Mbps upstream and downstream in coaxial cables. This is based on recommendation ITU-T G.993.1. VDSL2 can offer data rates up to 100 Mbps in both uplink and downlink directions. ITU-T G.993.2 is the standard for VDSL2. It has enabled operators and carriers to upgrade the xDSL infrastructure in a way to deploy triple play – voice, data and high-definition TV.

1.8.3 FTTN

FTTN or Fibre to the Node takes fibre optic cables into neighbourhoods. The end user is then connected to this node through the coaxial cable or twisted pair wiring. This permits the user to have access to high speed internet by the usage of protocols such as DOCSIS (Data Over Cable Services Interface Specifications) and data rates varying with the type of protocol used and the distance from the cabinet.

Cross talk is the most effective means to reduce the speed of the network. Vectoring is a technique based on the noise cancellation on a ADSL/VDSL line resulting in significant increase in the speeds attainable on such networks. The speed can be doubled in short lines while diminishing speed in longer lines where cross talk is weaker anyway. Vectoring technology measures and cancels interference across hundreds of lines over the full frequency spectrum they occupy. The interference is processed by subdividing the spectrum into narrow frequency bands, known as tones, and processing each tone independently. All copper lines deploying vectoring technology are processed simultaneously and the results are used in real time to develop antiphase compensation signals for each line, based on the actual signals transmitted on other lines in the bundle.

In FTTN, data is sent over the FTTN and then to the home over vector VDSL2 lines. The only requirement is that all the lines should be under the control of a single operator – no sub local loop unbundling. Cross talk elimination is not possible if lines belong to multiple operators and terminate on different nodes.

1.8.4 V+

Some key features of V+ include delivering speeds of more than 200 Mbps and up to 500 m. It also extends the frequency range used by VDSL2 17a to 35 MHz to achieve these higher speeds. V+ can be mixed with existing VDSL2 17a deployments to fill the gap between VDSL2 17a vectoring (100 Mbps aggregate at 700 m) and G.fast (500 Mbps+ aggregate at 100 m). It offers longer reach (higher bit rates beyond 250 m) and higher density (100–200 subscribers).

1.8.5   GPON

Passive Optical Network or PON is a technology where point to multipoint architecture is implemented by using the unpowered fibre splitters to enable single optical fibres to serve multiple end point customers without providing individual fibres between the hub and customer. It works on wavelength division multiplexing (WDM), using one wavelength for upsteam and one for downstream traffic on a single mode fibre. GPON or Gigabit Passive Optical Network is an enhancement of APON/ BPON supporting higher rates, enhanced security, and choice of layer 2 protocol (ATM, GEM, Ethernet).

1.8.6   GEPON

This is another standard of PON, known as Gigabit Ethernet Passive Optical Network or GEPON. Also called Ethernet PON, it uses Ethernet to send data. It uses 1 Gbps upstream and downstream data rates. BPON is based on ATM, GPON uses native Ethernet and GEPON supports ATM, Ethernet and WDM using a superset multi-protocol layer. A couple of other differences between GPON and GEPON are operational speed (1 Gbps asymmetrical in GPON vs. 2.5 Gbps symmetrical in GEPON) and protocol support for transport of data packets. GEPON is highly flexible and scalable. A broadcast method is used in the downstream and a TDMA (Time Division Multiple Access) protocol is used in the upstream.

1.8.7   G.Fast

The G.Fast acronym comes from Fast access to subscriber terminals with the G coming from the G series of ITU recommendations. It is a DSL standard for local loops smaller than 500 m. The technology uses time division duplexing unlike ADSL and VDSL that use frequency division duplexing. The main limitation in the system performance is due to the cross-talk phenomenon thereby making vectoring mandatory in the G.Fast systems. The latest developments make it possible to deliver 100 Mbps using G.Fast.

1.8.8   XG-PON1

Due to ever increasing demand on GPON, with requirements leading to higher line rates, optical extensions, and affordable migration conditions, the next generation GPON called XG-PON was developed. XG-PON, a 10 Gb capable passive optical network is based on ITU-T G.987. Asymmetric XG-PON is called XG-PON1 (10 Gbps upstream and 2.5 Gbps downstream) while symmetric XG-PON is called XG-PON2 (10 Gbps upstream). It uses TDMA in uplink and TDM in downlink as multiplexing schemes. However, the network operator needs to make significant investments to deploy XG-PON. (Note, the X in XG stands for the Roman numeral 10.)

1.8.9   XGS-PON

XGS-PON is the next generation development of PON with similarity to XG-PON but it is symmetrical in nature. The standard at the ITU aims to match GPON capacity for residential services and to add 6 Gbps of symmetrical capacity for new services. It is expected to have 10 Gbps in downstream and 10 Gbps in upstream directions.

1.8.10   NGPON2

Considering XG-PON where the capacity of the system increases fourfold (2.5–10 Gbps) versus investments needed, the solution, NGPON2 (Next Generation PON2), is actually relatively simple. The technology uses the time and wavelength division multiplexing technique and makes multiple NGPON1 systems work together. It provides a capacity of from 40 to 80 Gbps in downstream and 10–20 Gbps in upstream directions. It can coexist with the GPON and XG-PON1 systems.

1.8.11   NB-IoT

This is a Low Power Wide Area (LPWA) technology that is being developed for the IoT – to provide connectivity to devices in indoor locations in 3GPP release 13. This is not based on LTE but DSSS (Direct Sequence Spread Spectrum) modulations. Apart from coverage, the technology also focuses on long battery life, and a high number of devices connected in a small area. The standards for this are being developed. Some key applications are: smart metering, smart cities, in-building automation, etc. One negative aspect of NB-IoT is that it is difficult to deploy as it is not a part of LTE. Hence, it would need a side band using different software leading to an increase in cost or deployment will happen in a deprecated GSP spectrum.

1.8.12   LTE eMTC

eMTC stands for Enhanced Machine Type Communications. LTE eMTC is also known as LTE-M. MTC work started in 3GPP from release nine onwards, but significant progress was made in releases 12 and 13. LTE-M is more energy efficient because of its extended discontinuous repetition cycle (DRX). LTE-PSM from release 12 (power-saving mode) had a similar feature, but extended DRX was created specifically for LTE-M in release 13. A software upgrade to the LTE network and it is ready for MTC without any investment in antennas or other hardware. The data rates are higher than NB-IoT. Applications include wearables, energy management, utility, etc.

1.8.13   RFID

Radio frequency identification (RFID) refers to small electronic devices that contain a chip and antenna. The chip is capable of carrying 2 kB of data. This helps in providing unique identification for the object. The antenna enables the chips to transmit the information to the reader. The reader converts the information from the antenna signal into digital information and passes it on to computers for use. RFID systems use many different frequencies, but generally the most common are low frequency (around 125 kHz), high frequency (13.56 MHz) and ultra-high frequency or UHF (850–900 MHz). Microwave (2.45 GHz) is also used in some applications.

1.8.14   NFC

NFC is a short-range high frequency wireless communication technology that enables the exchange of data between devices over a distance of about 10 cm. NFC can be considered a development of RFID. This operates at a frequency of 13.56 MHz. The standards and protocols of the NFC format are based on RFID standards outlined in ISO/IEC 14443, FeliCa, and the basis for parts of ISO/IEC 18092.

1.8.15 ZigBee

ZigBee is a wireless technology developed as an open global standard to address the unique needs of low-cost, low-power wireless M2M networks. The ZigBee standard operates on the IEEE 802.15.4 physical radio specification and operates in unlicensed bands including 2.4 GHz, 900 MHz and 868 MHz. Sometimes, there is confusion between Bluetooth, Wi-Fi and ZigBee – both Bluetooth and Wi-Fi have been developed for the communication of large amounts of data with complex structures such as media files, software, etc. However, ZigBee has been developed looking into the needs of the communication of data with simple structures such as the data from sensors. The ZigBee protocol supports longer battery life, low latency, DSSS, approximately 65 000 nodes network apart from supporting multiple technologies (P2P, PMP, etc.).

1.8.16 Z-Wave

Z-Wave is a protocol used to communicate between devices that are used for home automation. Z-Wave is designed to provide reliable, low-latency transmission of small data packets at data rates up to $100 \, \text{kb s}^{-1}$. The throughput is $40 \, \text{kb s}^{-1}$ which is suitable for control and sensor applications with a communication distance of around 30 m. Communication between devices is possible even if they are not in line of sight with each other. Z-Wave is based on a proprietary design and a sole chip vendor.

1.8.17 LoRa

LoRa stands for Long Range Wide Area Networks. It targets key requirements in the IoT such as secure bidirectional communication, mobility and localisation services. This standard will provide seamless interoperability among smart things without the need for complex local installations and returns freedom to the user, developer, businesses enabling the roll out of the IoT. The communication between the end devices and gateways is spread out on different frequency channels and data rates. Date rates range from 0.3 to 50 kbs. The range is in the order of 15–20 km connecting millions of nodes with battery life of more than 10 years. In order to develop and promote the LoRa wireless system across the industry, the LoRa Alliance was set up.

1.8.18 SigFox

SigFox is a Low Power Wide Area Network. It is an operated network for the IoT. It is applicable for small messages and not for high bandwidth usage. The SigFox network operates on sub-GHz frequencies, on ISM bands: 868 MHz in Europe/ETSI and 902 MHz in the US/FCC. It uses an Ultra-Narrow Band (UNB) modulation, and with a 162 dB budget link SigFox enables long range communications, with much longer reach than GSM. The SigFox radio link uses unlicensed ISM radio bands. The exact frequencies can vary according to national regulations, but in Europe the 868 MHz band is widely used while in the US, 915 MHz is used. Some key characteristics of SigFox are: 140 messages per object in a day, with a payload size for each message of 12 bytes. The wireless throughput is up to 100 bps.

1.8.19 Li-Fi

Li-Fi or Light Fidelity is visible light communication. As the name suggests, it is based on light communication (coming from Optical Wireless Communications). It uses visible light, infrared, and the near violet spectrum. Practically it uses common LEDs for data transfer (and photodetectors for data reception) that can reach to more than 200 Gbps. Unlike Wi-Fi, Li-Fi cannot pass through walls, hence lights need to be switched on even during the day if connectivity through Li-Fi is required.

A part of the ITS or Intelligent Transport Systems, Dedicated Short Range Communications or DSRC are data systems only operating in the ISM bands. One of the main applications is for vehicle road side assistance. This is one- or two-way wireless communication. Operating in the licensed frequency bands (FCC: 75 MHz of spectrum in the 5.9 GHz band; ETSI: 30 MHz of spectrum in the 5.9 GHz band), it provides a secure wireless interface required by active safety applications. It supports high speed, low latency, and short range wireless communications while being immune in extreme weather conditions. It supports both Vehicle-2-Vehicle (V2V) and Vehicle-2-Infrastructure (V2I) communications.

1.8.20 V2V and V2I

V2V and V2I communication, an application for DSRC, is part of vehicular transportation systems (VTS). VTS is developed as part of ITS. This technology will allow vehicles to 'talk' to each other or to support infrastructure. V2V is also known as VANET (vehicle ad-hoc networks) which is a part of MANET (mobile ad-hoc networks). These are essentially enhanced versions of the Wi-Fi systems with the exception that they are optimised for moving vehicles with a range of up to 800 m. Each vehicle will have a designated ID and is connected to GPS that can locate the position of the vehicle to within a couple of metres. The speed, direction, and size of the vehicle is transmitted at a speed of $10\,\mathrm{s}^{-1}$ and receiving that same data from other V2V-equipped vehicles around. All information is then sent to onboard computers that will operate the electronic safety systems.

1.8.21 SRD

SRD or Short-Range Devices are radio devices that provide less interference to the other radio services due to less transmitted power. Some of the applications include CCTV cameras, LAN, remote controls, etc. As the power transmitted is in the range of 25–100 mW, the range of transmission is from a few metres to a few hundred metres, the user is free to operate such equipment but may require a licence in certain cases. This is based on ECC Recommendation 70–03.

1.8.22 RLAN

RLAN or Radio LAN is commonly called Wireless LAN or also Wi-Fi. It is used in home, office or small area public locations; it operates in 2.4 and 5 GHz frequency bands, based on IEEE802.11 standards.

1.8.23 PMR

PMR stands for professional mobile radio or private mobile radio. The most famous example is the TETRA (terrestrial trunked radio) systems that have been deployed with police/security agencies around the world. Some important features include point to multipoint, push to talk and close user group communications. Narrow band frequency modulation is used. In India, 3GPP band 27 is ear marked for police/PMRTS (public mobile radio trunking service).

Part I: 2G

GSM and EGPRS Network Planning and Optimisation

2

Radio Network Planning and Optimisation

Since the early days of GSM development, the GSM system network planning has undergone an extensive modification so as to fulfil the ever-increasing demand from operators and mobile users with issues related to capacity and coverage. Radio Network planning is perhaps the most important part of the whole network design owing to its proximity to mobile users. Before going into the details of the process, we first look at some fundamental issues.

2.1 Basics of Radio Network Planning

2.1.1 Scope of Radio Network Planning

The radio network is the part of the network that includes the base transceiver station (BTS) and the mobile station (MS) and the interface between them, as shown in Figure 2.1. As this is the part of the network that is directly connected to the mobile user, it assumes considerable importance. The base station has a radio connection with the mobile and this base station should be capable of communicating with the mobile station within a certain coverage area, and of maintaining call quality standards. The radio network should be able to offer sufficient capacity and coverage for the mobile users.

2.1.2 Cell Shapes

In mobile networks we talk in terms of 'cells'. One base station can have many cells. In general, the term cell can be defined as the area covered by one sector, i.e. one antenna system. The hexagonal shape of the cell is an artificial shape. This is the shape that is closest to the circular shape, which represents the ideal coverage of the power transmitted by the base station antenna. The circular shapes are themselves inconvenient as they have overlapped areas of coverage. But, in reality, their shapes look like the one shown in the 'practical' view in Figure 2.2. A practical network will have cells of nongeometrical shapes with some areas not having required signal strength for various reasons.

Fundamentals of Network Planning and Optimisation 2G/3G/4G: Evolution to 5G,
Second Edition. Ajay R. Mishra.
© 2018 John Wiley & Sons Ltd. Published 2018 by John Wiley & Sons Ltd.

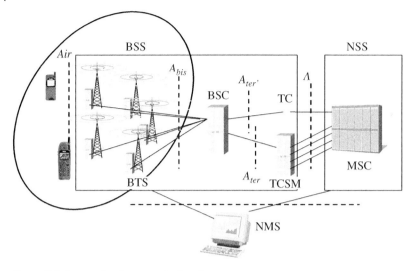

Figure 2.1 Scope of radio network planning.

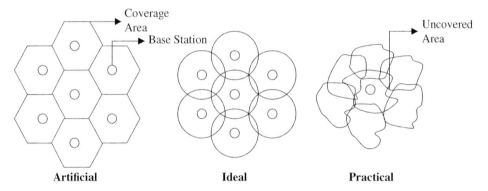

Figure 2.2 Cell shapes.

2.1.3   Elements in a Radio Network

2.1.3.1   Mobile Station (MS)

The mobile station is made up of two parts, as shown in Figure 2.3: the handset and the subscriber identity module (SIM). The SIM is personalised and is unique to the subscriber. The handset or the terminal equipment should have qualities similar to that of fixed phones in terms of quality apart from being user friendly. It also has functionalities like GMSK modulation and demodulation up to channel coding/decoding. It needs to be dual tone multi-frequency generation and should have a long-lasting battery.

The SIM or SIM card is basically a microchip operating in conjunction with the memory card. The SIM card's major function is to store data for both the operator and subscriber. The SIM card fulfils the needs of the operator and the subscriber as the operator is able to maintain the control over the subscription and the subscriber can

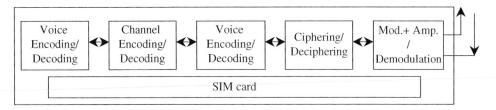

Figure 2.3 Block diagram of a GSM mobile station.

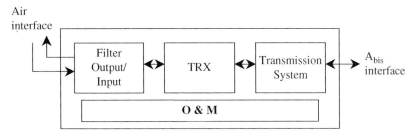

Figure 2.4 Block diagram of a base transceiver station.

protect his or her personal information. Thus, the most important SIM functions include authentication, radio transmission security and storing of the subscriber data.

2.1.3.2 Base Transceiver Station (BTS)

The base station is perhaps the most important element in the network, from the perspective of the radio network-planning engineer as it provides the physical connection to the mobile station through the air-interface. And on the other side, it is connected to the BSC via an A_{bis} interface. A simplified block diagram of a base station is shown in Figure 2.4. For a radio-planning engineer, the transceiver (TRX) is the most important part of the base station. The TRX basically consist of a low-frequency unit and a high-frequency unit. The low-frequency unit is responsible for the digital signal processing and the high-frequency unit is responsible for GMSK modulation and demodulation.

2.1.4 Channel Configuration in GSM

There are two types of channels in the air-interface: physical and logical. Physical channels are all the time slots of the BTS. There are again two types: Half-Rate (HR) and Full-Rate (FR). The FR channel is a 13 kbs coded speech or data channel with a raw data rate of 9.6, 4.8 or 2.6 kbs, while the HR supports 7, 4.8 or 2.4 kbs. The logical channel refers to the specific type of information that is carried by the physical channel. Logical channels can also be divided into two types: traffic (TCH) and control (CCH). Traffic channels are used to carry user data (speech/data) while the control channels carry the signalling and control information. The logical control channels are of two types: common and dedicated. Table 2.1 summarises the control channel types.

Table 2.1 Control channels.

Channel	Abbreviation	Function/application
Access grant channel (DL)	AGCH	Resource allocation, i.e. subscriber access authorisation
Broadcase common control channel (DL)	BCCH	Dissemination of general information
Cell broadcast channel (DL)	CBCH	Transmits the cell broadcast messages
Fast associated control channel (UL/DL)	FACCH	For user network signalling
Paging channel (DL)	PCH	Paging for a mobile terminal
Randon access channel (UL)	RACH	Resource request made by mobile terminal
Slow associated control channel (UL/DL)	SACCH	Used for transport of radio layer parameters
Standalone dedicated control channel (UL/DL)	SDCCH	For user network signalling
Synchronisation channel (DL)	SCH	Synchronisation of mobile terminal

2.2 Radio Network Planning Process

The main aim of the radio network planning is to provide a cost-effective solution for the radio network in terms of coverage, capacity and quality. The network planning process and design criteria vary from region to region depending upon the dominating factor, which could be capacity or coverage. The radio network design process itself is not the only process in the whole network design, and has to work in close coordination with the planning processes of the core and especially the transmission network. But for the ease of explanation, a simplified process just for radio network planning is shown in Figure 2.5.

The process of radio network planning starts with collection of the input parameters such as the network requirements of capacity, coverage, and network quality. These inputs are then utilised to make the theoretical coverage and capacity plans. The coverage definition would include defining the coverage areas, service probability and related signal strength. The capacity definition would include the subscriber and traffic profile in the region and whole area, availability of the frequency bands, frequency planning methods, and other information such as guard band and frequency band division. The radio planner also needs information on the radio access system and the antenna system performance associated with it.

Figure 2.5 Radio network planning process.

The pre-planning process results in preparation of the theoretical coverage and capacity plans. There are coverage driven areas and capacity driven areas in a given network region. The average cell capacity requirement per service area is estimated for each phase of network design to identify the cut-over phase where the network design will change from a coverage driven to a capacity driven process. While the objective of the coverage planning in the coverage driven areas is to find the minimum number of sites to produce the required coverage, the radio planners often have to experiment with both coverage and capacity, as the capacity requirements may have to increase the number of sites, resulting in a more effective frequency usage and minimal interference.

Site candidates are then sought and one of these candidates is then selected based on the inputs from the transmission planning and installation engineers. Civil engineers are also needed to do a feasibility study of constructing the base station at that site.

After site selection, the assignment of the frequency channel for each cell is done in a manner that causes minimal interference and the desired quality is maintained. The frequency allocation is based on the cell-to-cell channel to interference (C/I) ratio. The frequency plans need to be fine-tuned based on the drive test results and the network management statistics.

Parameter plans are made for each of the cell sites, which constitute a parameter set for each cell that is used for network launch and expansion. This set may include cell service area definitions, channel configurations, handover and power control, adjacency definitions and network-specific parameters.

The final radio plan would consist of the coverage plans, capacity estimations, interference plans, power budget calculations, parameter set plans, frequency plans, etc.

2.2.1 Radio Cell and Wave Propagation

Coverage in a cell is dependent upon the area covered by the signal. The distance travelled by the signal is dependent upon the radio propagation in the given area. Radio propagation varies from region to region and should be studied carefully, before the predictions for both coverage and capacity are made. The requirement from the radio planners is generally a network design that covers 100% of the area. But, as fulfilling this requirement is impossible, efforts are made so as to have a network that covers all the regions that may generate traffic and only the no-traffic zones are the holes in the network. In the radio network design, the land area is divided into three major classes: urban, suburban, and rural. The division of these areas is based on human-made structures and natural terrains. The cells (sites) that are constructed in these areas can be classified as outdoor cells and indoor cells. Outdoor cells can be further subclassified as macro-cellular and micro-cellular (Figure 2.6).

2.2.1.1 Macro-cell

When the base station antennas are placed above the average roof top level, the cell is called a macro-cell. As the antenna height is more than the average roof top level, the area covered is wide. A macro-cell range may vary from a couple of kilometres to 35 km. The area covered depends upon the type of terrain and the propagation conditions. Hence, this concept is generally used for suburban or rural environments.

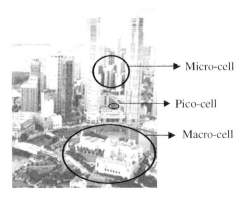

Figure 2.6 Macro-, micro- and pico-cells.

2.2.1.2 Micro-cell

When the base station antennas are placed below the average roof-level, then the cell is a micro-cell. As the antenna height is less than the average roof top level, the area covered by these kinds of cells is small making them micro-cells or small-cells. This kind of concept is applied for urban and suburban areas as the range of such cells is from a few hundred metres to a couple of kilometres.

2.2.1.3 Pico-cell

These are usually utilised for indoor coverage, i.e. coverage area is very small. They are defined as the same layer as micro-cells.

2.2.2 Propagation Effects and Parameters

The signal that is transmitted from the transmitting antenna (BTS/MS) and received by the receiving antenna (MS/BTS) travels a small and complex path. This signal gets exposed to a variety of man-made structures, passes through different types of terrains and gets affected by the combination of propagation environments from its journey between the transmitting and the receiving antennas. All these factors contribute to the variation in the signal level and thereby varying the signal coverage and quality in the network. Before we go into the propagation of the radio signal in urban and rural environments, let us consider some phenomena associated with the radio wave propagation itself.

2.2.2.1 Free Space Loss

Any signal that is transmitted by an antenna will have attenuation during its propagation journey in the free space. The amount of power received at any given point in space will be inversely proportional to the distance covered by the signal, i.e. power received would decrease as the distance between the transmitting and receiving point increases. This can be well understood by using the concept of isotropic antenna. An isotropic antenna is an imaginary antenna. It radiates power equally in all directions. As the power is radiated equally in all directions, hence, we can very well assume that a 'sphere' of power is formed, as shown in Figure 2.7.

The surface area of this sphere can be given as:

$$A = 4\pi R^2 \tag{2.1}$$

Figure 2.7 Isotropic antenna.

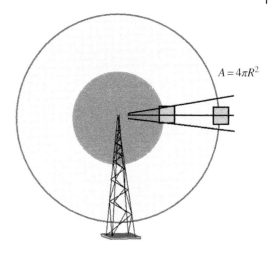

$$A = 4\pi R^2$$

The power density S at any point at a distance R from the antenna, can be then given as:

$$S = P^* G/A \tag{2.2}$$

where P is the power transmitted by the antenna, and G is the antenna gain.

Thus, the received power, P_r at a distance R, is given as:

$$P_r = P^* G_t^* G_r^* \left(\lambda/4\pi R\right)^2 \tag{2.3}$$

where G_t and G_r are the gain of the transmitting and receiving antennas, respectively. On converting this in dB,

$$P_r \left(\text{dB}\right) = P\left(\text{dB}\right) + G_t \left(\text{dB}\right) + G_r \left(\text{dB}\right) + 20\log\left(\lambda/4\pi\right) - 20\log d \tag{2.4}$$

The last two terms in Eq. (2.4) are called the path loss in free space or the free space loss. And, the first two terms, i.e. P and G_t combined are called the Effective Isotropic Radiated Power or EIRP. Thus,

$$\text{Free space loss}\left(\text{dB}\right) = \text{EIRP} + G_r \left(\text{dB}\right) - P_r \left(\text{dB}\right) \tag{2.5}$$

Thus, the free space loss can be given as:

$$L_{\text{dB}} = 92.5 + 20\log\left(f\right) + 20\log\left(d\right) \tag{2.6}$$

where f is the frequency in gigahertz and d is the distance in kilometres.

Equation (2.6) is the loss that would occur in the transmitted signal from the transmitting antenna to the receiving antenna.

2.2.2.1.1 Radio Wave Propagation Concepts

The propagation of the radio wave in the free space depends heavily on the interaction of the frequency of the signal and the obstacles in its path. There are some major effects on signal behaviour due to this. These are briefly described below.

2.2.2.1.2 Reflections and Multipath

When the radio wave is transmitted from the transmitting antenna of a base station, it travels in free space interacting with structures on the surface of the earth, both natural and man-made ones. Thus, the radio wave hardly ever arrives in one path to the receiving antenna, which also means that the transmission of the signal from the transmitting antenna to the receiving antenna is never in line of sight (LOS). Thus, the signal received by the receiving antenna is the sum of all the components of the signal transmitted by the transmitting antenna.

2.2.2.1.3 Diffraction or Shadowing

Diffraction is a phenomenon that takes place when the radio wave strikes a surface and changes its direction of propagation due to the inability of the surface to absorb it. The loss due to diffraction depends upon the kinds of obstruction in the path. This phenomenon is also called shadowing. In practice, the mobile antenna is at a much lower height as compared with the base station antenna. Moreover, there may be high buildings or hills in the surrounding the area. Thus, the signal travels through the diffraction phenomenon to reach the mobile antenna. The term 'shadowing' is used as the mobile is in the 'shadow' of these structures.

2.2.2.1.4 Effect of Building and Vehicle Penetration

When the signal strikes the surface of a building, it may be diffracted or sometimes, it may be absorbed. Due to the absorption, the signal strength is reduced. The amount of absorption is dependent on the type of building. The amount of the solid structure and glass on the outside surface, the propagation environment near the building, orientation of the building with respect to the antenna orientation, etc., contribute to the building penetration loss. This is an important factor in the coverage planning of a radio network. Vehicle penetration loss is similar to the building penetration loss except that the object in this case is a car (vehicle).

2.2.2.1.5 Propagation of Radio Signal over Water

Propagation over water is a big concern for radio planners. The reason is that as the radio signal can propagate over water, it may end up in creating interference with the frequencies of other cells. Moreover, as the water surface is a very good reflector of radio waves, there is a possibility of the signal getting reflected and causing interference to the antenna radiation pattern of some other cell(s).

2.2.2.1.6 Propagation of Radio Signal over Forest (Foliage Loss)

Foliage loss is caused by the propagation of the radio signal over vegetation. The variation of the signal strength depends upon many factors such as the type of trees, trunks, leaves, branches, their densities, and their heights relative to the antenna heights. Foliage loss occurs with respect to the frequency and varies according to the season. This loss can be as high at 20 dB in GSM 800 systems.

2.2.2.1.7 Fading of the Radio Signal

As the signal travels from the transmitting antenna to the receiving antenna, it loses strength. This 'fading' of the signal may be attributed to different reasons. It may be due to the phenomenon of path loss as explained above or may be due to the Rayleigh effect. Rayleigh (or Rician) fading is due to the fast variation of the signal level both in terms of amplitude and phase between the transmitting and receiving antennas when there is no

LOS (or LOS). Rayleigh fading can be characterised into two kinds: multipath fading and frequency selective fading.

Arrival of the same signal from different paths at different times and its combination at the receiver, causes the signal to fade. This phenomenon is called multipath fading and is a direct result of the multipath propagation. Multipath fading can cause fast fluctuations in the signal level. This kind of fading is independent of the downlink or uplink if the bandwidths used are different from each other in both directions. Frequency selective fading takes place due to variation in the atmospheric conditions. Atmospheric conditions may cause the signal of a particular frequency to fade. One point to remember at this stage is that atmospheric changes would react to only one particular frequency at any given time. When the mobile station moves from one location to another, the phase relationship between the various components arriving at the mobile antenna changes, thus changing the resultant signal level received. Here, the Doppler shift in frequency takes place due to the movement of the mobile with respect to the receiving frequencies.

2.2.2.1.8 Interference

Apart from the variation of the signal level due to above mentioned phenomenon and conditions, the signal at the receiving antenna can be weak due to interference from other signals. These signals may be from the same network or may be due to man-made objects. But, a major cause of interference in a cellular network is due to the radio resources in the network. There are many radio channels in use in a network that use common shared bandwidth. The solution to the problem is accurate frequency planning which is dealt with later in the chapter. The mobile station may experience a slow or rapid fluctuation in the signal level in a radio network. This may be due to one or more than one of the above-mentioned factors (as shown in Figure 2.8). These factors form the basis of cell coverage criteria.

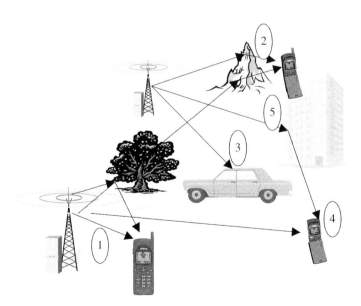

Figure 2.8 Factors effecting wave propagation: 1, direct signal; 2, diffraction; 3, vehicle penetration; 4, interference; 5, building penetration.

2.2.3   Network Planning Requirements: Dimensioning

In any network design, dimensioning constitutes the most important part. The more accurate the dimensioning is, the more efficient will be network rollout. The reason is that, in practice, network rollout very closely follows the output of network dimensioning/planning. For an efficient network rollout, the equipment is to be ordered well before the planning starts, i.e. after dimensioning. And, the equipment orders are made based on the dimensioning results. Thus, planning engineers should try to do a more realistic/accurate dimensioning. The output of the dimensioning exercise is to identify the equipment and the network type (i.e. technology employed) required in order to cater for the coverage and quality requirements apart from seeing that capacity needs are fulfilled for the next few years (generally three to five years). Thus, the accuracy of the immediate traffic requirements and the forecasted traffic requirements would lead to more accurate configuration prediction of the network in terms of capacity and coverage (by each cell site).

The inputs that are required for the dimensioning would contain:

- Geographical area to be covered and estimated traffic in each region.
- Minimum requirements of power in each region and blocking criteria.
- Path loss.
- Frequency band to be used and frequency re-use.

Thus, with the above parameters, the radio planner would be able to predict the number of base stations that would be required for coverage in the specified area to meet the individual quality targets. The radio planner would also predict the number of base stations that would be required to meet the expected increase in traffic in the next few years.

Thus, the network dimensioning would give the near to accurate network configuration and based on this, the network rollout plan for the immediate future and the near future could be made successfully.

2.3   Radio Network Pre-planning

Although in a real scenario, network dimensioning and pre-planning go hand in hand, they have been separated in this chapter for the ease of understanding. Pre-planning can be considered to be the next stage after dimensioning and it is at this stage that some concrete plans related to coverage, capacity and quality are made.

The major target of the radio planner is to increase the coverage area of the cell and decrease the amount of equipment needed in the network, giving the maximum coverage at minimum cost. Maximum coverage means that the mobile is connected to a given cell at a maximum possible distance. This is possible if there is a minimum signal to noise ratio (SNR) at both the BTS and MS. Another factor attributing to the path length between the two antennas (BTS and MS) is the propagation loss due to the environmental conditions.

Example 1 *Calculation of number of sites required in a region*

If there is a network to be designed that should cover an area of $1000\,km^2$.
The base stations to be used are three-sectored. Each sector (cell) covers a range of $3.0\,km$.
Thus, area covered by each site $= k * R^2$.
Where $k = 1.95$
$\Rightarrow$ *Area covered by each site $= 1.95 * 3^2 = 17.55\,km$.*
Thus, total number of sites $= 1000/17.55 = 56.98 \approx 57$ sites.

Capacity can be understood in simplest terms as the amount of mobile subscriber that a BTS can cater to a given time. The greater the capacity, the more mobile subscribers can be connected to the BTS at a given time, thereby reducing the amount of base stations in a given network. This reduction would lead to increase in the operation efficiency and thereby profits for the network operator. As the number of frequency channels in the GSM are constant, i.e. 125 for GSM 900 in either direction, the re-use of these frequencies would determine the number of mobile subscribers that can be connected to a base station. Thus, efficient frequency planning, which includes the assignment of given frequencies and its re-use, plays an important part in increasing the capacity of the radio network.

The quality of the network is dependent upon the parameter settings. Most of these are implemented during the rollout of the network, just before the launch. In some cases, these values are fixed, and in others they are based on measurements done on the existing networks. With the first GSM network to be launched in a given region/country, then it is helpful for the radio planners to plan these values beforehand for the initial network launch before they have the first measurement results. These may include Radio Resource Management (RRM), mobility management, signalling, handover, and power control parameters. Once there are some measurement results from the initial launch of the network, these values then can be fine tuned. This process becomes a part of the optimisation of the radio network.

2.3.1   Site Survey and Site Selection

When the pre-planning phase is nearing completion, the site search process starts. Based on the coverage plans, the radio planners start identifying specific areas for site location. Once the area is identified, the radio planner selects a few candidates, which can be prospective sites. Some points to remember during the process of site selection are:

- The process of site selection, from identifying the site to site acquisition, is very long and slow, which may result in delay of the network launch.
- The sites are a long-term investment and usually cost a lot of money.

Therefore, radio planners in conjunction with transmission planners, installation engineers and civil engineers should try to make this process faster by inspecting the site candidates according to their criteria and coming to a collective decision on whether the candidate site can be used as a cell site or not.

What is a good site for radio planners? A site that does not have high obstacles around it and has a clear view for the main beam can be considered to be a good radio site.

Radio planners should avoid selecting sites at high locations as this may cause problems with uncontrolled interference, apart from giving handover failures.

2.3.2 Result of Site Survey Process

There are two types of reports that are generated in the site survey process from radio planners. One is at the beginning of the site survey request and the other at the end, which is a report on the site selected. Both reports are very important and should have the desired information clearly given. The site survey request report should have the area where the site candidates should be sought. The report may contain more specific information such as the primary candidate for search and secondary site candidates. This gives the site selection team more specific information on where to place their priorities. Also, this report should contain the addresses, maps, and information in the local dialect if possible. The report made after site selection should have more detailed information. This may contain the height of the building/green-field, coordinates, antenna configuration (location, tilt, azimuth, etc.), maps, and top view of the site with exact location of the base station and the antennas (both radio and transmission).

2.4 Radio Network Detailed Planning

The detailed radio network plan can be subdivided into three sub-plans:

1) Link budget calculation
2) Coverage, capacity planning and spectrum efficiency
3) Parameter planning

2.4.1 Link Budget

This is also called the power budget. Before we go into the calculations, let us understand the results and constituents of the link budget calculations. The link budget calculation result gives the loss in the signal strength on the path between the mobile station antenna and base station antenna.These calcualtions help in defining the cell ranges along with the coverage thresholds. Coverage threshold is a downlink power budget that gives the signal strength at the cell edge (border of cell) for a given location probability. As the link budget calculations basically include the power transmission between the base station (including RF antenna) and the mobile station antenna, we shall look into the characteristics of these two pieces of equipment from the link budget perspective. Link budget calculations are done for both the uplink and downlink. As the power transmitted by the mobile station antenna is less than the power transmitted by the base station antenna, the uplink power budget is more critical than the downlink power budget. Thus, the sensitivity of the base station in the uplink direction becomes one of the critical factors as it is related to the reception of the power transmitted by the mobile station antenna. In the downlink direction, transmitted power and the gains of the antennas are important parameters. In terms of equipment loss, the combiner loss and the cable loss are to be considered. Combiner loss occurs only in the downlink calculations while the cable loss has to be incorporated in both directions.

For the other equipment, i.e. the MS, the transmitting power in the uplink direction is very important (as mentioned above, it is much less compared with the BTS). To catch the reception of the transmitted power from the BTS antenna even in remote areas, the sensitivity of the MS comes into play. The transmitting and the receiving antenna gains and the cable loss parameters are to be considered on the BTS side.

2.4.1.1 Important Components of Link Budget Calculations

- MS sensitivity: This factor is dependent upon the receive noise figure and minimum level of *Eb/No* (i.e. output SNR) needed. This is calculated by using the GSM specifications (ETSI GSM recommendation 05.05). The value of MS sensitivity given in these specification is according to the class of mobile being used. The recommended values of MS sensitivity in GSM 900 and 1800 are −102 and −100 dBm, respectively. However, when doing power budget calculations, value given by the manufacturer (or measured values) should be used.
- BTS sensitivity: Sensitivity of the base station is again specified by the ETSI's GSM recommendations 05.05 and is calculated in the same manner as the MS sensitivity. The recommended value of BTS sensitivity is −106 dBm. However, when doing power budget calculations, the value given by the manufacturer (or measured value) should be used.
- Fade margin: The difference between the received signal and receiver threshold is called the fade margin. Usually a fast fade margin is of importance in power budget calculations. Different values are used for different types of regions such as 2 dB for dense urban or 1 dB for urban.
- Connector and cable losses: As cables and connectors are used in power transmission, hence the loss incurred due to their usage should be taken into account. The cable attenuation figures are usually mentioned in loss (dB) per100 m. In such cases, the actual length of the cable should be multiplied by this value to get the actual (theoretical) losses taking place in the cable. Sometimes, these losses may exceed the desired values, thus pre-amplifiers also known as mast head amplifiers may be used to nullify the cable losses. Connector losses are usually much less in the order of 0.1 dB.
- MS and BTS antenna gain: The antennas used for MS and BTS have significantly different gain levels. For obvious reasons, the MS antenna has a lower gain, of the order of 0 dBi, while the BTS antenna gain can vary from 8 to 21 dBi depending upon the type of antenna (omnidiretcional to directional antennas) being used. This gain can be increased by using different techniques such as antenna diversity (both uplink and downlink).

Example 2   *Power budget calculation*

Consider a BTS and MS along with the parameters as shown in Figure 2.9.

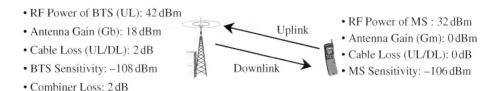

- RF Power of BTS (UL): 42 dBm
- Antenna Gain (Gb): 18 dBm
- Cable Loss (UL/DL): 2 dB
- BTS Sensitivity: −108 dBm
- Combiner Loss: 2 dB

Uplink

Downlink

- RF Power of MS : 32 dBm
- Antenna Gain (Gm): 0 dBm
- Cable Loss (UL/DL): 0 dB
- MS Sensitivity: −106 dBm

Figure 2.9 Power budget.

Uplink calculations

$$PLu\left(Path\ Loss\ in\ uplink\right)=EIRPm\left(Peak\ EIRP\ of\ mobile\right)-Prb$$
$$\left(Power\ received\ by\ the\ base\ station\right)EIRPm$$
$$=Ptm\left(Power\ transmitted\ from\ the\ MS\right)-Losses+Gm$$

$$Losses=Lcm\left(cable\ loss\ at\ mobile\right)+Lom\left(any\ other\ loss\right)$$

$$Prb=-Gb\left(antenna\ gain\right)+Losses+Bs\left(BTS\ sensitivity\right)$$

$$Losses=Lcb\left(cable\ loss\ at\ BTS\right)+Lob\left(any\ other\ loss\right)$$

$$PLu=EIRPm-Prb$$
$$=\left[\,Ptm-Lcm-Lom+Gm\right]-\left[-Gb+Lcb+Lob+Bs\,\right]$$
$$=\left[\,32-0+0+0\,\right]-\left[-18+2+0+\left(-108\right)\right]$$
$$=32+124=156\ dB$$

Downlink calculations

$$PLd\left(Path\ Loss\ in\ downlink\right)=EIRPb\left(Peak\ EIRP\ of\ BTS\right)$$
$$-Prm\left(Power\ received\ by\ the\ MS\right)EIRPb=Ptb\left(Power\ transmitted\ by\ BTS\right)$$
$$+Gtb\left(antenna\ gain\right)-Losses$$

$$Losses=Lcb\left(cable\ loss\ at\ BTS\right)+Lccb\left(combiner\ loss\ at\ BTS\right)$$

$$Prm=Ms\left(Mobile\ sensitivity\right)+Losses-Gm\left(mobile\ antenna\ gain\right)$$

$$Losses=Lcm\left(cable\ loss\right)+Lom\left(any\ other\ loss\right)$$

$$PLd=EIRPb-Prm$$
$$=\left[Ptb+Gtb-Lcb-Lccb\right]-\left[Ms-Lcm-Lom-Gm\right]$$
$$=\left[42+18-2-2\right]-\left[-106-0-0-0\right]$$
$$=56+106=162\ dB$$

As can be seen from the results, there is a difference in the uplink and the downlink power budget calculations, where the downlink path loss exceeds the uplink power loss. This is an indication that the area covered by the base station antenna radiations is more than the area covered by the mobile station antenna, thereby giving more coverage in the downlink direction. Reducing the power in the downlink direction can reduce this difference but this results in the loss of coverage. Another way is to introduce the diversity at the BTS or even by the introduction of low noise amplifiers (LNAs) at the BTS. Both will have a positive impact on the BTS receiver power levels. Another power budget calculation for the GSM 900 and 1800 systems using different classes of mobiles (A and D) is shown in Table 2.2.

Table 2.2 Simple power budget calculations.

RADIO LINK POWER BUDGET				MS CLASS 1
GENERAL INFO				
Frequency (MHz): **1800**			System:	**GSM**
RECEIVING END:		**BS**	**MS**	
RX RF-input sensitivity	dBm	−104.00	−100.00	A
Interference degrad. margin	dB	3.00	3.00	B
Cable loss + connector	dB	2.00	0.00	C
Rx antenna gain	dBi	18.00	0.00	D
Diversity gain	dB	5.00	0.00	E
Isotropic power	dBm	−122.00	−97.00	F=A+B+C − D − E
Field strength	dBμV/m	7.24	32.24	G=F+Z*
				$^*Z = 77.2 + 20^*log(freq[MHz])$
TRANSMITTING END:		**MS**	**BS**	
TX RF output peak power (mean power over RF cycle)	W	1.00	15.85	
	dBm	30.00	42.00	K
Isolator + combiner + filter	dB	0.00	3.00	L
RF-peak power, combiner output	dBm	30.00	33.00	M=K − L
Cable loss + connector	dB	0.00	2.00	N
TX-antenna gain	dBi	0.00	18.00	O
Peak EIRP (EIRP = ERP + 2dB)	W	1.00	79.43	
	dBm	30.00	49.00	P=M − N+O
Path loss due to ant./body loss	dBi	6.00	6.00	Q
Isotropic path loss	dB	146.00	146.00	R=P − F − Q
RADIO LINK POWER BUDGET				MS CLASS 4
GENERAL INFO				
Frequency (MHz): **900**			System:	**GSM**
RECEIVING END:		**BS**	**MS**	
RX RF-input sensitivity	dBm	−104.00	−102.00	A
Interference degrad. margin	dB	3.00	3.00	B
Cable loss + connector	dB	4.00	0.00	C
Rx antenna gain	dBi	12.00	0.00	D
Isotropic power	dBm	−109.00	−99.00	E = A + B + C − D
Field strength	dBμV/m	20.24	30.24	F = E + Z*
				$^*Z = 77.2 + 20^*log(freq[MHz])$
TRANSMITTING END:		**MS**	**BS**	
TX RF output peak power (mean power over RF cycle)	W	2.00	6.00	
	dBm	33.00	38.00	K

(*Continued*)

Table 2.2 (Continued)

RADIO LINK POWER BUDGET				MS CLASS 1
Isolator + combiner + filter	dB	0.00	3.00	L
RF-peak power, combiner output	dBm	33.00	26.00	M = K − L
Cable loss + connector	dB	0.00	4.00	N
TX-antenna gain	dBi	0.00	12.00	O
Peak EIRP (EIRP = ERP + 2dB)	W	2.00	20.00	
	dBm	33.00	34.00	P = M − N+O
Path loss due to ant./body loss	dBi	9.00	9.00	Q
Isotropic path loss	dB	133.00	133.00	R = P − F − Q

Placeholder Text How can there be an improvement in the power budget results? As seen above, apart from varying the power transmitted from the BTS antenna, these results can be improved by using enhanced planning techniques such as frequency hopping and/or by using some enhancements such as the receiver diversity, LNA for the uplink directions and boosters or filters for the downlink directions.

2.4.1.2   Output and Effect of Link Budget Calculations

- Path loss and received power: This is the main output of link budget calculations. The losses that occur in signal strength during the signal transmission from the TX antenna to the RX antenna are given by path loss, while the received power is the result of the path loss phenomenon. All the factors that contribute to the increases (e.g. antenna gains) and decreases (e.g. losses due to propagation) are taken into account during calculations. The higher the input data accuracy, the more accurate the results.
- Cell range: If the path loss is less, the signal from the TX (BTS) antenna would cover more distance thereby increasing the area covered by one BTS. Thus, the power budget calculations play a direct role in determing the covered area and thus, deciding the number of base stations that would be required in a network.
- Coverage threshold: Downlink signal strength at the cell border for a given location probability is known as coverage threshold. Although slow fade margin and MS isotropic power can be used to calculate this value, power budget calculations are utilised for this purpose. The effect of propagation models for the more accurate calculation of the cell range and coverage area as shown in Table 2.3.

2.4.2   Frequency Hopping

Frequency hopping (FH) is a technique that basically improves the C/I ratio by utilising many frequency channels. Employment of the FH technique also improves the link budget due to its effects: frequency diversity and interference diversity.

The frequency diversity technique increases the decorrelation between the various frequency bursts reaching the moving MS. Due to the increase in the decorrelation of the frequency bursts, the effects of fading due to propagation conditions reduces,

Table 2.3 Detailed radio link power budget.

	RADIO LINK POWER BUDGET				MS CLASS: 2
GENERAL INFO					**GSM**
Frequency (MHz):	900				
RECEIVING END:			System:	MS	
			BS	**MS**	
RX RF-input sensitivity	dBm		−106.00	−103.00	A
Interference degrad. margin	dB		2.00	2.00	B
Cable loss + connector	dB		2.50	10.00	C
Rx antenna gain	dBi		16.00	0.00	D
Diversity gain	dB		3.50	0.00	E
Isotropic power	dBm		−121.00	−91.00	F=A+B+C−D−E
Field strength	dBμV/m		8.24	38.24	G=ʼ+Z*
					*$Z = 77.2 + 20 \cdot \log(freq[MHz])$
TRANSMITTING END:			**MS**	**BS**	
TX RF output peak power (mean power over RF cycle)	W		3.00	33.66	
	dBm		34.77	45.27	K
Isolator + combiner + filter	dB		0.00	4.00	L
RF-peak power, combiner output	dBm		34.77	41.27	M=K−L
Cable loss + connector	dB		10.00	2.50	N
TX-antenna gain	dBi		0.00	16.00	O

(Continued)

Tabel 2.3 (Continued)

	RADIO LINK POWER BUDGET				MS CLASS: 2
Peak EIRP (EIRP = ERP + 2dB)	W		0.30	300.00	
Isotropic path loss	dBm		24.77	54.77	P=M–N+O
	dB		145.77	145.77	Q=P–F
CELL SIZES					
COMMON INFO		Region 1	Region 2	GENERAL	
MS antenna height (m):		1.5	1.5	1.5	
BS antenna height (m):		30.0	30.0	30.0	
Standard Deviation (dB):		7.0	7.0	7.0	
BPL Average (dB):		25.0	20.0	15.0	
BPL Deviation (dB):		7.0	7.0	7.0	
OKUMURA-HATA (OH)		Region 1	Region 2	GENERAL	
Area Type Correction (dB)		0.0	–4.0	–6.0	
WALFISH-IKEGAMI (WI)		Region 1	Region 2	GENERAL	
Roads width (m):		30.0	30.0	30.0	
Road orientation angle (degrees):		90.0	90.0	90.0	
Building separation (m):		40.0	40.0	40.0	
Buildings average height (m)		30.0	30.0	30.0	

INDOOR COVERAGE

	Region 1	Region 2	GENERAL
Propagation Model	OH	OH	OH
Slow Fading Margin + BPL (dB):	32.4	27.4	22.4
Coverage Threshold (dBμV/m):	70.6	65.6	60.6
Coverage Threshold (dBm):	-58.6	-63.6	-68.6
Location Probability over Cell Area(L%):	95.0%	95.0%	95.0%
Cell Range (km):	0.78	1.40	2.22

OUTDOOR COVERAGE

	GENERAL
Propagation Model	OH
Slow Fading Margin (dB):	7.4
Coverage Threshold (dBμV/m):	45.6
Coverage Threshold (dBm):	-83.6
Location Probability over Cell Area(L%):	95.0%
Cell Range (km):	5.92

thereby improving the signal level. There are again two types of frequency diversity techniques: random FH and sequential FH. Sequential FH is used more in practical network planning as it gives more improvement to the network quality.

If the number of frequency channels increases in the radio network, the number of frequencies used increases in the network, reducing the interference level at the MS. This leads to an increase in signal level indicating an improvement in the power budget.

2.4.3 Equipment Enhancements

2.4.3.1 Receiver Diversity

Diversity is the most common way in the mobile networks to improve the reception power of the receiving antenna. Major diversity techniques are space diversity, frequency diversity, and polarization diversity. Frequency diversity is also known as frequency hopping. Space diversity means installing another antenna at the base station. This means that there are two antennas receiving the signal at the base station instead of one and they are separated in *space* by some distance. There is no fixed distance of separation between the antennas, which depends upon the propagation environment. Depending on the environmental conditions, the distance between the main and the diversity antenna can vary from 1 to 15 λ. Polarisation diversity means the signals are received using two polarisations that are orthogonal to each other. It can be either vertical–horizontal polarisation or it can be ±45° slated polarisation.

2.4.3.2 Low Noise Amplifier (LNA)

Where the received power is limited by the long length of the cables, the LNAs can be used to boost the link budget results. As the name suggests, a LNA has a low noise value and can amplify the signal. Thus, the LNA is placed at the receiving end. The advantage of these is that they have a low noise figure, thus improving the received signal level by the amplification process. In the case where space diversity is being used, LNA should be used on both the main and the diversity antennas, thereby improving the diversity reception. As stated above, this is used for the uplink power budget improvement.

2.4.3.3 Power Boosters

Power in the downlink direction can be increased by the use of power amplifiers and power boosters. If the losses are reduced before the transmission by the use of amplifiers, which in turn increases the power, then the configuration is called a power amplifier. However, when the transmission power is increased, then it is done by using the booster. Power amplifiers are located near the transmission antennas while the boosters are located near the base station as shown in Figure 2.10.

2.4.4 Cell and Network Coverage

The cell and network coverage depend mainly upon natural factors such as geographical aspect/ propagation conditions and human factors such as the landscape (urban, suburban, rural), subscriber behaviour, etc. The ultimate quality of the coverage in the mobile network is measured in terms of location probability. For that, the radio propagation conditions have to be predicted with respect to the geographical aspects and landscape of the region as accurately as possible. There are two ways in which the radio planners

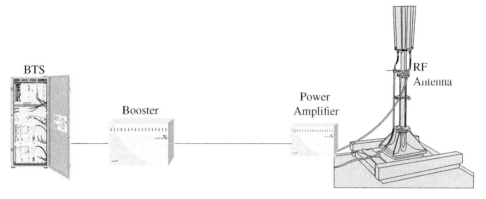

Figure 2.10 Power budget enhancement using a booster and power amplifier.

can use propagation models. They can either create their own propagation models for different areas in a cellular network, or they can use the existing standard models, which are generic in nature and are used for the whole area. The advantage of using their own model is that it will be more accurate but it would also be immensely time consuming. While usage of the standard models is economical from a time and money perspective, these models are standard and have limited accuracy. Of course, there is a middle option with the use of many generic models of urban, suburban and rural environments in terms of macro-cell or micro-cell structure.

2.4.4.1 Macro-cell Propagation Model

The Okumara–Hata model is the most commonly used model for macro-cell coverage planning. It is used for the frequency ranges of 150–1000 and 1500–2000 MHz. The range of calculation is from 1 to 20 km. The loss between the transmitting and the receiving station is given as:

$$L = A + B\log f - 13.82\log h_{bts} - a(h_m)$$
$$+ \left(44.9 - 6.55\log h_b\right)\log d + L_{other} \tag{2.7}$$

where f is the frequency (MHz), h is BTS antenna height (m), $a\ (h)$ is a function of the MS antenna height, d is the distance between BS and MS (km) and L_{other} is the attenuation due to land usage classes.

$$a(h_m) = \left[1.1\log(fc) - 0.7\right]h_m - \left[1.56\log(fc) - 0.8\right]$$

For a small or medium city:

$$a(h_m) = 8.25\left[\log(1.54 h_m)\right]^2 - 1.1; \text{for } fc \le 200\ MHz \tag{2.8}$$

For a large city:

$$a(h_m) = 3.2\left[\log(11.75 h_m)\right]^2 - 4.97; \text{for } fc \ge 400\ MHz \tag{2.9}$$

The value of the constants A and B varies with frequencies as shown below:

A = 69.55; B = 26.16 (for 150–1000 MHz).

A = 46.3; B = 33.9 (for 1000–2000 MHz).

The attenuation due to land usage types will vary with the type of terrain. This may include losses in an urban environment where small cells are predominant. Then there are foliage losses when forests are present in the landscape. Similarly, the effects of other natural aspects such as water bodies, hills, mountains, glaciers, etc. and the change in behaviour in different seasons have to be taken into account.

2.4.4.2 Micro-cell Propagation Model

The most commonly used micro-cellular propagation model is the Walfish–Ikegami model. This model is basically used for micro-cells in urban environments. This model can be used for the frequency range between 800 MHz and 2000 MHz, for heights up to 50 m (i.e. the height of building + height of the BTS antenna) for a distance of up to 5 km. This model talks about two conditions: line of sight (LOS) and no line of sight (NLOS). The path loss formula for the LOS condition is:

$$P = 42.6 + 26 \log(d) + 20 \log(f) \tag{2.10}$$

For the NLOS condition, path loss is given as:

$$P = 32.4 + 20 \log(f) + 20 \log(d) + L_{rds} + L_{ms} \tag{2.11}$$

where d is the distance (km), f is frequency (MHz), L_{rds} is the rooftop-street diffraction and scatter loss and L_{ms} is the multi-screen diffraction loss. The parameters in the equations above for the Walfish–Ikegami model can be understood from Figure 2.11.

The values of the rooftop-to-street diffraction loss are dependent upon the street orientation, street width and the frequency of operation. The multi-screen diffraction losses are dependent upon the distance and frequency.

Note: The Walfish–Ikegami model can also be used for macro-cells. However, it some radio planning engineers do use other models, such as ray tracing, for the micro-cellular environment.

2.4.4.3 Application of Propagation Models

The propagation models are usually not applied directly. The reason is that these models were developed taking particular cities into account and every city has its own

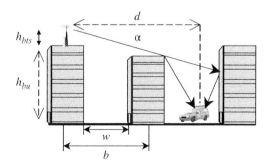

Figure 2.11 Walfish–Ikegami model. *d*, distance (km); *w*, road width; *b*, distance between the centres of two buildings; h_{bts}, height of the BTS antenna; h_{bu}, height of the building.

characteristics. The deviation or changes made in the propagation models are called correction factors. There are in fact no corrections to any error, but there are more accurate values that should be used, taking the conditions of the region for which the cellular network is to be designed. These values come from drive tests results. But if there is no existing cellular network, the radio network planning engineers install an omni-antenna at a location which would cover all or most of the type of regions, e.g. dense urban, urban, rural, etc. A drive test is performed and these values are determined. One such table of correction factors is shown in Figure 2.12a.

2.4.4.4 Planned Coverage Area

Based on the propagation models, drive tests and correction factors, the prediction for coverage areas is done. The sites are located according to the requirements of the network, and the coverage predictions are done as shown in Figure 2.12b. Usually some radio network planning tools are used for such an exercise.

2.4.4.5 Location Probability

As mentioned above, the quality of coverage is defined in terms of location probability. Location probability can be defined as the probability of the field strength being above the sensitivity level in the target area. For practical purposes, it is considered that a location probability of 50% is equal to the sensitivity of the receiver in the given region. As the received power at the receiver should be higher than the sensitivity, it is worth mentioning that location probability should be higher than 50%. Above, we have discussed the reason behind the fluctuation and fading of the signal strength. These fluctuations may be more or less than the sensitivity of the receiver. Hence, the design of the radio network incorporates a term called the fade margin. Planning is done in such a way that the field strength of the signal is higher than the sensitivity by this margin. So, when the fading is taking place, usually slow fading or shadowing being the most prominent, then the signal level after fading is way above the sensitivity of the receiver. A more accurate link budget calculation taking into account the propagation model effects is shown in Table 2.3.

2.4.5 Capacity Planning

Capacity planning is a very important process in the network rollout as it defines the number of base stations required and their respective capacities. Capacity plans are made in the pre-planning phase for initial estimations and are also done with coverage plans in a detailed manner. The number of base stations required in an area comes from the coverage planning and the number of transceivers required is derived from capacity planning as it directly associated with the frequency re-use factor. The frequency re-use factor is defined as the number of base stations that can be implemented before the frequency can be re-used again. An example of frequency re-use is shown in Figure 2.13. The maximum number of frequencies in a GSM 900 system is 125 in both the uplink and downlink directions. Each of these frequencies is called a channel. This means that there are 125 channels available in both directions. The minimum frequency re-use factor calculation is based on the C/I ratio. As soon as the C/I ratio decreases, the signal strength starts deteriorating, thereby reducing the frequency re-use factor. Another factor to keep in mind is the antenna height at the base station. If the antenna

(a)

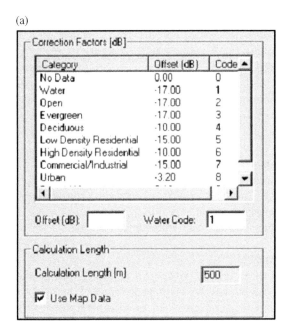

(b)

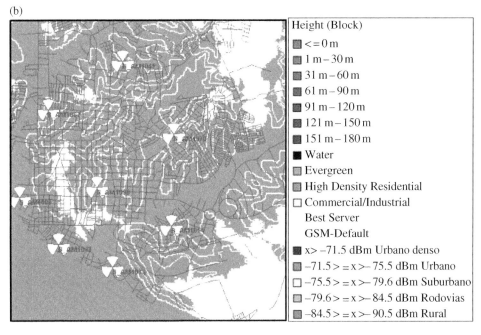

Figure 2.12 (a) Correction factors; (b) planned coverage area.

height at the base station is too high then the signal has to travel a greater distance, hence the probability that the signal causes interference becomes high. Thus, the average antenna height should be such that the numbers of base stations (fully utilised in terms of their individual capacity) are enough (minimum) for the needed capacity of the network. Of course, as seen above, this depends heavily on the frequency re-use factor.

Figure 2.13 Example of frequency re-use.

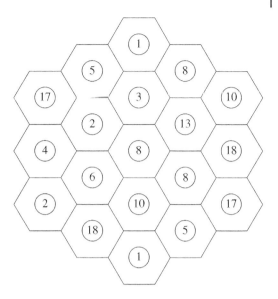

There are three essential parameters required for capacity planning: estimated traffic, average antenna height and frequency usage.

2.4.5.1 Traffic Estimates

Traffic estimation or traffic modelling is based on theoretical estimates or assumptions and on studies of existing networks (or experience). The traffic in the network (or the utilisation of the network by the subscriber) is dependent on the user communication rate and the user movement in the network. The user communication rate means how much traffic is generated by the subscriber and for how long. The user movement means an estimate of the user using the network in static mode and dynamic mode. Traffic estimation in the network is given in terms of 'erlangs'. One erlang is defined as the amount of traffic generated by the user when he or she uses one TCH for one hour (this one hour is usually the busy hour of the network). Another term that is frequently used in network planning is 'blocking'. Blocking is the situation when a user is trying to make a call and is not able to reach a dialled subscriber due to lack of resources. Generally, it is assumed that a user would be generating about 25 mErl of traffic during the busy hour and the average speaking (network usage) is about 2 minutes or 120 seconds. These figures may vary from network to network. Some networks use 35 mErl and 90 seconds as average usage time, respectively. Another factor is the user behaviour in terms of mobility. In the initial years of the GSM, the ratio of the static users to dynamic users was almost 0.7, but with rapid changes in time and technology, this ratio may soon change to 1.0! User mobility affects the handover rates, which in turn affects the capacity planning in the network. The actual traffic flowing in the network can be calculated by using the Erlang B tables that use the maximum traffic at a base station and the blocking rate. Commonly used Erlang tables are Erlang B and Erlang C. Erlang B takes into account that if the calls cannot go through, they get dropped (i.e. no queuing possible). Erlang C considers that if a call does not gets through, then it will wait in a queue (known as queuing). These Erlang tables are good enough for the circuit switched traffic but not for packet switching. We look into the packet switch aspect in later chapters. Erlang B tables are given in Appendix G.

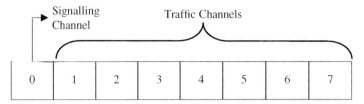

Figure 2.14 Time slot configuration for single TRX.

When the estimation of the traffic is done, it is important for the radio planner to know the capacity that can be offered by the base station equipment, which means that traffic handling capability of the transceivers takes precedence for capacity planning. Due to modulation, the modulated stream-of-bits are sent in *bursts* and these have finite duration. These bursts are generally called *slots* in GSM having a relatively fixed place in the stream and occurring after 0.577 ms. These slots have a width of about 200 kHz. This is known as one *time slot* (TS) in the GSM system. Due to the modulation schemes (i.e. TDMA), there are eight TSs at each frequency in each direction. All of these eight TSs can be used for sending the traffic or the signalling information. The channel organisation within the TS should be done in such a way that every time a burst is transmitted, it is utilised completely. A typical composition is shown in Figure 2.14.

Signalling requires one TS, e.g. TS0, and the remaining seven TSs can be used for traffic. In this configuration, the number of subscribers that can talk simultaneously is seven, hence called Traffic Channels or TCH. Now, when the number of transceivers (TRX) increase in the cell, the traffic and signalling channel allocation also changes. Generally, it follows that 2 TRX would have 15 TCH and 1 SCH (signalling channel). Four TRX will have 30 TCH and 2 SCH which means that the TCH increase by $7+8+7+8\cdots$ for every single increase in TRX while the SCH will have an increment at every alternate TRX (which means that a decrease in SCH in the TRX increases the TCH in it).

2.4.5.2 Average Antenna Height
The concept of average antenna height is important to understand, as it is the basis of the frequency re-use pattern determining the capacity calculations in a cellular network. Average antenna height is the basis of the cellular environment, i.e. whether it is macro-cellular or micro-cellular. Now, if the average antenna heights were lower, then the covered area would be smaller in an urban environment. This would lead to creation of more cells and hence increase the number of times the same frequency can be re-allocated. Exactly the opposite is the case in a macro-cellular environment. Here the coverage area would be more, so the same frequency can be re-allocated fewer times. All these calculations are based on the interference analysis of the system as well as the topography and propagation conditions playing an equally important role.

2.4.5.3 Frequency Usage
Frequency usage or frequency re-use is an important concept related to both the coverage and capacity usage. Frequency re-use basically means how often a frequency can be re-used in the network. If the average number of transceivers and the total number of frequencies are known, the frequency re-use factor can be calculated.

Example 3 *Frequency re-use factor*
If there are 3 TRX that are used per base station and the total number of frequencies available are 27, then the total number of frequencies available for re-use is 27/3 = 9.

2.4.6   Spectrum Efficiency and Frequency Planning

Spectrum efficiency is nothing but the maximum utilisation of the available frequencies in a network. In the radio planning process, this is known as frequency planning. Capacity and frequency planning go hand-in-hand, but the aspects described so far in this chapter provide the inputs for the frequency planning. A good frequency plan would mean utilisation of the frequency channels in a way that the capacity and coverage criteria are met without any interference. This is because the total capacity in a radio network in terms of number of sites is dependent upon two factors: transmission power and interference. The re-use of the BCCH TRX (which contains the signalling time slots) should be greater than that of the TCH, since it should be the most interference free.

We have already seen the factors affecting the transmission power flow. Now let us see the factors that should be kept in mind in order to keep the interference to a low or negligible level or, in other words, to create effective frequency plans.

2.4.7   Power Control

The power that is transmitted both from the mobile and the base station has a far-reaching effect on the efficient usage of the spectrum. Hence, the power control becomes a necessary feature in the mobile networks, both in the uplink and downlink direction. When a mobile transmits high power, it gives enough fade margin in the critical uplink direction, but it then can cause interference to other subscriber connections. Thus, the power should be kept at a level that the signal is received by the base station antenna above the required threshold without causing interference to other mobiles. Mobile stations thus have a feature such that their power of transmission can be controlled. This feature is generally controlled by the BSS. This is based on an algorithm that computes the power received by the base station and based on its assessment, it increases or decreases the power transmitted by the mobile station.

2.4.8   Handover

The automatic transfer of the subscriber during the call process from one cell to another, without causing any hindrance to the call, is called the handover. There are two main processes involved in this: the necessity to find a dedicated mode in the next cell as the mobile is on a call and the switching process being fast enough so as not to drop that call. So, how does the handover actually take place? There are many processes that may be used, but the most used is based on the power measurements. When a mobile is at the interface of two cells, the BSS measures the power that is received by the base stations of the two cells, and then the one that satisfies the criteria of enough power and least interference gets selected. As this kind of handover is directly related to the power control, this gives an opportunity to improve the efficiency of the spectrum.

2.4.9 Discontinuous Transmission

Discontinuous transmission or DTX is a feature that controls the power of the transmission when the mobile is in 'silent' mode. When the subscriber is not speaking on the mobile, a voice detector in the mobile detects this and sends a burst of transmission bits to the BSS, thereby indicating this inactivity. This function of the mobile is called voice activity detection or VAD. On receiving this stream of bits indicating DTX, the BSS asks the mobile to reduce its power for that period of time, thereby reducing the interference in the network and improving the efficiency of the network.

2.4.10 Frequency Hopping

Before we go into the concept and process of frequency hopping, let us understand the frequency assignment criteria in the GSM network. In GSM 900, the frequency band that is used is 890–915 MHz in the uplink direction and 935–960 MHz in the downlink direction, which means a bandwidth of 25 MHz in either direction. The whole of this band or some fraction of this band is available to the network operators. The central frequencies start at 200 kHz from the 'edge' of the band and are spread evenly in it. There are 125 frequency slots in this band. A major interference problem is between the adjacent bands because of the overlapping of the frequency at the borders of the individual channels. For this simple reason, the adjacent (and same frequency or co-channels) are not used on the cells belonging to same site.

Frequency hopping is a technique by which a frequency of the signal is changed with every burst in such a way that there is minimum interference in the network, and at the same time allocated channels are effectively utilised. This process in GSM is also known as SFH or slow frequency hopping. By using SFH, improvement takes place due to two phenomena: frequency diversity and interference diversity.

Every burst has a different frequency, so it will fade in a different way and time. Thus the decorrelation between each burst increases, thereby increasing the efficiency of the coding signal. This is called *frequency diversity*. The assignment of the frequency can be done in two ways: sequentially or randomly. In the former, the system follows a strict pattern of frequency assignment to each burst and in the latter, it assigns frequencies randomly.

If each mobile has one constant frequency, some mobiles may be affected by interference more than others. Thus, with the use of frequency hopping, the interference spreads within the systembeacuse the interfering signal's effect gets reduced. This is called *interference diversity* in averaging mode. As the interference becomes less, the frequency spectrum can be utilised better and so the capacity of the system increases.

Frequency hopping is of two types: base-band FH and RF FH. In the base-band FH, the calls are *hopped* between different TRXs. The number of frequencies that are used for hopping correlated to the number of TRXs and is thus constant. In the RF FH, the call stays on one TRX but the frequency change takes place with every frame. These frequencies are not included in the hopping sequence, thus effectively creating two layers in each cell, one FH and one non-FH. As the RF FH is not correlated to the number of TRXs, it is considered to be more robust and hence used in network deployment frequency planning.

2.4.11   Parameter Planning

The parameters that are used in the radio network are of two types: fixed and measured. These parameters include those related to signalling, Radio Resource Management or RRM, power control, neighbour cells, etc.

2.4.11.1   Signalling

Any flow of data in a network requires some additional information that helps the data to reach the destination in the desired fashion. This additional information is called signalling. Signalling in the GSM network is required at all the interfaces. Radio network planners deal mostly with signalling between the mobile station and base station (shown in Table 2.1). Signalling on all the interfaces except for the air-interface is done at 64 kbps. On the air-interface the signalling can be done either by using the Slow Associated Control Channels (SAACH) or by using the main channel itself wherein the signalling channel is sent instead of sending the data. This is called Fast Associated Control Channel (FAACH) signalling. The former is 'slow' and hence carries non-urgent messages, e.g. information containing handover measurement data, while the latter carries information that is more urgent, such as decisions leading to handover of the mobile. Signalling is also required in the air-interface for sending information about the mobile itself even when it is not on a call. Thus, signalling can be in the dedicated phase, i.e. when TCH have been allocated to the mobile and in the non-dedicated or IDLE mode when the mobile is not on a call but is camped on the network.

When the mobile first tries to get connected to the network, it requires the help of two channels: the Frequency Correction Channel (FCCH) and the Synchronisation Channel (SCH). These channels help the mobile to get synchronised and connected to the network. Each mobile once connected, keeps on receiving information from the base station and this is done through the Broadcast Channel or *BCCH*. Once a call is being initiated, the Paging Channel or PCH helps in transfer of information indicating that a dedicated channel will be allocated to the mobile. The allocation information comes via the Allocation Grant Channel or AGH. If the mobile needs to send the information to the base station, the request is made through the Random Access Channel or RACH. All these channels except RACH are downlink channels.

The above-mentioned channels are logical channels. There are three kinds of physical channels also: the traffic channels for full rate, half rate, and one-eighth rate also known as TCH/F, TCH/H, and TCH/8, respectively. TCH/F transmits the speech code at 13 kbps; TCH/H transmits speech code at 7 kbps. Although TCH/8 is a traffic channel, its rate is very low at almost one-eighth that of the TCH/F, thus its usage has been limited to signalling.

2.4.11.2   Radio Resource and Mobility Management

The management of radio resources, functions related to the mobility such as location update, communication management such as handover and roaming procedure handling, come under Radio Resource Management or RRM. For these management functions to happen, the information flow should take place to achieve the desired results. This information flow (traffic and signalling) takes place via three protocols, known as *link protocols*. Link protocols are from MS-BTS, BTS-BSC and BSC-MSC. LAPDm is present over the MS-BTS connection, LAPD over the BTS-BSC and MTP or Message Transfer Protocol for signalling transport over the SS7 network, as shown in Figure 2.15.

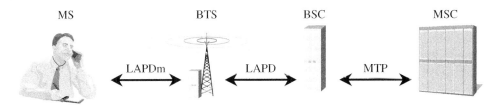

Figure 2.15 Link protocol in a GSM network.

RRM procedures basically relate to the processes taking place during the transition phase between different states of the mobile station, such as the IDLE state, the DEDICATED state, during mobility, during handover, when calls are getting re-established, etc. In the GSM system, there are two states that are defined for a mobile: IDLE or DEDICATED. The management of the radio resources is done for these two states of the mobile station. The radio network controls the access of the resources allocated to the mobile station. Thus, from an IDLE state, the mobile station enters a BTS access mode wherein 'access' is granted to the mobile based on whether or not the mobile station is allowed to 'use' the base station. Then the mobile station enters the DEDICATED mode and starts using the resources till it enters the release mode, i.e. the call ends.

Once a mobile is logged into the network, the transition procedure request always comes from the mobile. This is done through the Random Access Channel or RACH. As the timing of the request coming from a mobile(s) cannot be predetermined, this may cause some problems such as 'collision of call request'. Then factors such as congestion, call-repetition process, traffic increase, etc. come into the picture. The radio planner should keep the parameters such as the number of times the request for a channel can be sent, timing of re-requests which is usually kept 'random' and related effects to keep the throughput under control.

Paging is another function whereby when a request from the mobile reaches the MSC, then the MSC 'pages' the requested subscriber information to all the BSCs within a location area. The BSC in turn sends this information to the BTS to find the subscriber through the Paging and Access Grant Channel or PAGCH. The tasks assigned to the BSC and BTS may vary from network to network.

The subscriber should be able to use the network irrespective of his or her location, i.e. his or her movement should not affect connectivity to the network. The movement can be intra-region, inter-region, or even access to networks on moving to different countries. For this kind of flexibility, the subscription is associated with the concept of a Public Land Mobile Network or PLMN. Any operation network can be said to be a PLMN. There can be one or more than one PLMNs in one country. To give the user the immense flexibility to be connected even when changing between PLMNs (in the same or different countries) means that communication between these PLMNs should take place. The mobiles when entering a different PLMN search for their own home/serving cells. When no service is detected, then it can search in automatic mode or the desired PLMN can be selected manually.

2.4.11.3 Neighbour Cells
While in operation, a mobile is required to make evaluations regarding the quality and level of the neighbour cells. It has to decide which cells are better in terms of coverage and capacity. This is done by taking advantage of the TDMA scheme, whereby the

measurements are made during the uplink transmissions and downlink reception bursts. The evaluation is done with the help of the algorithms that are in the BSS. These algorithms make a decision and convey it to the mobile station on the basis of the measurement data that is sent by the mobile station to the BSS. One important thing to remember here is that every cell has its own identity code, known as the Base Station Identity Code (BSIC). Neighbour cells can have the same BSIC. The mobile in such cases identifies the neighbour cell by the 'colour code'. Since SCH and FCCH play an important role in logging the mobile to the network, these channels transmit this information on the frequency known as the *beacon* frequency. Neighbouring cells may transmit their beacon channels on the same frequency. And, for cases like this, BSIC helps in distinguishing channels of the same frequency.

2.5   Basics of Radio Network Optimisation

Optimisation of the radio network is part of the network planning process. The major task of this process is to monitor, verify and improve the performance of the radio network. It starts simultaneously somewhere near the last phase of the radio network planning, i.e. parameter planning. As the cellular network covers large areas, providing capacity to a large number of people, there are lots of parameters involved that are variable and have to be continuously monitored and corrected in order to maintain the coverage, capacity and the quality of the network. Apart from this, the network is always in the growth phase, i.e. always increasing subscriber number, traffic increase, creation of new shopping centres means that capacity requirement at these 'hot-spots' increases, etc. This means that the optimisation process should be conducted in the network at regular intervals, thus increasing the efficiency of the network leading to revenue generation from the network.

Radio network planners focus on three main areas: coverage, capacity, and frequency planning. Based on these three, they proceed to the rest of the tasks such as site selection, parameter planning, etc. In the optimisation process, focus is on the same aspects: coverage, capacity, and frequency planning. The only difference now is that sites are already selected and antenna locations are fixed but the subscriber is as mobile as ever, with continuous growth taking place. Optimisation tasks become more and more tedious with the increase in time gap between the launch and the radio network optimisation process.

Once a radio network is designed, a study of the performance of the network is done. This is done by monitoring the network and then comparing it against the chosen key performance indicators. After fine-tuning, the result (parameters) are then applied to the network to get the desired performance. Optimisation can be considered to be a separate process or as a part of the network planning process, as shown in Figure 2.16.

The main focus of the radio network optimisation is on areas such as power control, quality, handovers, subscriber traffic, and resource availability (and access) measurements.

2.5.1   Key Performance Indicators

For radio network optimisation (or for that matter any other network optimisation), it is necessary that the key performance indicators or KPIs be in place. These KPIs are parameters that are to be observed closely when the network monitoring process is

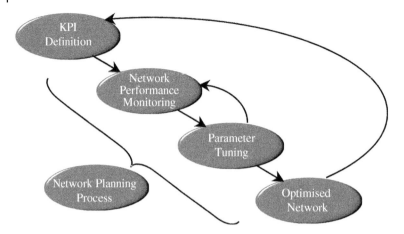

Figure 2.16 Radio network optimisation.

going on. Mainly, the term KPI is used for parameters related to voice and data channels. Network performance can be broadly characterised into coverage, capacity and quality criteria that cover the speech and data aspects also.

2.5.1.1 Key Performance Indicators for the Radio Network: Voice Quality

The performance of the radio network is measured in terms of KPIs related to voice quality, based on the statistics generated from the radio network. Drive tests and network management systems are the best source for generating the performance statistics. The most important of these from an operator's perspective are the Bit Error Rate (BER), Frame Error Rate (FER) and the Dropped Call Rate (DCR).

The BER is the measurement of the received signal bits before the decoding takes place, while the FER is a performance indicator after the incoming signal has been decoded. Correlation between the BER and FER is dependent upon various factors such as the channel coding schemes or the frequency hopping techniques. As the speech quality variation with FER is quite uniform, FER is generally used as the quality performance indicator for speech. The FER can be measured by using the statistics obtained by performing the drive test. Drive testing can generate both the uplink and the downlink FER.

The Dropped Call Rate or DCR as the name suggests is the measure of the calls dropped in the network. A dropped call can be defined as a call that gets terminated on its own after getting established. As the DCR gives a quick overview of the network quality and revenues lost, this easily makes it one of the most important parameters in the network optimisation. Both the drive test results and the NMS statistics are used to evaluate this parameter. At the frame level, the DCR is measured against the SACCH frame. If the SACCH frame is not received, then it is considered to be a dropped call. There is some relation between the number of dropped calls and voice quality. If the voice quality were not a limiting factor, perhaps the DCR would be very low in the network. Calls can drop in the network due to quality degradation, which may be due to many factors such as capacity limitations, interference, unfavourable propagation conditions, blocking, etc. DCR is related to the call success rate (CSR) and the handover

(HO) success rate. CSR indicates the number of calls that actually were completed after being generated while the HO rate gives the quality of the mobility management/RRM in the radio network.

The KPIs can be subdivided according to the areas of functioning, e.g. area level, cell level (including the adjacent level), TRX level. Area level KPIs can include SDCCH requests, dropped SDCCH total, dropped SDCCH A_{bis} failure, outgoing MSC control HO attempts, outgoing BSC control HO attempts, intra-cell HO attempts, etc. Cell level KPIs may include SDCCH traffic BH (av.), SDCCH blocking BH (av.), dropped SDCCH total and distribution per cause, UL quality/level distribution, DL quality/level distribution, etc. The TRX level includes UL quality distribution and DL quality distribution.

2.5.2 Network Performance Monitoring

The whole process of network performance monitoring consists of two steps: monitoring the performance of the key parameters and assessment of the performances of these parameters with respect to capacity and coverage. For the first step, the radio planners assimilate the information/parameters that they need to monitor. The KPIs are collected along with the field measurements such as drive tests. Tools are also used for this purpose. For the field measurements, the tools used are the ones that can analyse the traffic, capacity, and quality of the calls, and the network as a whole. For drive testing, a mobile is also used for this purpose that is known as a 'test mobile'. This mobile keeps on making calls in a moving vehicle that goes around in the various parts of the network. Based on the DCR, CSR, HO, etc., parameters, the quality of the network can be analysed. Apart from drive testing, the measurements can also be generated by the network management system. And finally, when the 'faulty' parameters have been identified and correct values are determined, the radio planner puts them in the network planning tool to analyse the change before these parameters are actually changed/ implemented in the field.

2.5.2.1 Drive Test

Drive test data constitutes the most significant data for radio network optimisation because it is the result from the field. The quality of the network is determined by the satisfaction of the ultimate user of the network, i.e. subscriber, and drive tests give the 'feel' of the designed network as they gives the results from the field. The drive test process starts with the selection of the 'live' region of the network where the tests need to be performed, and the drive test path. Before starting the drive tests, the engineer should have drive test kits that include mobile equipment (usually three mobiles), drive testing software (in a laptop), and a GPS (global positioning system) unit.

When the drive testing starts, two mobiles are used to generate calls with a gap of a few seconds (usually 15–20 s). The third mobile is usually used for testing the coverage. It makes one continuous call and if this call drops, it will attempt another call. The purpose of this testing to collect enough samples at a reasonable speed and in a reasonable time. If there are lots of dropped calls, the problem is analysed to find a solution for it and to propose changes. One such example of a drive test is shown in Figure 2.17.

Some typical drive tests results giving the received power levels from own cell and neighbour cells, FER, BER, MS power control, etc., are shown below in Figure 2.18.

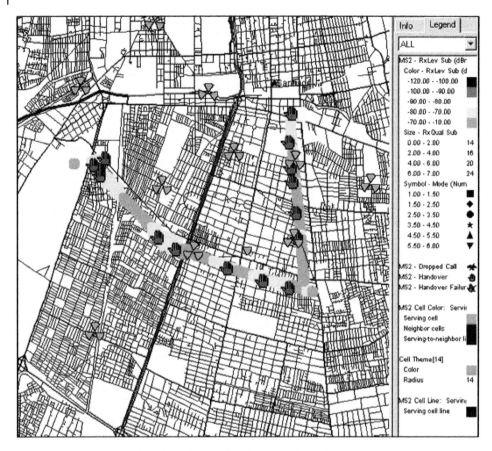

Figure 2.17 Drive test result analysis showing handovers on the path.

2.5.2.2 Network Management System Statistics

After the launch of the network, drive tests are performed periodically. But, the statistics are monitored on the NMS daily with the help of counters. The NMS usually measures the functionalities such as call set-up failures, dropped calls, handovers (success and failures). It also gives data related to traffic and blocking in the radio network, apart from giving data related to the quality issues such as frequency hopping, FER and BER in radio networks, field strength, etc. One such example of area level KPI statistics is shown in Figure 2.19.

2.5.3 Network Performance Assessment

The network performance can be assessed based on the performance of the key parameters with respect to coverage, capacity and quality. The performance indicators are:

- Amount of traffic and blocking
- Resource availability and access measurements
- Handovers (same cell/adjacent cell, success and failure) measurements
- Receiver level and quality measurements
- Power control measurements

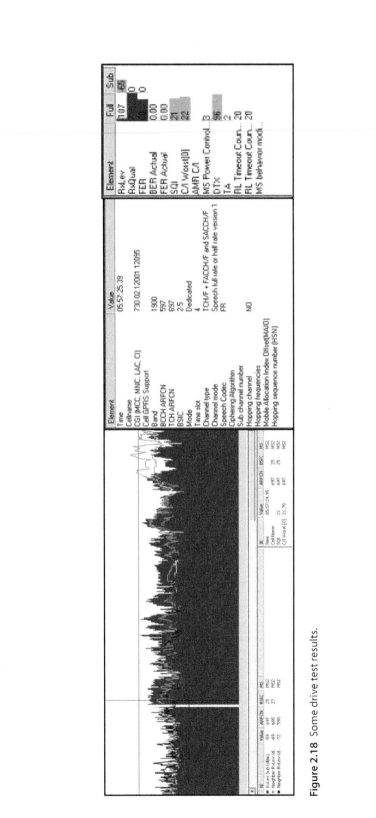

Figure 2.18 Some drive test results.

SDCCH

SDCCH availability ..../ava_4   81.30 %

Dynamic SDCCH allocation (TCH reconfigured to SDCCH):

. Attempts ..../c1154   0

SDCCH requests ..../c1000   6617

. HO in../c1006   0   (0.00 %)

. blocked ..../blck _15 (blck_5a)   33   (0.50 %)

. To FACCH call setup ..../c1099   0   (0.00 %)

. LU ..../c3019   4102   (61.99 %)

. MTC (incl. SMS) ..../c3012   145   (2.19 %)

. MOC (incl. SMS,SS) ..../c3013   343   (5.18 %)

. supplementary service request (S9) ..../c3044   4   (0.06 %)

. IMSI detach (S7) ..../c3033   114   (1.72 %)

. call re-establishment......................../c3020   0   (0.00 %)

. emergency call ..../c3021   255   (3.85 %)

. other (fails, ghosts) ..../sd_1a   1625   (24.56 %)

SDCCH usage ..../trf_7b   0.02 %

Average SDCCH seizure length ..../trf_4   2.05 sec

SDCCH failures

SDCCH seizures ..../trf_54   6584

SDCCH drop ratio ..../sdr_1a   27.70 %

. SDCCH_RADIO_FAIL............................/c1003   166   (2.52 %)

. SDCCH_A_IF_FAIL_CALL......................./c1078   0   (0.00 %)

. SDCCH_ABIS_FAIL_CALL........................./c1075   1658   (25.18 %)

. SDCCH_USER_ACT............................./c1037   0   (0.00 %)

. SDCCH_BCSU_RESET........................../c1038   0   (0.00 %)

. SDCCH_NETW_ACT............................/c1039   0   (0.00 %)

. SDCCH_BTS_FAIL............................./c1036   0   (0.00 %)

. SDCCH_LAPD_FAIL.........................../c1035   0   (0.00 %)

. SDCCH_RF_OLD_HO........(HO drop)............/c1004   0   (0.00 %)

. SDCCH_ABIS_FAIL_OLD....(HO drop)............/c1076   0   (0.00 %)

. SDCCH_A_IF_FAIL_OLD....(HO drop)............/c1079   0   (0.00 %)

Figure 2.19 Example of area level KPI statistics from the NMS.

2.5.3.1 Coverage

Drive test results give the penetration level of signals in different regions of the network. These results can then be compared with the network plans made before the network launch. In the urban areas, coverage is generally found to be less at the far-end of the network, in the areas behind high building and inside buildings. These issues become more serious when important areas and buildings are not having the desired level of signal even when care has been taken during the network-planning phase. This leads to an immediate scrutiny of the antenna locations, heights and tilt. The problems are usually sorted out by moving the antenna locations and the tilting of the antennas. If optimisation is being done after a long time, new sites can also be added.

Coverage also becomes critical in rural areas, where the capacity of the cell sites is already low. The populated areas and the highways usually form the region that should have the desired level of coverage. A factor that may hinder the signal level could be propagation conditions, so study of link budget calculations along with the terrain profile becomes a critical part of the rural optimisation. For the highway coverage, addition of new sites may be one of the solutions.

2.5.3.2 Capacity

The data collected from the network management system is usually used for assessment of the capacity of the network. As coverage and capacity are interrelated, data collected from the drive tests is also used for capacity assessment. The two aspects of this assessment are drop calls and congestion. Generally, capacity-related problems arise when the network optimisation is taking place after a long period of time. Radio network optimisation also includes providing capacity to the new hot spots or indoor coverage planning aspects. Once the regional/area coverage is planned and executed in the normal planning phase, the optimisation should take into consideration providing as much coverage as possible to the places that would expect high traffic such as inside office buildings, inside shopping centres, tunnels, etc.

2.5.3.3 Quality

The quality of the radio network is dependent upon its coverage, capacity and frequency allocation. Most of the severe problems in a radio network can be attributed to interference. For uplink quality, BER statistics are used and for downlink, FER statistics are used. Once it is found that interference exists in the network, the source of the interference needs to be found (i.e. if the source is internal or external). The entire frequency plan is checked again to verify if the source is internal or external. The problems may be caused by inaccuracy in the frequency plans, configuration plans (e.g. antenna tilts), inaccurate correction factors used in propagation models, etc.

2.5.3.4 Parameter Tuning

The end of the assessment process sees the beginning of the tedious and complex process of fine-tuning of parameters. The main parameters that are fine-tuned are signalling parameters, radio resource parameters, handover parameters and power control parameters. The concepts that are discussed in the radio planning process and the KPI values should be achieved after the process is complete. The major complexity of this process is the non-homogeneity of the radio networks. The radio network planning takes into account the standard propagation models and corrections are made based on trial and error methods that may be valid for some parts of the network and invalid for others. During the network deployment, some more measurements are done and the parameters are fine tuned again. Once the network goes 'live', the drive test and NMS statistics help in further fine-tuning of the parameters. It is at this point that a set of default parameters is created for the whole network. These parameters are decided based on the analysis done in some critical parts of the network (by drive testing and NMS). However, as the network is non-homogeneous, these default parameters may not be so accurate in the other regions, thereby bringing down the overall quality leading to a reduction in revenue inflow for the network operator. For this one main reason, the radio network optimisation is a continuous process in the network that begins during the pre-launch phase and continues throughout the existence of the networks.

2.6 GPRS Network Planning and Optimisation

GSM can provide a data rate of 9.6 kbps on a single timeslot. With the advent of High Speed Circuit Switched Data or HSCSD, the capability of the network increased eight-fold, i.e. 115.2 kbps. But in practice, it was only 64 kbps due to the limitation of the A- interface and the core network. The main benefit in the implementation of HSCSD was that with limited upgrade, i.e. with minimum investment, the capacity of data transfer was increased to up to four TSs on the receiving side and two TSs on the transmitting. However, the nature of traffic was still circuit switched, which meant that the access time to the network was long. As charging is proportional to the logging time, this means that the subscriber ends up paying more. This led to evolution to the packet switched network. Using this technology, the access time to the network reduces and charging is done solely on the usage of the network, i.e. even when the connection is there and not being used, the subscriber would not be charged. The usage of the network resources becomes more dynamic and efficient. They are no longer reserved for a user logged to the network, even when the user is not using the resources. This system was called a General Packet Radio System or GPRS.

GPRS is an addition to the existing GSM system, enabling packet switched transmission in the network keeping the existing value-added services like SMS, etc., in the network. Due to this, data rates also increase substantially. For the very reason behind the development of this technology, the user now can log in to the GPRS network, make use of all eight TSs dynamically and get charged only when using the resources. The packet data can also be sent in the idle times, i.e. between speech calls, making effective use of the network resources and saving money for the subscriber.

2.6.1 GPRS System

GPRS technology is an addition to the existing GSM technology. Hence, there are a few additions to the GSM system for transformation to a GPRS system. As the packet switching is now there, the new network elements are the ones that are capable of performing packet switching. The main ones include the Serving GPRS Support Node or SGSN and Gateway GPRS Support Node or GGSN.

The present GSM system design is oriented towards providing a voice service. So, apart from the addition of new elements such as SGSN and GGSN, there are a few minor changes required in the GSM network elements in the BSS and HLR. These are both hardware and software related changes. These minor changes are due to the higher-level coding schemes that are used in GPRS technology. The most important change is the addition of the PCU (Packet Control Unit) at the base station controller. The GPRS system with all these elements looks the same as the GSM except with the addition of the packet handling core part, as shown in Figure 2.20.

   As all the network elements of the GSM have already been explained, only new elements, such as SGSN, GGSN and PCU, etc., will be discussed here. But first let us see the changes in the mobile station.

2.6.1.1   GPRS Mobile Station

The fundamental difference between a GSM mobile and GPRS mobile (shown in Figure 2.21) is that the GPRS mobile is able to handle the packet data at a higher speed.

   The GPRS mobile stations have been classified into three classes, A, B and C, based on their ability to handle the cellular networks. Class A mobiles are connected to both the GSM and GPRS networks and can use them simultaneously. Class B mobiles are

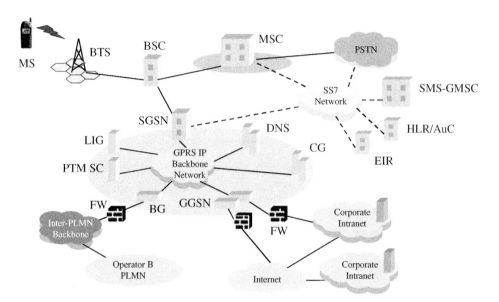

Figure 2.20 GPRS network.

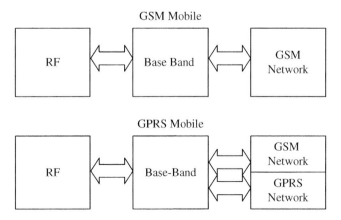

Figure 2.21 GSM and GPRS mobile stations.

connected to both the networks, but can use only one at one time. Class C mobiles can be connected to either one of the networks.

2.6.1.2   Serving GPRS Support Node (SGSN)

SGSN is the most important element in the GPRS network. It is the service access point for the mobile station. Its main functions include mobility management and registration and authentication. It also interacts with mobile packet data flow and functions related to it like compression and ciphering. These are handled by protocols such as the SNDCP (Sub-Network Dependent Convergence Protocol) and LLC (Logical Link Control). SGSN is also responsible for the GTP tunnelling to other GSNs.

2.6.1.3   GPRS Gateway Support Node (GGSN)

GGSN is connected to the SGSN on the network side and to the outside world external networks such as the Internet and X.25. As it is a gateway to external networks, its main function is to act as a 'wall' for these external networks in order to protect the GPRS network. When data comes from the external network, after verification of the address, the data is forwarded to the SGSN. If the address is found invalid, data is discarded. On the other hand, the SGSN also routes the packets it receives from the mobile to the correct network. Thus, for outside networks, the SGSN acts as a router.

2.6.1.4   Border Gateway (BG)

This interconnects different GPRS operators' backbones, thereby facilitating the roaming feature. It is based on standard IP router technology.

2.6.1.5   Legal Interception Gateway (LIG)

This performs a 'legal' functions in the network. Subscriber data and signalling can be intercepted by using this gateway, enabling the authorities to track criminal activities. LIG is required when launching the GPRS service.

2.6.1.6   Domain Name System (DNS)

The DNS does the translation of the IP host names to the IP addresses, thereby making the IP network configuration easier. In the GPRS backbone SGSN uses DNS to obtain GGSN and SGSN IP addresses.

2.6.1.7   Packet Control Unit (PCU)

This is a new card that is implanted in the BSC to manage the GPRS traffic. The PCU has limitations in terms of the number of transceivers and base stations it can manage, thereby creating a bottleneck for the network design usually in terms of capacity. Increase in the capacity of the network would lead to increase in the PCU capacity, thereby increasing the hardware costs of the network.

2.6.2   Interfaces in a GPRS Network

Due to the addition of some network elements, some new interfaces are added in the GPRS network. All these new interfaces are called G interfaces as shown in Figure 2.22. A brief description of these interfaces is given below.

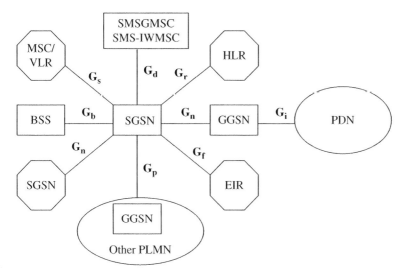

Figure 2.22 GPRS interfaces.

- Gb interface: This lies between the BSS and SGSN. It carries the traffic and signalling information between the BSS (of GSM) and the GPRS network, thus easily making it the most important interface in network planning.
- Gn interface: This is present between the SGSN and SGSN/GGSN of the same network. This provides data and signalling for intra-system functioning,
- Gd interface: This is present between the SMS-GSMC/SMS-IWMSC and SGSN, providing a better use of the SMS services.
- Gp interface: This is the interface between the SGSN and the GGSN of other PLMNs. Thus, it is an interface between the two GPRS networks. This interface is highly important considering its strategic location and functions that include security and routing.
- Gs interface: This is present between the SGSN and MSC/VLR. Location data handling and paging requests through the MSC are handled via this interface.
- Gr interface: As this is an interface between the SGSN and HLR, all the subscriber information can be assessed by the SGSN from the HLR.
- Gf interface: This interface gives the SGSN the equipment information that is present in the EIR.
- Gi interface: This is an interface between the GGSN and external networks. This is not a standard interface, as the specification would depend upon the type of interface that would be connected to the GPRS network.

2.6.3 Protocol Structure in a GPRS Network

The protocol structure in GPRS is quite different from that in the GSM as the interworking between different network elements takes place on the basis of this protocol. The protocol structure is shown in Figure 2.23.

2.6.3.1 MS Protocols

There are two kinds of physical layers that have been defined in the GSM 05 specifications. They are the *physical RF layer* and the *physical link layer*. The physical RF layer performs modulation/demodulation of the input signal (information) apart from FEC

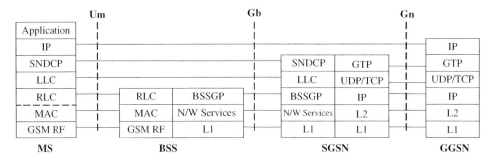

Figure 2.23 GPRS protocol stack.

coding, interleaving, and congestion detection, while the physical link layer performs the function of providing services that are required to transfer this information over the air-interface. These include coding, framing, synchronisation, monitoring of radio quality, power control procedures and transmission error detection and correction. This layer can support multiple mobile stations that share the same physical channel.

The *RLC/MAC layer* is present over the physical RF layer. RLC or Radio Link Control is responsible for data transmission over the physical layer of the GPRS radio interface. Also, it provides a link to the upper layers. The MAC or Medium Access Control layer, as the name suggests, is responsible for the control of the access functions of the mobile over the air-interface. These functions include the allocation of channels and multiplexing of the resources.

The *LLC layer* or Logical Link Control layer is present over the RLC layer and, as the name suggests, is responsible for the creation of a logical link between the mobile and the SGSN. Thus, it is responsible for the transfer of the signalling and data transfer. It is independent of the lower layers.

The *SNDCP layer* or Sub-Network Dependent Convergence Protocol performs mapping and compression functionalities between the network layers and lower layers. The compression of both the control and data information is done by this layer. It also performs the segmentation/desegmentation of the information to/from the lower LLC layer.

The above-mentioned are responsible for information flow between the MS and SGSN. The IP layer is responsible for the formation of the backbone of the GPRS network and a direct interface between the MS and GGSN.

2.6.3.2 BSS Protocols

The BSSGP or Base Station Subsystem GPRS Protocol layer is responsible for the information transfer between the SGSN and the RLC/MAC layer. Thus, its primary function is to create an environment for the data flow between these two entities.

2.6.3.3 SGSN Protocols

The *GTP* or GPRS Tunnelling Protocol runs over the UDP layer and is responsible for the data and signalling information between the BSC and the GSN nodes. This protocol forms a tunnel for each subscriber and each tunnel is identified by a tunnel endpoint identifier.

The *UDP/TCP* or User Data Protocol/Transport Control Protocol is the backbone network protocol that is used for routing the network data and control signalling. While the TCP is a connection- oriented protocol providing reliable data transmission service, UDP is a connectionless protocol providing data transmission services that are unreliable.

The *IP* or Internet Protocol is the datagram-oriented protocol. Both UDP and TCP interface with the IP directly. The major function of the IP is as a routing protocol that provides the means for devices to discover the topology of the network as well as detect changes of state in nodes, lines and hosts.

2.7 Network Planning in a GPRS Network

The main difference between the GSM and the GPRS network is the addition of the packet data handling capability of the GPRS. All the changes in the network planning are due to this additional aspect of the GPRS network. Radio network planning and core network planning are most affected by the change compared with the transmission network planning that experiences minimal changes.

2.7.1 GPRS Radio Network Planning

The radio part of the network planning process remains the same as for the GSM described previously. Some of the aspects change due to the introduction of packet data and planning engineers need to take them into account. This would mainly affect the dimensioning, detailed planning which directly affects the coverage and capacity planning leading to an impact on the radio network quality. In GSM, radio network quality meant voice quality but in GPRS network quality would include both voice and data quality. This factor would lead to changes in the KPIs and hence some new aspects to radio network planning. In the coming sections, only the areas that would need to change with respect to the radio planning in GSM are explained.

2.7.1.1 Basics of Radio Network Planning in GPRS Networks
There are some concepts that the network planning engineer needs to know before planning a GPRS radio network. These fundamental concepts include the following:

- Logical channels in GPRS
- Coding schemes
- Management: RRM and MM (Radio Resource Management and Mobility Management)
- Power control
- Resource allocation

2.7.1.2 Logical Channels in a GPRS Network
Due to the change in the scope of the functionality of the network, i.e. involvement of the data packets, some new channels are used. We already have seen the logical channels used in GSM, now we look into the logical channels that are used in GPRS. A separate set of channels is allocated for the packet data allowing more flexibility in the

Table 2.4 GPRS logical channels.

Channel	Abbreviation	Function/application
Packet broadcast control channel (DL)	PBCCH	Broadcast system information specific to packet data
Packet common control channel	PCCCH	Contains logical channels for common control signalling
Packet data traffic channel	PDTCH	Channel temporarily used for data transfer
Packet associated control channel	PACCH	Used for signalling information transfer for a given mobile
Packet access grant channel (DL)	PAGCH	Notifies that mobile about resource assignment before actual packet transfer
Packet notification channel (DL)	PNCH	Used for sending information to multiple mobile stations
Packet paging channel (DL)	PPCH	Pages a mobile station before packet transfer process begins
Packet random access channel (UL)	PRACH	Used by the mobile station for initialisation of the uplink packet transfer

Table 2.5 Coding scheme in GPRS.

Coding Scheme	Code rate	Data rates (kbs)	Data rates (kbs) (excl. Headers: RLC/MAC)
CS-1	1/2	9.05	8
CS-2	~ 2/3	13.4	12
CS-3	~3/4	15.6	14.4
CS-4	1	21.4	20

signalling. These channels are illustrated in Table 2.4 and are mapped into physical channels, i.e. PDCH (Packet Data Channel). Different logical channels can find their place on one single physical channel.

2.7.1.3 Coding Schemes

The radio block consists of header, data, and control information, which are basically the MAC header, RLC data block and MAC/RLC control information. RLC data is encoded for protection purposes. In GPRS systems, there are four coding schemes that are used for the packet data CS-1, CS-2, CS-3, and CS-4. Table 2.5 shows the coding schemes and related parameters.

In coding Scheme CS-1, half-rate convolution code is used for forward error correction (FEC). It has a data rate of 9.05 kbps. Coding Schemes CS-2 and CS-3 are the same as CS-1 but in punctured format. Puncturing is done so as to increase the data rate but it comes at the cost of reduction in redundancy. Coding Scheme CS-4 has 'no coding', i.e. no FEC, further increasing the data rates to 21.4 kbps.

2.7.2 Mobility and Radio Resource Management

In the GSM system, IDLE and DEDICATED are the two states of the mobile station. In the GPRS system, there are three states: IDE, STANDBY, and READY. In IDLE mode, the subscriber is not attached to the GPRS network, while in the STANDBY mode, the GPRS network knows the routing area location of the mobile station. The mobile station enters the READY state by sending a service request to the network. In this state, the mobile station gets attached to the GPRS mobility management and is known by the network on the call basis. Once the call is finished, the mobile enters the STANDBY mode again. Thus, when moving from the STANDBY to READY state, SGSN receives and processes *GPRS attach request* and for READY to STANDBY movement, SGSN receives and processes *GPRS detach request*.

For the movement of data between the MS and GPRS network, the PDP (Packet Data Protocol) context needs to be activated. Either the MS or the GPRS network can generate the PDP context. These requests are sent and processed by the SGSN. A PDP context generally contains the activated information such as addresses (required for data traffic), QoS related information, protocol types, etc. The MS uses the PDP contexts in the STANDBY and READY states. The number of PDP contexts is one of the factors that have a direct impact on the SGSN capacity.

Radio time slot allocation also changes in the GPRS air-interface. It becomes more dynamic. A GPRS mobile is capable of using the network both for the voice (CS) and data (CS and PS). GPRS traffic is managed by the BSC as it does the allocation of resources for the CS and PS data. The timeslots that handle the CS traffic fall under the *CS territory* and the timeslots that handle PS traffic fall under the *GPRS territory*. Some of the timeslots are in DEDICATED mode and some in default. Each group of timeslots is known as a *territory*. Consider a base station with two transceivers TRX1 and TRX2. There are eight radio timeslots in each of the two transceivers. In TRX1, two timeslots are allocated to the signalling, i.e. BCCH and PBCCH. Of the remaining six, three fall under dedicated territory and three under default territory. The timeslots that are to be only used for the packet data and cannot be used for the CS traffic are known as dedicated timeslots and fall under dedicated territory. The remaining three timeslots can be used both for voice and data and fall under default territory.

The eight timeslots that are used for the voice traffic come under the CS territory. Consider a case when all eight timeslots in TRX2 are occupied and there is a ninth call, then one timeslot from the default territory is assigned to the CS traffic. Once the traffic in TRX2 decreases, the call on default territory would be switched back to CS territory leaving the default timeslots for the possible PS traffic.

2.7.3 Power Control

Power control in GPRS networks is more complicated because of the addition of PS traffic. Power control takes place in both the uplink and downlink directions. The UL power control is used to reduce the interference and helps in increasing the battery life of the mobile station. In the downlink direction, the power control feature is used to control the output power of the base station and subsequently help in decreasing the interference in the network. As in the GSM system, the mobile station performs the measurements and based on these measurements the output power of the base station is controlled.

2.7.4 Concept of Temporary Block Flow

A new concept is introduced in the GPRS network called Temporary Block Flow or TBF. The network and mobile establish a connection for data flow. This connection is unidirectional in nature and is maintained for the duration of the call. As it is established for the packet data (i.e. blocks) and is not permanent (i.e. temporary), it is called Temporary Block Flow. This can be uplink, downlink or simultaneously uplink and downlink.

2.8 GPRS Detailed Radio Network Planning

The planning process for the GPRS radio network would be similar to that for the GSM radio network. Both the GSM and GPRS radio networks have the same radio wave propagation principles. While the system planning remains the same for the GPRS radio system, the one important consideration would be the addition of data traffic. Pre-planning and nominal planning steps such as site-survey and site selection remain the same as for GSM as explained previously. Detailed planning of the GPRS network would be different from that of the GSM network due to the addition of data traffic and the new equipment added to the network for this. The detailed radio network planning would again be focusing on the same aspects as that of the GSM radio network, i.e. coverage, capacity, frequency and parameter planning. Thus, detailed planning can be subdivided as:

1) Coverage planning (including the link budgeting)
2) Capacity planning
3) Frequency planning
4) Parameter planning

2.8.1 Coverage Planning

The two important aspects on which coverage in a GPRS network depends on are the SNR and the data transmission rates. Interference can be a limiting factor for the maximum amount of data rate in the GPRS network. Each of the coding schemes (e.g. CS-1, CS-2, etc.) works for a certain range of C/I ratio for a given value of Block Error Rate (BLER). The coverage plans are made with the objective of providing a balanced link budget for both the uplink and downlink directions. The link budget in GPRS radio network is similar to that of the GSM radio network. However, the threshold requirements change in GPRS networks, thereby giving a different value of the covered area by a cell compared with that of the GSM network. The two main parameters that are required for link budget are the transmitted power (from mobile station and base station) and the receiver sensitivity. As the coding schemes in the GPRS radio network have different SNR requirements, hence the area covered would be different. Coding Scheme CS-1 covers a larger area compared with Coding Scheme CS-4. One important change that is seen in the link budget calculations is the removal of body loss value for CS-2 thereby giving the GPRS service a 3 dB advantage compared with GSM services. The SNR requirements for CS-3 and CS-4 are quite high, thereby reducing the area covered by them. CS-1 and CS-2 are usually used for the GPRS radio network coverage planning while CS-3 and CS-4 are used for the call centre. As the coverage in the GPRS network is interference limited (rather than noise limited), the C/I ratio distribution would become a detrimental factor in coverage area predictions.

2.8.2 Capacity Planning

The capacity planning of a GPRS network may be subdivided into two parts: capacity planning for the radio interface and capacity planning for the G_b interface. In this section, we deal with the former.

The GPRS network has three kinds of traffic: voice, CS-data and PS-data. All of these have to be considered when doing capacity planning for the radio interface. CS traffic always has a higher priority over the PS traffic. But, due to the delay sensitive nature of some PS services, some timeslots are *dedicated* to carry PS traffic only. CS-traffic calculations are the same as that of the GSM radio interface that predominantly includes the Erlang B tables, blocking, and C/I thresholds. Assume the case shown in Figure 2.24. There is one cell that has two TRXs. In ideal conditions (i.e. without blocking), 14 (voice) users can use the timeslots continuously, i.e. a traffic of 14 Erl would be generated, if there is no blocking. If the number of voice users is reduced to eight, then the remaining six timeslots could be used for data. It should be however noted that data, which is not delay sensitive, could always be sent through the *gaps* in the air-interface. Only the data that is delay sensitive would need uninterrupted availability of timeslots. When an existing GSM network is upgraded to a GPRS network, the available capacity would fall short for the PS-data. In such cases, increasing the number of TRXs and the timeslots in GPRS territory (dedicated + default) would be one effective way to tackle the capacity problems. Quality of service (QoS) has a deep impact on the capacity planning in GPRS networks. An increased load would decrease the quality of the call. For critical applications, minimum QoS should be met which means that the load increase can take place only until minimum QoS is met. Thus, frequency planning is important in achieving a desired QoS level in a GPRS radio network.

2.8.3 Frequency Planning

Coverage and capacity planning go hand-in-hand. Coverage planning is related to frequency planning. An effective frequency plan would increase the coverage areas significantly. We have seen before that in the GPRS network, coverage is interference limited rather than noise limited. There are two methods to reduce interference: frequency planning and power control.

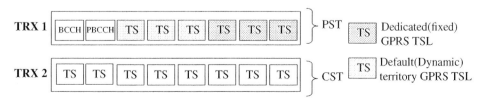

Figure 2.24 GPRS timeslot allocation to CS and PS traffic.

The principles and methods of frequency re-use (the same as that used in GSM radio networks, e.g. frequency hopping) are used extensively so that the spectrum is used effectively. Power control is more necessary in the downlink direction. Using the BCCH layer provides this. The BCCH layer has an important characteristic, the burst transmission in the DL is constant and has full power. This means that the variation in the throughput is due to user multiplexing over the same timeslot, thereby making the timeslot capacity constant (and independent of the GPRS traffic load). Timeslot capacity is also interference limited. With increase of traffic, the number of users per timeslot decreases due to the increase in interference level. Thus, in the GPRS network, the interference reduction becomes an important aspect of the whole network-planning scenario.

2.8.4 Parameter Planning

Parameter planning in the GPRS network can be considered to be an extension of the GSM parameter planning. The parameters still remain the same, i.e. signalling, RRM, power control and handover, etc. However, packet data related parameters are added to the original GSM parameters.

The major enhancements are the signalling parameters. As seen in the Table 2.1, there is whole list of parameters associated with the signalling of packet data transfer. One important parameter that is to be planned is whether or not the GPRS traffic goes on the BCCH timeslot or not. Then there are parameters that are related to defining the GPRS territory. Lastly, there are parameters related to the routing and location area codes that need to be planned, just to ensure enough capacity is available for paging.

2.9 GPRS Radio Network Optimisation

As the GPRS traffic may increase tremendously, the existing GPRS network may not be able to cater for this increased volume of traffic with the required QoS and would need optimisation. Steps in the optimisation process of the GPRS network would remain the same as those for the GSM network, i.e. KPI definitions, network performance monitoring, and parameter tuning. However, GPRS network optimisation is more complicated than GSM network optimisation. This is due to that fact that the GPRS network is an extension of GSM network optimisation, i.e. it uses resources such as frequency of the existing GSM network.

Thus, the optimisation process in a GPRS network would focus on improvement in the following areas:

- Accessibility to the network: The mobile subscriber should be able to get the desired service when it is requested.
- Quality: Once the user gets access to the requested service, it should come with the desired quality.
- Utilisation of the resources: On the air, A_{bis} and Gb interface.
- Improvement in both the CS and PS quality levels along with improvement in coverage and capacity.
- Improvement in the security aspects of the network.

The whole of the optimisation process can be subdivided into three main parts: radio network optimisation, transmission network optimisation and core network optimisation. Only radio network and packet core network optimisation processes are covered in the following sections as transmission network and CS-core network optimisation processes are covered in Chapters 3 and 4.

2.10   EDGE Network Planning and Optimisation

GPRS networks are able to achieve higher bit rates than GSM networks but still fall short of data rates that could make existing GSM networks deliver services that are still of a higher speed, something comparable with that of the characteristics of the third-generation networks. The delay in the third-generation system deployment also added to a faster emergence of a technology known as EDGE or Enhanced Data rates for GSM Evolution that was capable of giving a similar kind of service to that of third-generation networks yet with implementation on the existing networks, i.e. second generation (e.g. GSM).

EDGE or Enhanced Data rates for GSM Evolution, as the name suggests, indicates that the existing GSM network becomes capable of having higher data rates. The enhancement from GSM was GPRS, i.e. voice and packet, while enhancement of GPRS led to EDGE networks as shown in Figure 2.25. In EDGE networks, the fundamental concept remains the same, i.e. voice, CS data and PS data being carried in the network with the network architecture remaining the same as in the GPRS network. Enhancement of HSCSD is known as ECSD (Enhanced Circuit Switched Data), while enhancement of GPRS is known as EGPRS.

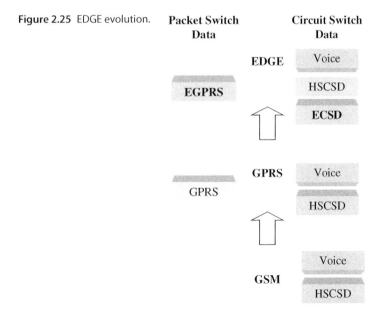

Figure 2.25 EDGE evolution.

2.10.1   EGPRS and ECSD

EGPRS is enhancement of the GPRS network, while ECSD is enhancement of the HSCSD network. EGPRS implementation would lead to a major effect on the protocol structure, e.g. on layer 1 or 2. The modulation and coding schemes are quite different in EGPRS compared with GPRS (this is explained later in the chapter). In ECSD, though the user data rates do not go beyond 64 kbps, less timeslots are required to achieve this compared with HSCSD. Also, the architecture of ECSD is based on HSCSD transmission and signalling, thus having minimal impact on the existing specifications.

2.10.2   EDGE System

As shown in Figure 2.26, the EDGE system is quite similar to the GPRS system. When some changes and modifications are made to this network, it becomes capable of handling higher data rates. The most important change is the new modulation scheme. In GSM and GPRS, the GMSK modulation scheme was used. In GMSK modulation, only one bit per symbol was used. However, in the EDGE network, the octagonal phase shift keying or 8-PSK modulation scheme is used which allows a three times higher gross data rate of 59.2 kbps per radio time slot by transmitting three bits per symbol with the existing symbol rate. GMSK is a constant amplitude modulation while 8-PSK has variations in amplitude. This variation in amplitude changes the radio performance characteristics making hardware changes in the base stations mandatory.

2.11   EDGE Radio Network Planning Process

The radio network planning process is similar to the one used for the GSM and GPRS networks. However, due to minor hardware and software changes in the existing network that lead to major changes in the network performance, the network planning parameters change quite a bit. Due to increased bit rates, transmission planning undergoes a major change with the introduction of the dynamic A_{bis} concept. However, core network planning remains almost the same as in the GPRS network.

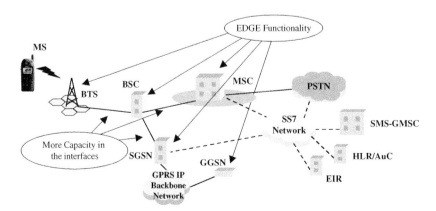

Figure 2.26 EDGE network (simplified).

2.11.1 EDGE Radio Network Planning

The major change in the radio network planning in the EDGE network is due to the introduction of the new modulation scheme, i.e. 8-PSK and new coding schemes.

2.11.1.1 Coding Schemes

The EDGE system is an enhancement to the existing GPRS system. There are nine modulation and coding schemes (MCS-1 to MCS-9) that provide different throughputs as shown in Table 2.6. The MCS carry data from 8.8 to 59.2 kbps on the existing GSM carrier of 270.833 kbps. For coding schemes MCS-1 to MCS-4, modulation is still GMSK while for MCS-5 to MCS-9 it is 8-PSK.

Based on these coding schemes and their capability to carry data, a data rate of 473 kbps (8 bits × 59.2 kbps) can be achieved. Though GMSK is a more robust scheme, 8-PSK gives more data throughput. However, increased data rate comes with a decreased sensitivity in the system. This has an impact on the network planning as the higher the data rates, the more limited the coverage. One more advantage of EDGE networks is that switching between different coding schemes can take place easily, which was not possible in the GPRS network. If data transmission takes place in a coding scheme, it was not possible to switch the coding scheme on reception failure and the retransmission would then take place with the same protection as the initial transmission. In EGPRS, it is possible to change the MCS, i.e. the data block can be sent again but with better protection than for its initial transmission. This is done through the Link Adaptation process. One more observation on the coding schemes is that the new set of GMSK coding schemes is being used, e.g. CS-1 is equivalent to 8 kbps in GPRS networks as seen previously; similarly the changes are also there in CS-2, CS-3, and CS-4. These changes are done to provide the incremental redundancy support.

2.11.1.2 Link Adaptation and Incremental Redundancy

As the propagation conditions change with time and region, the quality of the signal changes. Due to this, the modulation and coding schemes change all the time. Link adaptation or LA is used for maximising the throughput per channel and changing the coding

Table 2.6 Modulation and coding scheme in the EDGE system.

MCS	Modulation	User rate (kbps)
1	GMSK	8.8
2	GMSK	11.2
3	GMSK	14.8
4	GMSK	17.6
5	**8-PSK**	**22.4**
6	**8-PSK**	**29.6**
7	**8-PSK**	**44.8**
8	**8-PSK**	**54.4**
9	**8-PSK**	**59.2**

scheme depending upon the channel conditions does this. Basically, this leads to providing the highest throughput possible with the minimum amount of delay. This gives better link quality and makes EDGE a more efficient system. The algorithms that are responsible for the link adaptation to take place are known as LA algorithms. These LA algorithms activate the LA feature based on BEP or Bit Error Probability measurements. Incremental redundancy or IR improves the throughput and it is done by automatically adapting the total amount of transmitted redundancy to the radio channel conditions. This is done using two techniques: ARQ (Automatic Repeat reQuest) and FEC (Forward Error Correction). In the GPRS system, when errors are detected in the RLC blocks, retransmission is requested and provided till the correct information reaches the destination. FEC provides the redundant information to user information that is used by the receiver to correct errors caused by the radio channel disturbances. However, in the EDGE system, all the redundant information is not sent immediately. Only a small amount is sent at first. If decoding is successful, then it ends up saving high capacity, but if unsuccessful, then the normal ARQ process takes place. LA selects the amount of redundancy for each individual transmission, so this process basically reduces the number of retransmissions and subsequent delays. The re-transmission mechanism in EDGE is more efficient (than in the GPRS networks) due to the IR phenomenon.

LA operates always on a first-sent block or a retransmitted block and upon receiving BEP measurements will change the MCS according to the condition of the network. IR is a specific retransmission algorithm also known as Hybrid ARQ II (it includes puncturing, storage and soft combining at the receiving end) that if enabled together with LA will allow the change in MCS within the family. In order to change MCS during retransmission, LA should be enabled. (Note: ETSI specifications makes IR mandatory only for MS as the receiving side, not for the BTS.)

2.11.1.3   Channel Allocation in the EDGE Network

The channel allocation in EDGE networks is nearly the same as in GPRS networks. BCCH, PCH, RACH, and AGCH are the signalling channels. PACCH is the only associated channel when the physical resources are assigned. Channel allocation algorithms are responsible for the assignments of the channels to the mobile stations. The EDGE base station should be capable of being synchronised with the existing GSM base stations. This would maximise the efficiency as both base stations could be configured as one sector instead of two, thereby making only one BCCH necessary for the operation. Moreover, an increase in data rates would lead to an increased signalling requirement for a given traffic and applications.

2.11.1.4   Smart Radio Concept

Implementation of the smart radio concept enhances the performance of the radio link, both in the uplink and downlink directions by the use of diversity methods described later in the chapter.

2.12   Radio Network Planning Process

The basic process of radio network planning remains the same as in GPRS networks. However, due to changes in the modulation and coding schemes, there are some changes in the coverage, capacity, and parameter planning of EDGE radio networks.

2.12.1 Coverage Planning

Link budget has a direct impact on the coverage. As the EDGE network focuses more on the (PS) data, the delay tolerance becomes higher, thereby improving the quality level of the system. Some factors that affect the link budget calculations specifically in EDGE are discussed in the following.

2.12.1.1 Incremental Redundancy

Use of IR (in coordination with LA) not only makes retransmission more efficient but also optimises the performance of the system. IR reduces the required C/I ratio by at least 3 dB. Link budgets are calculated for a given modulation and coding scheme for specific BLER. The value of BLER directly affects the gain due to IR; the higher the BLER, the higher the IR gain.

2.12.1.2 Body Loss

No body loss is taken into account for packet data services in the EDGE network.

2.12.1.3 Diversity Effects

Usage of diversity schemes generally has a positive impact on the link performances, thereby increasing the amount of area covered by individual sites. Both the uplink and downlink diversity schemes are possible. However, smart radio concepts can be used to increase the coverage performance of the EDGE network drastically. In uplink diversity, multiple antennas are used so as to cancel the correlated noise received by the antennas. Reduction of noise would lead to a gain in the signal level. Where there is no noise signal, the system would then allow the signal to flow without noise reductions. In the downlink direction, two transmitters are used and the signal is transmitted through two uncorrelated paths in bursts with slight delays. The transmitted power increases substantially with the use of two transmitters compared with a single transmitter. The transmission of the signal through two different paths with the introduction of some delays reduces the effects of fast fading. Thus, the link performance and coverage can be increased substantially in EDGE radio networks compared with GSM or GPRS radio networks.

2.12.1.4 RX Signal Strength

Signal strength in the radio networks is expressed in relation to the interference signals. There are three parameters that specify this: E_b/N_o, E_s/N_o, and C/N. Both E_s/N_o and C/N can be expressed in terms of Eb/No. The ratio of available bit energy to the noise spectral density is E_b/N_o: also known as the SNR. Energy per symbol to noise ratio is E_s/N_o. For the GPRS system (GMSK modulation), this is unity, while for the EDGE systems (8-PSK modulation), this is $E_b/N_o + 4.77$ dB (as three bits is one symbol). While the ratio of total received power to total noise is given by C/N. C/N can be expressed as:

$$\frac{E_b}{N_o}(\text{dB}) = \frac{E_s}{N_o} - 4.77 \tag{2.12}$$

$$\frac{C}{N}(\text{dB}) = \frac{E_b}{N_o} + 6.07 \tag{2.13}$$

Link budget calculations can be done by using MCS and BLER. However, received signal strength can be calculated for some specified data rates as well. E_s/N_o can be used for throughput (per timeslot) calculations in a cell. As seen previously, the received signal strength is dependent upon the transmitted signal strength, losses, antenna heights and gains (TX and RX) and distance travelled by the signal. Loss in the signal strength would give the cell range. Thus, the relation of energy per symbol limited by noise can be known and plotted with respect to the cell range. Link level simulations with the above calculations can give the throughput per timeslot for each coding scheme. The timeslot capacity is directly dependent upon the interference levels apart from the L2 protocol performances.

2.12.1.5 Capacity Planning

Capacity planning in EDGE networks is quite similar to that for the GPRS networks. However, the fundamental differences in the two technologies and an increase in the throughput per radio timeslot in EDGE changes some aspects of the capacity planning. A brief overview of these concepts and their effect on throughput per radio timeslot is described in the following sections.

The territory aspects explained earlier, i.e. dedicated, default, CSW, etc., are the same for the EDGE networks. Dedicated territory is specifically for PS traffic, CSW territory for CS traffic, and default can be used for PS traffic if the CS traffic is not utilising it (i.e. CS traffic has priority over PS traffic for default territory). The number of timeslots that are assigned to each of these territories can be changed dynamically based on the load conditions.

The concept of frequency re-use is similar to that of GSM/GPRS radio networks. The re-use pattern defines the number of cells that could be used within a cluster in the manner that there are no two neighbour cells having the same frequency. A frequency re-use of 3/9 would mean that each frequency is used only once in three sites/cluster, wherein each site is three-sectored. The frequency re-use of 1/3 will have a higher value of interference and thus would degrade the throughput per radio timeslot. Thus, a higher frequency re-use value will give a higher throughput and less delay. However, the spectral efficiency is higher in cases of lower frequency re-use as fewer frequencies are being used. Timeslot capacities have a larger dynamic range compared with GPRS radio networks. The number of timeslots available is less compared with the number of users in a cell (or network). This would mean that several users would be using the same timeslot, reducing the throughput per user. Thus, the higher the number of users (per timeslot), the lower the throughput (per user) and the higher the delay. PS traffic can be allocated to the BCCH TRX or non-BCCH TRX. Since the spectrum efficiency is usually the same for BCCH and non-BCCH cases, the TSL capacity remains constant on the BCCH layer making the BCCH layer more suitable for achieving high throughput than the non-BCCH layer TRXs. Usually, the frequency re-use patterns are found to be more stringent on the non-BCCH TRXs, so BCCH TRXs give higher throughput. Enabling frequency hopping does not have a major impact on capacity or quality of the EGPRS radio network.

Capacity planning dimensioning would require inputs related to cell configuration and traffic behaviour. This would mainly include the number of TRXs, EDGE territory definitions (number of timeslots in CSW, dedicated and default territories), CS and PS traffic, etc. The outputs of capacity planning would mainly include the amount of PS traffic, maximum, minimum and average PS load, available timeslots of CS traffic, blocking rate, etc. Capacity planning based on this dimensioning is done on a cell basis to make sure that the required capacity is available for CS and PS traffic apart from signalling in an EDGE network. Traffic types determine the signalling loads in the EDGE network. Unlike in GPRS networks where short messages increase the resource (PRACH/PAGCH) requirements for channel set-up, EDGE networks do not face such a problem because of enhanced data rates. However, in EDGE networks, signalling requirements may be greater (i.e. number of subscribers supported is less). This is because as more and more users are able to get attached to the network using the same timeslot even when the net load is constant (i.e. higher number of users/timeslot would mean less throughput/user), a decrease in TCH utilisation would take place during a TBF. If this TCH utilisation remains constant, more signalling channels would be required, decreasing the number of users getting connected to the network.

The steps involved in the dimensioning of the EDGE radio network can be summarised as:

1) Territory parameter definition: CS, PS territory should be identified. With the PS territory, the number of timeslots in the default and dedicated modes should be defined.
2) Total traffic load including CS, PS, and combined to be defined.
3) Delay versus load factor should be studied.
4) Rate reduction factor/parameter to be calculated.
5) Calculation of number of TRXs required for supporting the traffic.

Outputs of dimensioning can be:

1) Number of timeslots required for voice traffic.
2) Number of timeslots required for data traffic.
3) Number of timeslots in default and dedicated territory.
4) Average and maximum PS load.
5) Average and minimum user throughput.
6) Number of TRXs required to support the above parameters.

2.12.1.6 Parameter Planning

Apart from the parameters that have been discussed earlier, there are a few additional parameters in the EDGE radio networks. The most important ones are related to the LA and IR feature. Once EGPRS is enabled, the initial coding scheme is selected. LA parameters are dependent upon the modulation and coding scheme selections, both for initial transmission and re-transmissions. Although the MCS selections are based on the BTS parameters, the MCS used for the transmission is based on the BLER limits. However, MCS for both transmission and re-transmission can be affected by the mobile station memory. Then the parameters related to multi-BCF and common BCCH become important in the EDGE radio network. As mentioned before, the EDGE capable and non-EDGE capable TRXs in one sector can be configured to having only one BCCH. The TBF parameter settings make it possible for TBFs of GPRS and EDGE radio networks to be multiplexed dynamically on one timeslot. However, this scenario should be avoided as the performance suffers in both the UL and DL. In UL, GPRS performance suffers due to the large amount of 8-PSK re-transmissions taking place while in DL, performance suffers due to the GMSK modulations being used where 8-PSK is carrying higher data rates for EDGE. Parameters related to delay and throughput assume importance due to higher subscriber expectation (from the EDGE network) of higher throughput with minimum delay.

2.13 EDGE Radio Network Optimisation

The process of radio network optimisation is similar to that described before, i.e. taking the measurement from the drive tests and NMS, analysis and parameter planning. However, focus in the EDGE radio network optimisation would be on throughput. In the following, we also look into throughput improvement as that is one factor in the EDGE networks that would affect the capacity and coverage directly. One important aspect to remember is that since EDGE implementation would be generally on top of the GPRS network, it would be essential that throughput improvement is measured and analysed with respect to that of the GPRS networks.

2.13.1 Key Performance Indicators

In the EDGE network, there are a few more parameters in addition to the parameters in the GSM/GPRS networks. The major ones relate to the definitions of the EDGE territory, BLER (Block Error Rate), MCS schemes used, BCCH usage, etc. These parameter settings define the initial throughput. Adjustments to the parameter settings (apart from the ones in the GSM/GPRS networks) would enable throughput to be maximised.

2.13.2 Performance Measurements

The tools that are part of the performance measurements include drive test tools, reports from the network management system, and network-planning tools. However, the focus will be on throughput in EDGE networks, apart from the general optimisation of coverage, capacity, and quality. The key areas of performance measurement would be the air-interface and the A_{bis} interface. The air-interface measurements would as usual focus on capacity, coverage and quality while the A_{bis} interface measurements are the variation of the throughput with respect to the modulation and coding scheme variation. One such example of performance based on the accessibility, retainability, quality (ARQ) and peak traffic is shown in Figure 2.27.

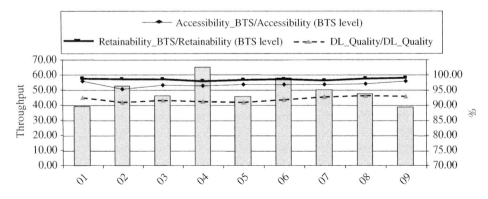

Figure 2.27 Test results (ARQ and peak traffic: BTS1).

Table 2.7 BTS performance measurements.

Date	BTS ID	Accessibility (BTS level)	Retainability (BTS level)	UL_Quality (dB)	DL_Quality (dB)	TCH Erlang
04/06/03	BTS1	98.61	99.05	92.73	72.01	0.55
04/07/03	BTS1	98.21	96.49	91.35	91.23	0.68
04/08/03	BTS1	66.18	100.00	99.11	99.33	2.06
04/09/03	BTS1			100.00	95.62	0.11

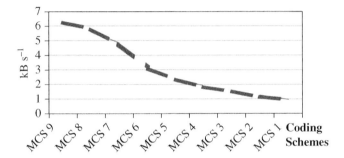

Figure 2.28 MCS versus throughput.

As stated before, the performance indicators, such as the ability of the mobile to get access to the cell (accessibility), DCR (retainability), and call quality in the downlink (DL-Quality), need to be measured. They are stated as percentages in Table 2.7.

Measurement of the A_{bis} throughput capacity with respect to the coding schemes used by the mobile subscribers needs to be done as this would be very critical in adjusting the boundaries of the dynamic pool. The example in Figure 2.28 shows the variation of the pool with respect to the modulation and coding schemes.

In most of the networks, the performance of the EDGE network would be a measure of the backdrop of the GPRS network. The achieved mean throughput values and delay values in the EDGE network are better than in the GPRS network, even when the load is higher. The measurement information serves as input to not only increase the coverage and capacity but also the QoS (experienced by end users), which will be closely observed in EDGE networks. The optimum value of QoS would be ultimately dependent upon the traffic requirements. As will be seen later in the book, the interactive traffic has more stringent requirement than the background traffic. Spectrum efficiency will play a big role in achieving this. The trade-off between the number of TRXs and interference level, i.e. higher throughput or higher number of users, would be a delicate issue in these networks. Final optimisation plans would have similar solutions to that of GSM/GPRS networks, but with stringent QoS requirements guiding the results.

2.13.3 Improvement of Throughput in EDGE Networks

Though the QoS in a network is service dependent, higher throughput is what will be the main 'QoS filling' requirement in the EDGE network. Throughput on the air-interface and coverage (both at hotspots and cell-edges) will be the main focus area in the

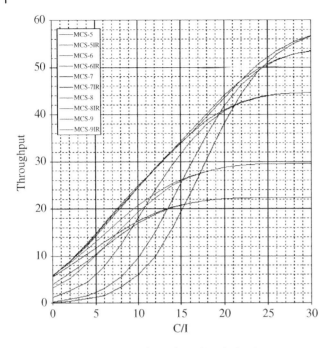

Figure 2.29 C/I ratio versus throughput (simulations).

planning as well as the optimisation of the radio network. Throughput is dependent upon a number of factors such as C/I, handovers, MCS changes, etc. An increase in C/I ratio will result in increased throughput (shown in Figure 2.29). Measurement at the cell-edges would be critical and performance improvement can be made by using concepts such as antenna tilts, power increase/decrease, etc. Increase in the number of handovers will also decrease the throughput. Optimisation of the neighbours to reduce the unnecessary handover can be one possible solution to increase the throughput. Throughput decreases with the number of MCS changes from higher throughput MCS to lower throughput MCS. Optimisation can be achieved by parameter auditing after observing the C/I ratio and receiver level statistics against MCS changes.

2.14 Conclusions

The traffic loads (both CS and PS), and the applications used affect the performance of the EDGE network. C/I criteria are stricter in the EDGE network than in the GPRS networks. There are two new aspects in the measurement, i.e. related to coding schemes and dynamic A_{bis} (both in UL and DL). The spectrum efficiency in the EDGE radio network is higher than in the GPRS networks, which is mainly because of the higher throughout and better link performance. Though the performance measurements and subsequent results are usually based on the data throughput, in the real network optimisation both voice and data traffic performance need to be assessed. Thus, with numerous kinds of applications running on the network, there would be lot of interaction between the applications with applications affecting each other.

3

Transmission Network Planning and Optimisation

3.1 Basics of Transmission Network Planning

3.1.1 Scope of Transmission Network Planning

In the GSM system, the transmission network is generally considered to be the network between the base stations and the transcoder sub-multiplexers (TSCM). The transmission network connects the radio network to the mobile switching centre (MSC), and hence, by virtue of its position and functionality, it acquires an important position in the mobile network infrastructure (see Figure 3.1).

3.1.2 Elements in a Transmission Network

3.1.2.1 Base Station Transceiver (BTS)

An overview of the BTS is provided in Chapter 2. Here we consider the BTS from the transmission planning perspective, i.e. the transmission system within a base station, as shown in Figure 3.2. The base station consists of a transmission unit or TRU. This unit interacts with the A_{bis} and A_{ter} interfaces. It also reallocates the traffic and the signalling channels to the correct transceiver or TRX. These units may be capable of making cross-connections at the 2 Mbps level and drop-and-insert functionalities at the 8 kbps level. Transmission units are capable of holding branching tables. These tables guide the transmission units on what to do with the traffic coming in from the PCM lines, e.g. should it just let it pass for the next BTS or should it drop some traffic meant for its own TRX?

3.1.2.2 Base Station Controller

The base station controller or BSC is capable of many functions. Radio channel management is an integral function of the BSC and includes tasks such as channel management and release. It is responsible for the handover management function that is based on an assessment of signal power, signal quality at the MS in both the uplink and downlink directions, and adjustments related to the minimisation of the interference in the radio network. It is also responsible for the measurements related to the implementation of frequency hopping (FH) within the network. Other important functions include interaction with the base station and MSC on either side, encryption management such as storing the encryption parameters and forwarding the same to the base stations,

Fundamentals of Network Planning and Optimisation 2G/3G/4G: Evolution to 5G,
Second Edition. Ajay R. Mishra.
© 2018 John Wiley & Sons Ltd. Published 2018 by John Wiley & Sons Ltd.

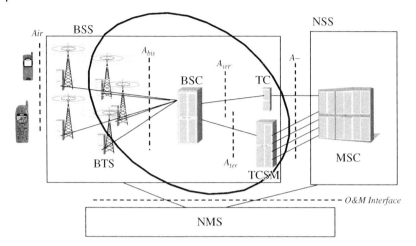

Figure 3.1 Scope of transmission network planning.

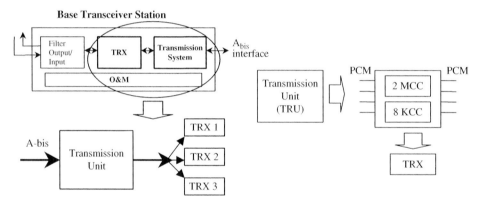

Figure 3.2 Transmission system in BTS.

traffic and channel management, etc. Also, the BSC-BTS signalling, which includes the LAPD, transceiver signalling (TRXSIG), and BSC-MSC signalling which is CCS7, is handled by the BSC itself. BSC is also responsible for carrying the O&M information to the BTS.

Apart from these functions, the BSC also performs one more function of interest to transmission planners, which is traffic concentration between itself and the BTS. This means that even if there were a channel available on the air- and A_{bis} interface, the subscriber would not be able to make a call because of the blocking on the A_{ter} interface.

3.1.2.3 Transcoder and Sub-Multiplexer (TCSM)
Transcoders and sub-multiplexers are used between the BSC and MSC. Although they are present in the BSS system, as shown in Figure 3.1, physically they may be present at the MSC location. The TCSM is capable of two functions: transcoding of speech signal and sub-multiplexing of the PCM signals. The transcoder does the speech coding in the downlink direction, which is decoded in the MS, and decodes the speech signal in the

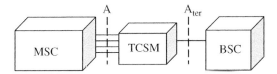

Figure 3.3 Transcoder-submultiplexer (TCSM).

uplink direction. Sub-multiplexers are located at the same place as the transcoders and are responsible for multiplexing the PCM links between the BSC and the MSC. As shown in Figure 3.3, three PCM signals (this figure may be different depending upon the equipment manufacturer) coming from the MSC are multiplexed to one PCM signal towards the BSC.

3.2   Transmission Network Planning Process

The transmission network planning process should generate a network plan for the transmission network that would provide a high-quality network with an immense amount of spare capacity at very low cost. In practice, a transmission planning engineer has to use his or her expertise in balancing these three factors, i.e. cost, quality and capacity.

The process of transmission network planning includes five main phases before the final plan is generated, as shown in Figure 3.4. Though the process looks simple in theory, in practice it is much more complicated because there could be many iterations at each step before a final plan is ready.

The process begins with data collection. This includes requirements for capacity and quality that lay the foundations for the whole process. Other essential information, including what transmission equipment is available, including capacities, quality targets, tools for transmission planning including link budget calculations, topologies that would be used, etc., is all part of the initial process.

The pre-planning phase focuses on the dimensioning aspects of the transmission network. The pre-planning starts with the inputs from the radio planning engineers. Once the number of base stations and their capacities are known, transmission planners would start with defining the quality targets for the network. According to the transmission products available such as radios, base stations, base station controllers, information on the media, an imaginary network topology is defined or even the topology principles that would be used in the transmission network are formed. This gives a better idea on where the transmission network should be heading. The final network design after the whole process is quite different to the one at this stage. All along the media of transmission is considered. If microwave transmission is to be used, then the

Figure 3.4 Transmission network planning process.

link budget aspects have to considered as well, so that the link design meets the availability and quality targets.

The site survey, line of sight (LOS), and site selection process starts at the very beginning of the network rollout. The site selection is basically a radio planning issue but the transmission planning engineers should get involved in the early phase, so that it will be easier to find sites that have the desired LOSs with other sites. Thus, there is connectivity with the other sites. The LOS process should be done very professionally because once a site is *constructed* and there is no LOS with another site, then this is a big loss for the network owner.

When microwave is the chosen medium of transmission, link budget calculations should be done such that the availability and quality targets are met. These calculations are generally based on the ITU recommendations. Tools used for the link budget and availability calculations should be based on the ITU/ETSI, etc., recommendations.

The detailed planning phase consists of frequency allocation/frequency planning, defining the time slot allocation plans, routing of the PCM signals, synchronisation principles and network management planning.

This whole process is usually an iterative one at different points, as the transmission planners try to achieve a balance between the network capacity, quality, and costs.

3.3   Transmission Pre-planning

Based on the capacity, coverage and quality targets, radio network planners arrive at the number of base stations that will be required in a given region. The task of a transmission planning engineer will be to be to connect these sites. In doing so, they would need to know the existing network infrastructure, if any, in terms of existing and spare capacity. The site locations of the BTS, BSC and MSC sites should be known as well, along with the present and future capacity requirements for each base station.

Once the above information is known, the transmission planner needs to determine the required capacity of each link. This process is dependent upon the traffic generated by each base station, the ability to cross-connect and groom each site so as to make complete use of the PCM links. The transmission media, future capacity and topology have their own individual impacts on the number of PCM links required to inter-connect the base stations, and subsequently the base stations to the mobile.

3.3.1   One PCM Connection

One PCM line consists of 32 timeslots as shown in Figure 3.5. Timeslot 0 (TS0) is used for the link management while TS1–TS31 are used for traffic and signalling. The bit rate of each TS is 64 kbps and based on the type of signalling, the TCH can be 16, 32, or 64 kbps. The PCM shown in Figure 3.5 is the PCM on the A_{bis} interface. It contains TCHs that carry the traffic from the transceiver (TRX) and also the signalling associated with it. This PCM also contains the O&M signalling for the base station. Depending upon the topology, this PCM signal would also contain the bits for synchronisation and loop control.

	Bits →			
	1 2	3 4	5 6	7 8
0	LINK MANAGEMENT			
1	TCH .1	TCH .2	TCH .3	TCH .4
2	TCH .5	TCH .6	TCH .7	TCH .8
3	TCH .1	TCH .2	TCH .3	TCH .4
4	TCH .5	TCH .6	TCH .7	TCH .8
5	TCH .1	TCH .2	TCH .3	TCH .4
6	TCH .5	TCH .6	TCH .7	TCH .8
7	TCH .1	TCH .2	TCH .3	TCH .4
8	TCH .5	TCH .6	TCH .7	TCH .8
9	TCH .1	TCH .2	TCH .3	TCH .4
10	TCH .5	TCH .6	TCH .7	TCH .8
11	TCH .1	TCH .2	TCH .3	TCH .4
12	TCH .5	TCH .6	TCH .7	TCH .8
13	TCH .1	TCH .2	TCH .3	TCH .4
14	TCH .5	TCH .6	TCH .7	TCH .8
15	TCH .1	TCH .2	TCH .3	TCH .4
16	TCH .5	TCH .6	TCH .7	TCH .8
17	TCH .1	TCH .2	TCH .3	TCH .4
18	TCH .5	TCH .6	TCH .7	TCH .8
19	TCH .1	TCH .2	TCH .3	TCH .4
20	TCH .5	TCH .6	TCH .7	TCH .8
21	TCH .1	TCH .2	TCH .3	TCH .4
22	TCH .5	TCH .6	TCH .7	TCH .8
23	TCH .1	TCH .2	TCH .3	TCH .4
24	TCH .5	TCH .6	TCH .7	TCH .8
25	TRXSIG 1	OMUSIG 1	TRXSIG 2	OMUSIG 2
26	TRXSIG 3	OMUSIG 3	TRXSIG 4	OMUSIG 4
27	TRXSIG 5	OMUSIG 5	TRXSIG 6	OMUSIG 6
28	TRXSIG 7	OMUSIG 7	TRXSIG 8	OMUSIG 8
29	TRXSIG 9	OMUSIG 9	TRXSIG 10	OMUSIG 10
30	TRXSIG 11	OMUSIG 11	TRXSIG 12	OMUSIG 12
31	x	x	x	x

Time slots (vertical axis)

Figure 3.5 One PCM.

3.3.2 PCM Requirements on the A_{bis} and A_{ter} Interface

Chapter 2 defines the blocking for the air-interface. Blocking parameters need to be defined for the A_{bis} and A_{ter} interface. Typically, the value of the air-interface blocking is 1–2% and that on the A_{ter} interface is 0.1–0.5%. It is considered that the A_{bis} interface should have no blocking apart from no blocking between the TCSM and MSC. Radio planning engineers use the air-interface for dimensioning the radio network, while transmission planning engineers use the A_{bis} and A_{ter} interfaces. The transmission capacity is determined by factors such as traffic generated in the air interface, the grooming ability of the equipment, topologies and the locations of the BSCs and TCSMs. Spare capacity plays an important part in taking the future requirements of the network into account.

Radio planning engineers determine the number of BTS and TRXs in the network. Transmission planners start with information about the radio planning dimensioning, i.e. number of TRXs per base station. As we see from Figure 3.5, one TRX needs 2×64 kbps timeslots while for the signalling, 16, 32, or 64 kbps timeslots may be required. A lower signalling rate means that more subscribers can use the network at one time. The object of the dimensioning for the A_{ter} is to find the number of traffic channels required. Once that is done, then based on the number of channels that can be multiplexed on the A_{ter}, the capacity of the A_{ter} interface can be found.

Example 1   *Capacity requirements on the A_{ter} interface*
Assume, that radio planners have decided that there are five sites of 2 + 2 configuration under a single BSC. Air-interface blocking is 2% and A_{ter}

Five sites of 2 + 2 configuration = 10 cells, each having 15 TCH.
Air-interface blocking = 2%.
Using Erlang B tables, 15 TCH support = 9.01 Erl of traffic.
Traffic offered to the BSC = 10 × 9.01 = 90.1 Erl.
If A_{ter} blocking probability is 0.1%, then the number of traffic channels supported = 117 (approx.).
If the number of traffic channels that can be multiplexed on the A_{ter} = 120, then A_{ter} interface capacity would be = 117/120 = 0.975 ≈ 1 E1.

3.3.3   Equipment Location

The base station and TCSM location needs to be decided during the nominal planning phase. As shown in Figure 3.3, the transmission capacity required between the TCSM and MSC is more compared with the connection towards the BSC. Thus, if the TCSMs were placed physically near the MSC, it would save transmission costs. When it comes to BSC locations, they can be located far from the MSC or co-located with the MSC. Both the remote and co-located BSC locations have their own advantages and disadvantages. The remote BSC location criterion is chosen to cater for the needs of the huge number of BTSs that are located far from the MSC. In this way, transmission costs are saved, though there may be heavy usage of the A_{bis} links. This is usually preferred for rural regions. A co-located BSC location is generally chosen for metropolitan areas where there is a heavy concentration of traffic.

3.3.4   Network Topology

Selection of proper network topology may lead to huge savings in the transmission networks. There are four main topologies, as shown in Figure 3.6. They are discussed here with their benefits.

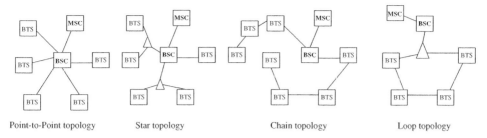

Figure 3.6 Transmission network topologies.

1) Point-to-point topology: As the name suggests, it is a point-to-point connection between the base stations and the BSC. This is quite prevalent during the early phase of network evolution as the implementation is easy and quick. But, the capacity may or may not be fully utilised. Also, protection methods are generally equipment protection in such cases, which is an additional expense.
2) Star topology: As Figure 3.6 suggests, the topology looks like a star, hence the name. The capacities can be utilised in a better way by using cross-connect equipment. This topology is also simple to implement, but if the node fails due to any reason, the traffic coming through it towards the BSC also is lost. Again, protection is expensive in this kind of topology.
3) Chain topology: This topology is generally used to connect the sites that extend from one region/city to another region/city. A typical example is the connection of sites on a highway. The capacity of the links is usually fully utilised. However, the last link of the chain can be vulnerable.
4) Loop topology: This is an extension of the chain topology with the last link of the chain being connected back to the BSC. Although this requires an additional link in the chain, it then provides protection without additional hardware redundancy. Also, there is requirement for an additional node, which is able to control loop protection.

In practice, a transmission network would consist of all of the above-mentioned technologies. Transmission planning engineers should balance the technological advantages with the costs involved keeping in mind the strategic importance of the site, before deciding which topology should be implemented.

3.3.5　Site Selection and Line-of-Sight Survey

3.3.5.1　Site Selection
The site selection process has been described in Chapter 2. This is predominantly a radio network planning engineer's domain with the active involvement of transmission planning engineers. The types of site that are required by radio and transmission planning engineers have opposite characteristics. Radio planners prefer sites that are not too high so as to prevent interference problems from other sites while transmission planning engineers need sites that are very high so that connectivity to a large number of other sites is possible. However, both agree that a site needs to have connectivity. Hence, there can be some compromise on the height of the site to be selected in a given

region. Sometimes, the important sites chosen by the radio planners are low or between high building structures, and in such cases repeaters are used by transmission planners. These repeater sites are usually high buildings without a base station, having only transmission equipment.

3.3.5.2 Line-of-Sight Survey

For transmission planning engineers (and radio planning engineers as well), the connectivity of the site (to other sites) is a very important criterion for site acceptance. The transmission planners make sure that one site is visible from the other. This means that it is not just sufficient for one site to be merely visible from the other site but should *clear* the LOS criteria before acceptance. This process may also lead to construction of the path profile, especially in the case of longer hops. This clearance criterion is based on the concept of a *Fresnel zone*. A typical Fresnel zone and path profile is shown in Figure 3.7.

3.3.6 Radius of the Fresnel Zone

Typically, the Fresnel zone can be defined as an area covered by an imaginary ellipsoid (a three-dimensional ellipse, as shown in Figure 3.8) drawn between the transmitting and receiving antennas in such a way that the distance covered by the ray being reflected from the surface of the ellipsoid and reaching the receiving antenna would be half a wavelength longer than the distance covered by the direct ray travelling from the transmitting to the receiving antennas, i.e. $d_1 + d_2 = d + \lambda/2$.

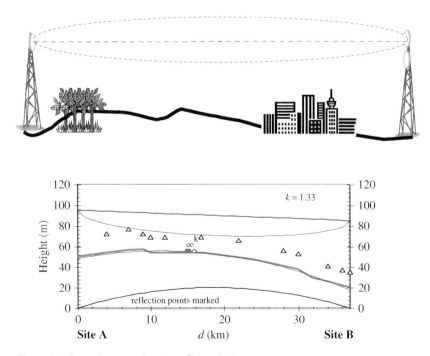

Figure 3.7 Fresnel zone and path profile analysis.

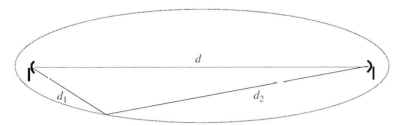

Figure 3.8 Radius of Fresnel zone.

The Fresnel zone is dependent upon two factors: frequency of transmission and the distance covered. Mathematically, the radius of the Fresnel zone can be calculated as:

$$F_1 = 12.75 \sqrt{\frac{d_1^* d_2}{f^* D}}$$ (3.1)

where F_1 is the radius of the first Fresnel zone, F is the frequency of the transmitting signal in GHz, and D, d_1, and d_2 are distances in km (as shown in Figure 3.8).

In most of the cases, the radius of the first Fresnel zone should be kept 100% clear. This means that there would be no reduction in signal level due to the terrain. If this first Fresnel zone were blocked, then the percentage of losses would be proportional to the percentage blocking in the Fresnel zone. In practice, transmission planning engineers try to keep some 'clearance' between the Fresnel zone and the highest obstacle in the radio path.

3.3.7 Microwave Link Planning

This constitutes a very important part of the transmission planning process in networks that use microwave transmission. The aim of microwave link planning is to achieve the desired performance at the lowest cost. The two major aspects of the link planning process are the link budget and propagation phenomena.

3.3.7.1 Link Budget Calculations

The aim of link budget calculations is to find out what path length would be suitable for getting the desired signal level. This received signal level is then used for the calculation of the fade margin. The fade margin can be defined as the margin required for countering the effects of decrease in the signal level (also known as 'fading') due to atmospheric conditions, etc.

The first factor required in the calculation of the the fade margin is the free space loss, which is same as the one introduced in Chapter 2. The phenomenon of free space loss or FSL stands true for microwave planning, except that now the distance changes from a few hundred metres to a few kilometres. FSL is dependent upon the frequency of the microwave signal and the distance covered.

$$\text{FSL}\left(L_{fs}\right) = 92.5 + 20\log d + 20\log f$$ (3.2)

The second factor is the gain of the antenna. The antenna should have sufficient gain so as to receive the signal at a desired level. The gain of the microwave antenna can be given as:

$$G = 10 \log \left(\frac{4\pi \, \overset{\cdot}{A} \, e}{\lambda^2} \right) \tag{3.3}$$

where G is the gain of the antenna in dB, A is the area of the antenna aperture, e is the efficiency of the antenna, and λ is the wavelength (same units as A).

Equation (3.3) can also be written as (for parabolic antennas):

$$G = 20 \log (D_a) + 20 \log f + 17.5 \tag{3.4}$$

where G is the gain of the antenna in dB, D_a is the diameter of the antenna in metres, and f is the frequency of operation in GHz.

Based on these, the hop loss Lh can be calculated as:

$$L_h = L_{fs} - G_t - G_r + L_{ext} + L_{atm} \tag{3.5}$$

where G_t is the gain of the transmitting antenna, G_r is the gain of the receiving antenna, L_{ext} is the extra attenuation (due to radome, etc.), and L_{atm} stands for the atmospheric losses due to water vapour and oxygen.

The received signal level P_{rx} can be given as:

$$P_{rx} = P_t - L_h \tag{3.6}$$

Mathematically, fade margin can be described as the difference between the received signal power and the receiver threshold (R_{xth}).

$$FM = P_{rx} - R_{xth} \tag{3.7}$$

Example 2 *Calculation of fade margin*
Calculate the fade margin of a microwave link whose dimensions are as follows:

Hop Length = 10 km
Frequency = 15 GHz
Antenna diameter = 0.6 m
Transmit power = 20 dBm
Extra attenuation = 0 dB
Atmospheric attenuation = 0 dB
Receiver threshold = −75 dB

Using Eqs. (3.2)–(3.7):

$$FSL \left(L_{fs} \right) = 92.5 + 20 \log (10) + 20 \log (15) = 136.02 \, dB$$

$$\text{Gain of antenna} \left(G_t \text{ and } G_r \right) = 20 \log (0.6) + 20 \log (15) + 17.5 = 36.58 \, dBm$$

$$Hop\ loss\left(L_h\right)=136.02-36.58-36.58-0-0=62.86\ dB$$

$$Received\ signal\ level\left(P_{rx}\right)=20-62.86=-42.86\ dB$$

$$Thus,\ fade\ margin\left(FM\right)=-42.86-\left(-75\right)=32.14\ dB.$$

3.3.7.2 Propagation Phenomena

In this section, we try to look into the factors that would affect the received signal level at the receiving antenna. As seen above, there are losses in the received power between the transmitting and the receiving antennas and this loss has been termed 'free space loss'. Apart from this, there are many other factors such as weather conditions, rain, terrain profile, etc., that need to be considered before the link design can reach the implementation stage. But before that, the phenomenon of propagation should be understood.

When a signal passes from one medium to another, the trajectory of the signal changes along with changes in velocity. This is due to the refraction phenomenon that takes place in the atmosphere. The ratio of the velocity of the signal in the free space to the velocity of the signal is known as the refractive index (η) of that medium, i.e. it signifies the refractivity of the medium.

$$\eta = \frac{V_{fs}}{V_m} \tag{3.8}$$

where V_{fs} is the velocity of the signal in free space and Vm is the velocity of the signal in the medium through which it is passing.

Now, the atmosphere is a mixture of gases. The refractive index of the atmosphere is dependent upon three factors: pressure, humidity, and water vapour. As we move up into the atmosphere, all these three change. On a mountain, you can feel the change in pressure and humidity and, of course, it becomes cooler, i.e. a change in temperature. Thus, the refractive index of the atmosphere, which is considered to be close to unity, has to be calculated taking these three effects into account. Mathematically this value is approximately 1.00045. Another term that is used is radio refractivity, N, which is defined as:

$$N =\left(\eta-1\right)^{\cdot}10^6 \tag{3.9}$$

Up to 300 GHz, this equation is also written in terms of pressure, humidity and temperature as:

$$N = 77.6\frac{P}{T}+3.73^{\cdot}10^{6^{\cdot}}\frac{e}{T^2} \tag{3.10}$$

where e is the partial pressure of water vapour (millibars), T is the absolute temperature in degree Kelvin, and P is pressure (millibars).

Due to changes in the pressure, temperature and humidity at every single point in the atmosphere, there would be a change in the radio refractivity at every stage. Thereby, we

(a)

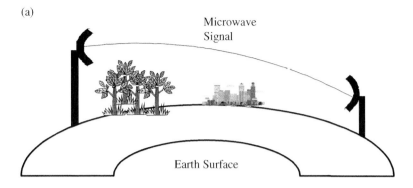

(b)

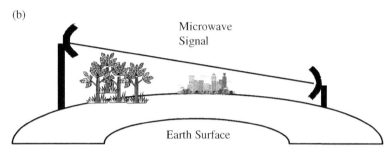

Figure 3.9 Microwave signal trajectory.

can well assume that small change in height, dh, would bring small change in radio refractivity, dN. Thus, the path of the signal is not a straight *line* but is in the shape of an arc. Hence, in reality, the trajectory of the signal with relation to the surface of the earth is as shown in Figure 3.9a. This kind of situation, i.e. two arcs, where one is the microwave signal and one is the Earth's surface, is complicated to analyse from a mathematical perspective. To make the analysis easy, one of the two arcs is made into a straight line as shown in Figure 3.9b, where the microwave signal is made into a straight line, thus changing the earth's radius in the process.

Transmission planning engineers have to consider another parameter in the process known as the k-factor. This k-factor is the one that is used to calculate the antenna heights and is described as the ratio of the *effective* radius of the earth k_a to the actual radius of the earth a.

Thus:

$$k = \frac{k_a}{a} \tag{3.11}$$

This k-factor can also be derived using the above analogy and using Eq. (3.10), because this factor is dependent upon the curvature of the microwave signal which is in turn is dependent upon the pressure, humidity and temperature at that point on the Earth's surface. As these factors are varying with height, it can be said that the k-factor also varies with height.

$$k = \frac{157}{157 + \dfrac{dN}{dh}} \tag{3.12}$$

During standard conditions of temperature and pressure, the value for dN/dh is considered to be $-40\,\text{N}\,\text{km}^{-1}$. Thus, the k-factor turns out to be 4/3. This is the reason why transmission planning engineers plan the microwave links using a k-value of 4/3. From the k-value, it is easy to predict the trajectory of the microwave signal if the path profile is known. Using the concept of path profile and the trajectory of the microwave signal along with the concept of the Fresnel zone, the minimum distance between the highest obstacle and the Fresnel zone can be determined. This would lead to the calculation of antenna heights on two sites.

The pressure, humidity and temperature, all change in the 24-hour cycle of the day. This leads to a continuous variation of radio refractivity, N, and thus the k-factor during the day. Transmission planning engineers should make themselves familiar with the climatic changes in the city/region that they are planning the microwave links for. In practice, the earth's curvature remains constant and only the trajectory varies with the value of k, but it is easy to understand and analyse the phenomenon if the trajectory is kept constant (i.e. a straight line) and variation is made to the surface of the Earth. The variation in the values of k and subsequent variation in the Earth's curvature is shown in Figure 3.10. The condition $k = 1.33$ is for the standard atmosphere variability (dN/dh) of $-40\,\text{N}\,\text{km}^{-1}$. Under this condition, the Fresnel zone should be 100% clear.

When the variability increases beyond $-40\,\text{N}\,\text{km}^{-1}$, then the path will experience sub-refractive conditions. Due to this increase and the subsequent variation in the k-factor, the Earth's surface would appear flat, thereby taking the signal closer to the surface of the Earth. This would mean that the obstacles that were 'clear' before the occurrence of this condition now start to obstruct the Fresnel zone. This condition is shown as $k = 0.8$.

When the variability reduces substantially from the $-40\,\text{N}\,\text{km}^{-1}$ value, then the path experiences super refractive conditions which means that the Earth's curvature increases substantially. If the value of dN/dh can be reduced to $-157\,\text{N}\,\text{km}^{-1}$, then $k = \infty$. This

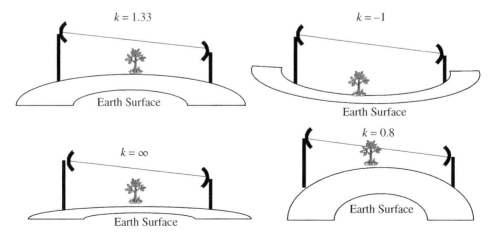

Figure 3.10 Variation of k-factor.

would mean that the Earth would appear to be flat and so would the microwave signal. Further reduction would mean a negative value of k, thereby increasing the probability of ducting taking place in the microwave link.

3.3.7.3 Multipath Propagation

When the microwave signal leaves the transmitting antenna, it would take different paths to reach the receiving antenna. One of the paths taken by the signal is a straight line or the Fresnel zone. Other paths may include the reflections from the atmosphere or the surface of the Earth. This is known as multipath propagation and results in more than one signal reaching the receiving antenna. This may result in reduction of the signal level if the phase difference is an odd multiple of half a wavelength. The severity of the reduction of the signal value will depend upon the reflected signal level. Thus, the signal level received at the receiving antenna is the sum of all the signals that are arriving at the receiving antenna (Figure 3.11).

As stated above, the multipath signals will deteriorate the signal level at the receiving antenna, with the reduction level being dependent upon the signal level of the reflected signal. Thus, the distance covered by the reflected or indirect signal and the surface of reflection play an important role in this. Generally, most of the surfaces reflect the signal, but the losses during the reflection makes the reflected signal so weak that it is either not able to reach the receiving antenna or it is not able to make any impact on the received signal level.

The term multipath fading is used generally for the fading of the signal caused by (Earth) surface reflections, while fading caused by the reflection/refractions from the atmosphere is called selective fading.

The layers formed by the atmosphere that reflect the microwave signal, will cause fades in the received signal that can range from tens of decibels to $100\,\mathrm{db\,s^{-1}}$, but the duration of these fades lasts only for a fraction of a second. This phenomenon is also known as the layering phenomenon. The layering of the atmosphere can be caused by phenomena such as radiation nights, advection, subsidence, frontal systems, etc. *Radiation nights* occur over land when a cold night follows a warm day, resulting in the surface of the Earth cooling and the air at the surface cooling down faster than the air higher up in the atmosphere, causing a temperature inversion. *Advection* is caused by the high-pressure system moving land-based warm dry air over the cooler moist air above the sea, causing a temperature inversion. *Subsidence* is the descent of dry air in a high-pressure system that becomes heated by compression and spreads out over the cooler moist air causing a temperature inversion. Localised inversion can also be caused

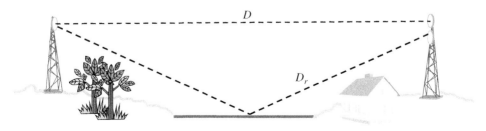

Figure 3.11 Multipath propagation.

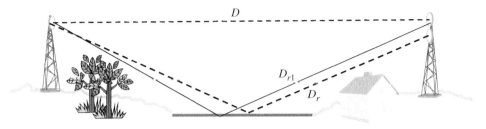

Figure 3.12 Multipath fading.

by the *frontal system* driving cold air beneath warm air. Thus, the layering phenomenon is usually caused in hot climatic regions and land near large water bodies.

Multipath fading has substantial impact on the signal strength below a frequency of 10 GHz. As the frequency increases, the impact of this kind of fading diminishes and signal fading gets more prone to rain effects. The effects of multipath fading can be counted by using diversity. Usually space diversity is used, i.e. two receiving antennas separated from each other in space.

When $D_{r1} - D$ is an odd multiple of wavelength, multipath fading will occur. Thus, the difference between the two diversity (at the receiving site) antennas should be an odd multiple of the wavelength (Figure 3.12).

3.3.7.4 Fading
Fading or the deterioration of the signal level can be broadly classified into two kinds:

1) Flat fading
2) Frequency selective fading

3.3.7.4.1 Flat Fading
Flat fading is 'flat' across the transmitted spectrum, which means that in the absence of any other source that would reduce the signal level, the fading would be a function of thermal noise only. There are two forms of flat fading: ducting and fading due to rain attenuation.

Ducting   In certain conditions, the atmosphere acts as a duct and thereby, becomes capable of trapping the signal within itself. In such cases, the signal will leave its original direction and move in the direction of the duct, as shown in Figure 3.13.

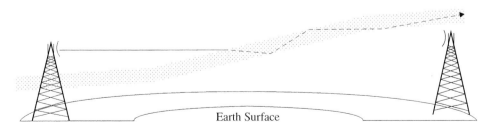

Figure 3.13 Ducting.

Once the signal is trapped in the layer, it propagates as if travelling inside an optical fibre cable and going in the direction of the waveguide. The signal gets trapped only when the critical angle is exceeded, i.e. the signal should enter the duct at an angle less than the critical angle. The ducting phenomenon also depends upon the intensity of the duct, dimensions, and the frequency of operation. It is not necessary for ducts to travel in an upward direction (as shown in Figure 3.13) but they may also travel downwards, i.e. towards the surface of the Earth.

Rain Attenuation   The rainfall attenuates a microwave signal. The level of attenuation is dependent upon factors such as frequency of operation, polarisation of the signal, path length, and rainfall rate. Other factors that influence the attenuation include drop size distribution, terminal velocity, temperature and shape of the raindrops. Rain starts having an impact over 10 GHz.

Attenuation due to rain becomes significant beyond a frequency of 10 GHz and critical beyond 15 GHz. The higher the frequency of operation, the more significant would be attenuation due to rain. For links working at frequencies of 12–15 GHz, depending upon the rainfall rate, the attenuation may reach $10-12\,\mathrm{dB\,km^{-1}}$, which is significantly high. Transmission planning or microwave link planner engineers should remember that rainfall rate varies with time and location. Thus, the accuracy of the calculations depends upon the how the rainfall was recorded and how stable the cumulative distribution remains over a given period of time. In general, it has been seen that one year of data has 0.1% of error, while two years of data contains 0.2% of error and if the data is collected for a period of 10–15 years, the error rate reduces to 0.001%. In the absence of any data, it is recommended to use the maps given in the ITU recommendations.

The attenuation (A) due to rain is given by:

$$A = \alpha \cdot R^{\beta} \tag{3.13}$$

where A is the attenuation due to rain, R is the rainfall rate, and α and β are constants.

Both parameters α and β are defined for the spherical drops and are polarisation independent. The value of these constants depends upon the drop size distribution, canting angle, terminal velocity, and the properties of water. During network designs, these values are not easily available hence it is recommended that transmission planning engineers use ITU recommendation 838 as it gives the values for these parameters for both horizontal and vertical polarisations.

The attenuation due to rain is dependent upon the rain rate and polarisation. Attenuation also increases with distance. As seen in Figure 3.14, as the frequency increases, the attenuation due to rainfall increases. Thus, in summary, attenuation increases with increase in distance, frequency and rain rate.

So how does the attenuation take place due to rain? Energy is lost from the microwave signal when the raindrops scatter or absorb it. Scattering occurs when a wave impinges on an obstacle without being absorbed and the effect of the scattering is to change the direction of travel. If the particle is small compared with the incident wavelength the process is described by the Rayleigh scattering phenomenon (<10 GHz) but when the particle is of similar size to the incident wavelength, the scattered wavelength, the process is described using the Mie scattering theory (>10 GHz).

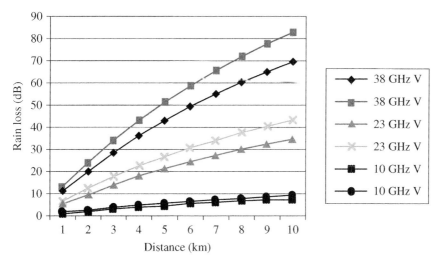

Figure 3.14 Attenuation due to rain.

Due to the rainfall, wet radomes may cause additional losses that need to be accounted for in the link budget calculations. These losses can be caused by reflection absorption and/or scattering. If the antenna (and its radome) is quite old, the losses may be even greater than a couple of decibels.

Thus, attenuation due to rain can be summarised as:

- As the distance increases, the attenuation increases for a given rain fall rate.
- With a constant distance, the rainfall rate increase would increase the attenuation in the microwave link.
- Vertically polarised links experience less degradation in the signal level than the horizontally polarised links.
- Extra attenuation losses, e.g. radome losses, need to be taken into account.

3.3.7.4.2   *Frequency Selective Fading*

In the case of ducting, the entire signal is trapped and (usually) lost but in the case of frequency selective fading, only a part of the signal may be lost. In this case, the atmosphere acts in such a way that the beam is not trapped completely, but only deflected. The signal may still reach the receiving antenna but then this reflected signal rarely reaches the receiving antenna in phase with the direct signal.

Thus, the total fading (*PTF*) in a system can be considered to be a sum of flat and selective fading.

$$P_{TF} = P_{FF} + P_{SF} \tag{3.14}$$

where P_{TF} is the Probability of fading (combined effect of flat and selective fading), P_{FF} is the probability of fade due to flat fading, and P_{SF} is the probability of fade due to selective fading.

Apart from flat fading and frequency selective fading, there are some other types of fading that transmission/microwave planning engineers should be aware of.

3.3.7.4.3 *k-Fading*

As explained and shown above (Figure 3.10), the fading of the microwave signal due to the variation of the atmospheric conditions, i.e. pressure, temperature and humidity, which results in the variation of the k-factor is called k-fading. As object which could be considered as clear from the first Fresnel zone, may well start acting as an obstacle with the changing atmospheric conditions, i.e. with the changing value of the k-factor. For this reason, transmission/microwave planning engineers should plan the link for clearance not only for standard conditions (i.e. $k = 4/3$), but also for non-standard conditions such as $k = 0.6$, etc. The variation of the received signal power (with diversity antenna) versus the k-value is shown in Figure 3.15.

3.3.7.4.4 *Diffraction Fading*

Diffraction occurs in microwave signals when they graze an object and in the process lose (disperse) energy. The amount of energy lost is dependent upon the type of surface/object. The loss of signal due to the surface/objects in the microwave path is known as diffraction fading. These objects may act as obstacles, i.e. interfere in the first Fresnel zone under the standard atmospheric conditions (i.e. $k = 4/3$), or may act as obstacles under the changing atmospheric conditions, i.e. during k-fading. Fading due to diffraction primarily is of two types: smooth sphere (or earth) and knife edge, as shown in Figure 3.16.

Smooth-sphere Diffraction   This takes place when the microwave signal passes over the surface of the Earth, e.g. water. Microwave links having path lengths into tens of kilometres, passing over vast plain land or a water body may display this when there are changes in the atmospheric conditions (i.e. k-fading). The losses due to smooth sphere diffraction may be as high as 12–15 dB.

Variation in k value

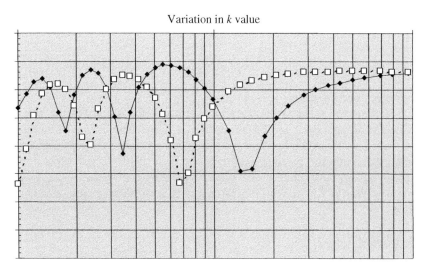

Figure 3.15 Received field (dB) at station as a function of k.

(a) 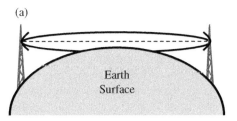   (b)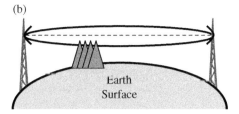

Figure 3.16 Types of diffraction: (a) smooth sphere; (b) knife edge.

Knife-edge Diffraction   The losses that occur due to grazing of the microwave signal over bushes, trees or forests are called knife-edge diffraction. This may occur during normal functioning of the links and due to variation in atmospheric conditions (k-fading). There are a few methods to calculate the losses due to this type of diffraction; the major ones are the Epstein–Peterson method, the Bullington method, and the Deygout method. Each of these methods have their own advantages, so transmission planners are advised to see their benefits in particular conditions before applying them. In general, it is assumed that diffraction loss due to knife-edge obstacles can be approximately 6 dB.

3.3.7.4.5   Atmospheric Fading
Fading due to the atmosphere may include fading due to gases (e.g. oxygen) and water vapour. The losses in the signal level due to oxygen and water vapour are predominant beyond a frequency of 15 GHz. The attenuation due to oxygen peaks at a frequency of about 60 GHz, and the losses can be as high as $12-15\,\text{dB}\,\text{km}^{-1}$. For water vapour, the peaks are below 45 GHz, and attenuation can be as high as $0.5-2.0\,\text{dB}\,\text{km}^{-1}$.

3.3.7.5   Other Factors Affecting Fading
Transmission/microwave planning engineers should take into account the effect of some other factors such as snow, sand, dust, etc. Snow affects the performance of the link in two ways: the water content of the snow degrades the signal level; and snow on the radome degrades performance of the antenna, thereby degrading the signal level. Losses due to snow can go as high as three times the equivalent amount due to rainfall, depending on the water content present in the snow. Dry snow causes relatively less attenuation. Hail, which is a mixture of air, water and ice, can cause a loss of up to $6\,\text{dB}\,\text{km}^{-1}$. The effect of sand and dust is relatively less, with the signal degrading by $1.37\,\text{dB}\,\text{km}^{-1}$ at 37 GHz.

3.3.8   Design Principles for Microwave Link

Climatic conditions cannot be predicted accurately for all times. If the climatic conditions of a few years are observed, the microwave links can be designed with fairly good accuracy. Some design rules/principles that will help as guidance for the engineers are:

- The fade margin should be sufficiently large. It should be able to handle degradation in signal level due to rain, multipath fading, k-fading, etc.
- The clearance should be checked for various fade margins, e.g. 1.33, 0.6, 0.5, etc.

- The first Fresnel zone should be free of obstacles for $k = 1.33$.
- For hops under 15 km and for hops over 15 km, the clearance should be zero for k-factors of 0.3 and 0.5, respectively.
- Over-water hops should be avoided. If unavoidable, then the antenna heights should be chosen such that the reflection point is not falling over water. Another method to avoid the reflection point falling over water is to place the antenna such that the reflected ray is blocked (e.g. placing the antenna on the far side of the terrace so that the reflected ray is blocked by the roof of the building itself).
- Choose higher antenna heights if the probability of k-fading is high in the region.
- In regions where ducting phenomenon is higher, choose higher antenna heights, as it is easier to move the antenna down than move them up on the tower (especially if the antennas are placed at the top of the tower).

3.3.9 Error Performance and Availability

Once the link is designed, its performance is measured against error performance and availability targets. The error performance objective defines the parameters to be measured against certain standard values, while the link availability is characterised on the basis of the error performance parameters. These objectives are defined in ITU recommendations ITU-R G.826.

Example 3 *Error performance calculations for radio relay system*

Calculated hop from		Site A	To	Site B
Radio frequency	8.1 GHz vertical			
Hop length	35.0 km			
Latitude	6N			
Longitude	80E			
Percentage pL	50.0			
Geoclimatic factor	1.41E-003			
Rain rate (0.01%)	120.0 mm h−1			
Station heights (ref. level)		200.0 m		200.0 m
Antenna heights (above ref. level)		50.0 m		50.0 m
Feeder lengths		60.0 m		60.0 m
Feeder loss/100 m		6.1 dB		6.1 dB
Feeder type		EW 77		EW 77
Antenna diameters		3.7 m		3.7 m
Transmitter output power (P_{tx})		27.0 dBm		
Free space loss (Lo)		141.6 dB		
Additional terrain loss (Lad)		0.0 dB		
Antenna branching loss(Lbr)		5.0 dB (SD, HSB)		
Feeder losses (L_{c1})		4.2 dB		
Connector Loss (L_{c2})		4.2 dB		
Antenna gains (G_{a1} and G_{a2})		47.3 dBi		
Hop loss in nonfaded state (L_{ho})		60.3 dB		
Received unfaded power (P_{rx})		−33.3 dBm		
Receiver threshold power (P_{rxth})		−76.5 dBm		
(at BER 10−3)				

Flat fading margin (M)	43.2 dB
Calculated flat outage time (p_{fm})	0.0283%
Total outage time (nondiversity) p	0.0283%
Annual unavailability due to rain	0.0003%
With space diversity:	
Diversity antenna diameter (D_d)	3.7 m
Antenna gain (G_d)	47.3 dBi
Antenna spacing (s)	10.0 m
Diversity improvement (I_{fm})	34.9

3.4 Detailed Transmission Network Planning

3.4.1 Frequency Planning

A microwave link is rarely situated in an isolated environment. There are usually many microwave links in a given region, of the same and/or different network operators. Then there are links that are not related to cellular networks radiating frequencies that may cause problems in these networks, e.g. radar systems. Owing to the presence of so many different sources of radiations in a region, it is very important to use the correct frequency plans for one's own network. The process usually starts with application for a frequency being made to the local governing bodies. Due to congestion, the number of frequency bands given to a cellular operator is usually less than desired making the need for an efficient, i.e. interference free, frequency plan even more necessary.

Before we go into frequency planning, let us see the various kinds of interferences and sources of interference that may exist in a network:

- intra-system interference (noise, imperfections, etc.)
- inter-channel interference (adjacent/co-channel, etc.)
- inter-hop interference (front-to-back, over-reach)
- external interference (other systems, radar, etc.)

For frequency planning in the microwave network, inter-channel interference, and inter-hop interference are of utmost important for transmission planning engineers. Intra-system interference is caused by noise generated within the system and this kind of interference can be removed by good system design. Inter-channel interference consists of interferences caused by co-channel and adjacent channels. Co-channel interference means that the same signals from the same frequency bands are reaching the microwave antenna. However, co-channel interference can only be cross-polar, i.e. both the signals will be 90° with respect to each other (e.g. one is horizontally polarised while the other is vertically polarised). Adjacent channel polarisation means that both the frequency bands of both the signals are next to each other, e.g. one can be 18.1–18.4 and the other is 18.2–18.6. Unlike co-channel, adjacent channel interference can be both co-polar and cross polar. These are shown by signals 1, 3 and 4 in Figure 3.17. Using non-consecutive frequency bands in the same or adjacent links can prevent this kind of interference.

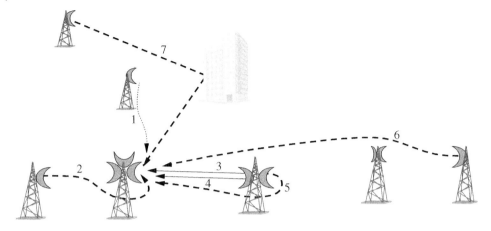

Figure 3.17 Sources and types of interference. 1, Co-channel/adjacent channel from different hop; 2, opposite hop front to back reception; 3, adjacent channel, same hop; 4, cross-polarisation, same hop; 5, front to back radiation; 6, over-reach; 7, terrain reflection.

Inter-hop interference is caused due to the undesired signal arriving at the receiving antenna. Using different frequency spots at potential sources of interference can prevent this kind of interference. Techniques such as using different polarisation in subsequent links and tilting of antennas are the remedies for these kinds of interference. Sometimes interferences are detected in a network, even though frequency plans are perfect. This may be due to the possible reflection that might be taking place in the network. In these kinds of situations, again, techniques such as antenna tilting and power reductions of the transmitting antenna (interfering source) can be used.

Example 4 *Interference calculation (over-reach): The example calculates the signal to interference ratio (SIR) and threshold degradation due to over-reach interference.*

The two links are assumed to be operating at the same frequency. The links, however, may belong to the same or a different network, as shown in Figure 3.18.

The following aspects are considered:

1) *Site 4 will receive interference signals from site 1 and* vice versa. *The interference may cause degradation of the receiver threshold to an extent dependent upon the received SIR at the site.*
2) *Since the antenna at sites 1 and 4 are oriented towards sites 2 and 3, respectively, an antenna discrimination would result, depending upon the off-axis angles θ and ϕ.*
3) *Also contributing to this discrimination could be a difference in polarisation of the two antennas at sites 1 and 4.*
4) *Furthermore, since the antennas at sites 1 and 4 are mounted at heights dependent upon the path between sites 1–2 and sites 3–4, respectively, the path between sites 1 and 4 may cause additional attenuation over and above the free space loss, because of obstructions. As a worst-case analysis, however, the additional loss may be assumed to be zero.*

(Note: Along with the above form of interference, other sources may also contribute to interference as shown in Figure 3.17)

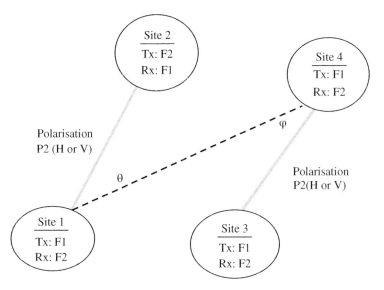

Figure 3.18 Interference calculation.

The following factors should be taken into account:

- Antenna discrimination
- Cross polarisation discrimination
- Diffraction loss due to obstructions in the interfered path

Inputs required:

System details	Site 1	Site 4	Units
Transmitter output	Ptx1	Ptx4	dbm
Feeder loss	Lf1	Lf4	db
Connector loss	Lc1	Lc4	db
Received carrier power	Prx1	Prx4	dbm
Polarisation	P1	P4	H, V
Threshold degradation	Y1	Y4	db
Receiver threshold	Th1	Th4	dbm
Antenna gain	G1	G4	dBi

Geographic details	Site 1	Site 4	Units
Distance	D	—	Km
Off axis angle	θ	ϕ	deg

Polarisation	Site 1	Site 4
HH	HH1	HH4
VV	VV1	VV4
VH	VH1	VH4
HV	HV1	HV4

Interference level at site 4 (I4):

$$= Received\ signal\ level\ at\ site\ 4, receiving\ from\ site\ 1$$
$$-combined\ antenna\ discrimination\ at\ both\ ends$$
$$= TX\ power\ level\ of\ 1 + antenna\ gain\ at\ 1 - wave - guide\ loss\ at\ 1$$
$$-attenuation\ over\ path\ 1 - 4\left(FSL + additional\ terrain\ loss\right)$$
$$+ antenna\ system\ gain\ at\ 4$$
$$- wave - guide\ loss\ at\ 4 - combined\ antenna\ discrimination\ at\ both\ ends$$

or

$$I_4 = P_{tx}1 + G_1 - \left(Lf_1 + Lc_1\right) - FSL - additional\ terrain\ loss$$
$$+ G_4 - \left(Lf_4 + Lc_4\right) - combined\ antenna\ discrimination\ at\ both\ ends$$

Combined antenna discrimination can be derived from radiation pattern details of sites 1 and 4 antennas, depending upon the polarisation combination (HH, HV, VH or VV) between sites 1 and 4.

Example 5 *Frequency plan*
Taking a cue from Figure 3.17, an example of a possible frequency plan is shown in Figure 3.19. Interferences 1, 3, and 4 (from Figure 3.18) can be cured by using different frequency bands and polarisations. While interference from 2, 5, and 6 can be removed by the use of a polarisation altering technique, i.e. altering the horizontal and vertical polarisations. Interference due to signal 7, which is detected only (usually) after the network is live, can be removed by either using different frequency bands or polarisation. As mentioned previously, other techniques such as antenna tilting can also be used. However, as this requires the link to be non-functional, other techniques such as power reduction can be used.

As seen above, it is always better to have more frequency bands for better frequency planning, but then due to congestion of the frequency bands, especially at lower frequencies, it is usually not possible to obtain a large number of frequency spots.

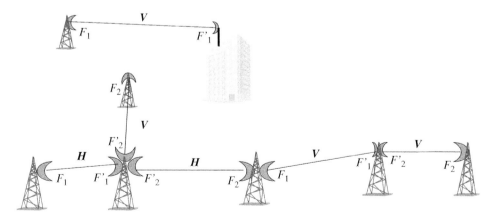

Figure 3.19 Frequency plan; frequency spots used [F_1, F_1'] and [F_2, F_2'].

3.4.2 Timeslot Allocation Planning

The understanding of timeslot allocation principles and methodology is very crucial for the efficient utilisation of the PCM lines available. These timeslot allocation principles are also crucial for the traffic distribution at the base station.

First let us understand the movement of one PCM or 2 Mbps from the MSC to the MS. One PCM is shown in Figure 3.5. On the A interface, one 2 Mbps consists of 32 timeslots. TS0 is used for link management. Each of these timeslots is a 64 kbps channel. The next stop for the 2 Mbps signal is the transcoder/sub-multiplexers, which convert these 64 kbps signals into 16 kbps traffic channels. The common channel signalling passes through unchanged. The sub-multiplexer then maps them into a single 2 Mbps channel and sends them to the base station. The traffic at the base station determines the number of TRXs there. Traffic channels remain at 16 kbps on the A_{bis} interface along with the addition of some signalling channels that are required for the TRXs and the base station signalling. Beyond the base station, i.e. on the air-interface, the 16 kbps channels are converted into the 13 kbps traffic and 3 kbps inband signalling channels.

As seen before, each TRX needs two timeslots of 64 kbps and one signalling channel of 16 kbps for each TRX. One signalling channel is required for each base station, known as the base station control or BCF signalling. One TRX has a handling capacity of 12 TRX (this figure may vary from equipment to equipment).

Basically, timeslot allocation planning involves the allocation of the traffic channels (and timeslots) for traffic and signalling apart from the pilot bits (for synchronisation) and loop control bits. There are a number of ways that the transmission planner can allocate the timeslot and signalling channels. The allocation of the timeslots should be done in such a way that there is a possibility for an upgrade from, say, $1 + 1 + 1$ configuration to $3 + 3 + 3$. While doing such an upgrade, there should be minimal changes made in the allocation table and the branching tables. There should always be space for the management bits for the loop protection, synchronisation, etc. Timeslots can be grouped either in linear fashion or in block fashion.

Linear allocation of timeslots means that the next free timeslot is allocated for the new TRX. The main advantage of this technique is that it is simple to plan. However, the network starts varying, and a few changes are being made, it becomes difficult to keep track. This is not very suitable for big and complicated networks.

Block allocation of the timeslots in the whole of PCM is divided into standard blocks and each of these blocks is then assigned to the respective TRX. This scheme of allocation is easy to upgrade. As the spare capacity is distributed along the whole 2 Mbs signal, it is difficult to add a whole new base station.

One way of doing timeslot allocation is shown in Figure 3.5, where all the traffic channel are allocated timeslots 1–24 while all the signalling is confined to the lower part of the 2 Mbps signal. Another way of doing timeslot allocation planning is shown in Figure 3.20. The example shows $1 + 1 + 1$ configuration. The timeslots for traffic and signalling for each TRX are allocated together. Timeslots 1–3 contain the traffic for TRX1 and signalling for TRX1 and TRX 2. Subsequent timeslots 4–7 are empty assuming that the capacity for the cell would increase from one TRX to three TRXs. The same pattern is followed for other cells. This scheme can also be used for a condition where four base stations are using the same 2 Mbps signal. The synchronisation bits and management bits can be placed in either TS0 or TS31. These bits can also be placed in other timeslots as well.

3.4.3   2 Mbps Planning

The routing of the 2 Mbps is part of timeslot allocation and detailed planning. This involves the planning of the routes from where the 2 Mbps line would pass and where it would terminate. Planning engineers have to pass on these plans to the commissioning engineers and so these plans should be very carefully and clearly made. Apart from this, planning engineers during the course of 2 Mbps planning would calculate the amount of 2 Mbps ports that would be utilised at the BSC. The example shown in Figure 3.21 is for two sites that are connected between two BSCs. The sites are connected to each other by microwave radios. They are path protected, i.e. in a loop, and thus 2 Mbps are terminating at cross-connect (here it is SXC or SDH cross-connect). Each site or base station is using one PCM and terminating at the ET ports at the BSC. Each of the PCM requires one port at the base station. In some cases, one PCM is shared between two or more base stations, as shown in Figure 3.20, but the number of ET ports required at the BSC would be still one. Transmission planning engineers should remember to reserve some ports for the 2 Mpbs connections between the BSC and MSC.

In big upcoming networks, the 2 Mbps plans keep on changing. Planning engineers need to keep track of how the 2 Mbps paths are getting routed as this directly affects the ET port allocation plans and the complete utilisation of the PCM signals.

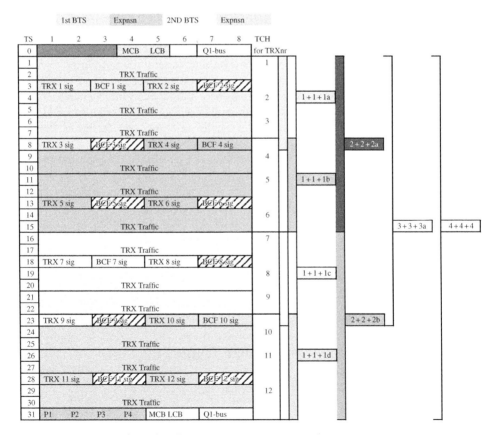

Figure 3.20 Example of timeslot allocation.

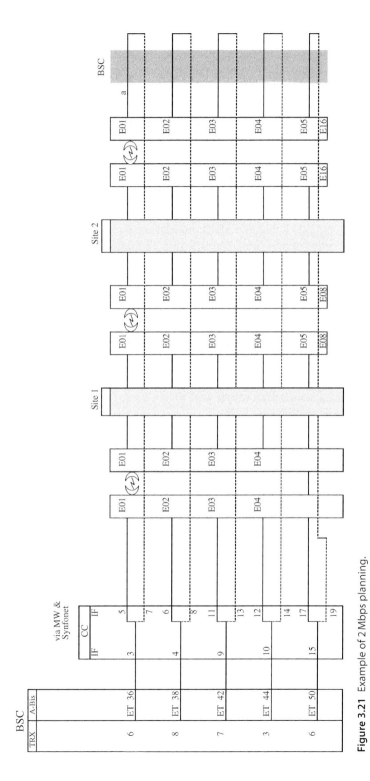

Figure 3.21 Example of 2 Mbps planning.

3.4.4   Synchronisation Planning

Another important aspect of detailed transmission planning is the synchronisation plans. Usually networks are designed using mixed technologies, i.e. SDH and PDH. In such mixed network that do not have proper synchronisation planning, 'clicks' are heard due to bit errors in the data service (Figure 3.22). SDH networks themselves do not require any synchronisation for proper operation. The reason is that the SDH payloads are easily transferred within the network while the frequency differences can be taken care of by the pointer mechanism. Pointers provide the mechanism for dynamically accommodating the variation in the phase of the multiplex section into which they are to be multiplexed. Thus, when both the PDH and SDH equipment is present in the network, the degradation takes place due to the pointer movements. If the synchronisation planning is done properly, then these 'clicks' due to bit errors disappear.

Another reason for the synchronisation to be present is that there are other equipment/systems other than the SDH and PDH in the network. These may include the likes of base stations and other transmission equipment. The internal clock of this equipment is not very accurate, so external synchronisation is needed for movement of 'click free' call through the network. The required accuracy for the GSM networks is of the order of 10^{-8}.

Thus, for having a good quality network, synchronisation is required. For good network synchronisation, clock distribution should be there. The quality of the clock plays an important role in determining the quality level of the synchronisation within the network. There are three clock levels that are recognised by the ITU-T G.803. They are:

1) PRC or the primary reference clock
2) Slave clock (synchronisation supply unit or SSU)
3) SEC or the SDH equipment clock

The PRC usually provides the clock for the whole network, thereby acting as the master. The accuracy of the PRC is very high usually of the order of 10^{-11}. The specification

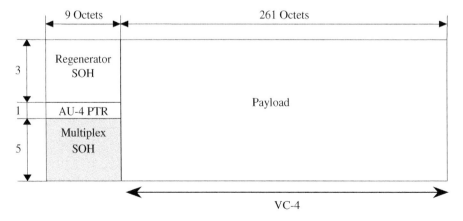

Figure 3.22 SDH frame for STM-1 interface. The frame size of an SDH frame is 260 bytes $\times$ 8 bits $\times$ 9 rows and the frame is transmitted 8000 s^{-1}, so the payload bit rate is 149 76 Mbps. The first column in the virtual container is VC-4 path overhead and the header consists of regenerator and multiplex section overhead (SOH) and AU-4 pointer fields.

of such clocks, which can act as PRC, has been given in ITU-R recommendation G.811. As PRC is the master, it is advisable to have a redundant source, i.e. two PRCs, with one clock in standby mode. A typical example is the caesium clock. The main advantage of these kinds of clocks is their high degree of accuracy but the high cost of the clock may act as a disadvantage in small networks.

Slave clocks are so called because they are used in conjunction with the master clocks. These clocks are always locked to the clocks of higher accuracy (e.g. caesium clocks) through the network. They are also known as *refresher clocks* as they may be used in long chains to refresh the timings. They also have the capability to hold the clocks, when the reference (master) clocks are lost temporarily. The accuracy of such clocks is of the order of 10^{-8} based on the ITU-T recommendation G.812. The major advantage of such clocks is their low cost but lifetime may be limited.

SEC or the SDH equipment clock has accuracy of the order of 10^{-4} based on ITU recommendation of G.813. Use of such clocks is recommended for SDH networks. Mixed networks (that have PDH equipment) should not use this clock for timing distribution as it may degrade the signal quality.

3.4.4.1 Synchronisation Planning Principles

As the cellular networks are *mixed* networks, they generally follow the master–slave technique of distributing the clock signal. In this method, a higher-level clock synchronises the lower level clock. This lower level clock further distributes the timing signal as shown in Figure 3.23. MSC receives the clock from the PRC and distributes it to the BSC. The BSC then further distributes it to the base stations, which further distribute the timing signal.

Topology also plays an important role in the synchronisation planning. Long chains of timing distribution should be avoided. There should not be any more than 10 SSUs in the chain and the number of the SECs between two subsequent SSUs should not be more than 20. This implies that the maximum number of G.813 clocks in the synchronisation chain is 60, and the maximum number of G.812 clocks is 10.

If there are loops (loop topology) in the network, then the engineers should make sure that there is no loop of the timing signal to avoid a timing loop.

Synchronisation should be protected. There should be more than one source of timing signal for each network element. The clocks should be derived from reliable sources such as SDH equipment rather than 2 Mbps signals from the leased lines whose accuracy is not known.

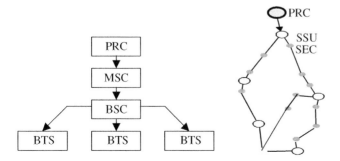

Figure 3.23 Clock distribution in a typical cellular network.

3.4.4.2 Implementation of the Synchronisation

The PRC is generally the sourse of timing signal in a cellular network. The clock should be re-traceable back to the source. For loop protection, MCB and LCB bits in TS31 (or any other timeslot) are used. The MCB or master control bit indicates whether or not the clock timing is based on the master, with '0' indicating that it is based on the timing sent by the master and '1' indicating that the timing is not based on the master. If the synchronisation is based on the master–slave technique, then MCB '1' also indicates that there is some fault with the flow of the timing signal and the equipment is not receiving the timing from the master. The LCB or loop control bit would indicate '0' when there is possibility of a timing loop, which means that if there is a loop topology in the network, LCB would be '0' indicating that the timing loop is not there (based on the principle of synchronisation planning explained above). As every element should have more than one clock source, so the LCB indicates from where the equipment should take its timing signal. So:

- MCB: 0, Signal is based on the master clock.
- MCB: 1, Signal is based on some other clock, e.g. internal clock.
- LCB: 0, Synchronise slave from master, no possibility of timing loop.
- LCB: 1, Do not synchronise slave from master (e.g. there may be the possibility of a timing loop).

3.4.5 Transmission Network Management Planning

Transmission network management planning is quite important in the cellular networks and engineers should take into account the capability of the transmission network elements and the future expansions. The network management system or NMS can be of many types. They may be able to manage the whole networks as such or may be specifically for the transmission element management in the network. Generic applications of the NMSs include network element or NE (and configuration) management, security management, alarm management, performance and fault management.

NE and configuration management functions may include controlling the NE, collecting and providing the data to the NEs. This function also makes possible the integration of the NE to the network. The alarm management function enables the user to collect the alarms from the NEs. These alarms indicate the status of these NEs. Fault management function detects the failures and schedules the correction of these faults apart from testing and bringing the faulty NE back into working condition. The generation of the performance behavioural reports of the network and its elements is done by the performance management function of the network. And finally, prevention and detection of the misuse of the network from breaches of security is taken care of by the security management function of the NMS.

The network management functionality is based on the master–slave protocols. The planning of the NMSs starts with choosing the master. The NMS can act as the master and the NEs as the slave. Sometimes, NEs such as the base station controller or the base stations can also be the master. Once the master is defined, the definition of the management bus and its transfer method is decided. The next step is to decide the parameters for each of the NEs that the master would control.

Table 3.1 Example of parameter setting for microwave radio.

Group	Setting	Value/Operation	Comments
Identification	Equipment type	MWR	
	Equipment number	2233	
	Equipment name station ID	According to the transmission plan/station identification	
Service options	Baud rate	1200 bit s^{-1}	Default: 9600 bit s^{-1}
	Address	1 ... 4000 (Individual address)	Default: 4095
Equipment settings	Mgmt channel loop protect	OFF	Default: OFF
	RX Mgmt Channel	Disconnected	Default: Normal

Each master that is chosen has its own capability of managing a number of NEs. The capacity may vary from managing a few NEs to hundreds of NEs. Apart from that, each master has its own speed of managing the elements, i.e. the speed of the management buses may vary in the network. These management buses may have a speed of 1200 bps, 9600 bps or even higher.

Definition of the management bus includes deciding the NEs that would be engaged by a particular NE. Also, the topology of the management bus and the protection scheme that is associated with it are decided. One thing to remember is that the higher the number of elements associated with the management bus, the slower will be the speed of the bus.

NEs can be managed by attaching the slave to masters through a cable. Other techniques involve sending the managements bits through the PCM signal from TS0 to TS31 and through the auxiliary data channels of the frame overhead, e.g. radio frames.

The most important part of the transmission management planning is to define the parameters for management of the NEs. Engineers need to define the parameters such as the addresses of the NEs along with their identifications, and the speed to the management buses. These parameters depend upon the type of management system that is being used in the network and its capability to handle the amount and types of NEs. One typical example of parameter settings is shown in Table 3.1. There may be more parameters than shown depending upon what type of master is being used and how the management channels are being transferred and controlled.

3.5 Transmission Network Optimisation

3.5.1 Basics of Transmission Network Optimisation

In the past, the term network optimisation meant solely radio network optimisation. Transmission planning itself was considered to be a line-of-sight job. Any problems in the network coverage capacity or quality were attributed to the radio part of the network. With more and more studies being conducted on the mobile networks, it was

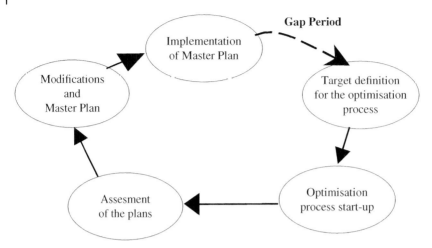

Figure 3.24 Cellular transmission network optimization cycle.

found that transmission and the core network also play an equally important role in contributing to a highly efficient network. Thus, both the transmission and the core network optimisation processes gained importance. Moreover, some networks that came during the evolutionary years of GSM, were designed in a hurry, without much consideration for future capacity requirements and the knowledge that the quality of transmission network would affect the quality of the whole network.

In Chapter 2, we saw that radio network optimisation is a process that starts almost simultaneously with the radio planning process. This is not the case with transmission network optimisation. The main reason for this is the static nature of the transmission network. The Transmission network once up and running should not be touched unless a problem is identified and verified, e.g. if one microwave link (near the BSC) in a chain is 'down' for a few minutes, the network traffic and hence revenue generated from all the sites getting connected to that BSC would be lost. It is for this very reason that the transmission networks are not touched during the network optimisation.

The process of transmission network optimisation is shown in Figure 3.24. Once it has been identified that the transmission network needs to be optimised, the target areas of optimisation are defined. The process starts with making plans based on the data available and the targets to be achieved by the process. These plans are then assessed and optimised, to achieve a balance between the costs, quality and required time frame for the process. The process of transmission network optimisation ends up with a few plans and these plans are then analysed. Again, based on the requirements and the costs, a master plan is prepared. This master plan is then implemented. Unlike in the radio network, where the optimisation is an on-going process, there is a gap period before the next optimisation cycle can begin in the transmission network.

3.5.2 Transmission Network Optimisation Process

3.5.2.1 Process Definition

There are usually two major reasons (problems) behind the transmission network optimisation activity: either the capacity is less or the quality is not up to the desired standards. The optimisation definition and process, analysis of the network problem

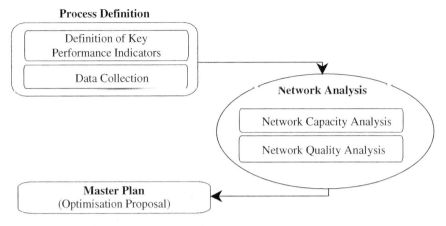

Figure 3.25 Cellular transmission network optimisation process.

and solutions creation are based on either one or both of these problems. Once the main problem is identified, the key performance indicators (KPIs) should be decided upon. The main KPIs in a transmission network are related to the quality of the microwave links such as ESR and SESR. Apart from this, some parameters like slip frequency can be observed during a synchronisation study (Figure 3.25).

3.5.2.2 Capacity Analysis

During the initial network *rollout*, the benchmarks are usually on how many sites are *on air*. For going *on air*, the base stations need to be connected to the BSC. When the deadlines are fast approaching, the usual way to connect these sites are through leased lines. The topology design, its effects and benefits may somehow get neglected. These temporary solutions may result in the under utilisation of the media that is connecting them. Another major factor is the congestion in the network that may be due to the high increase in the number of mobile users. This may lead to increase in the number of TRXs at each site, thereby forcing the transmission planning engineers to come up with solutions to cater for the traffic increase without an increase in the number of base stations. Another reason could be the need to upgrade the network technologically, e.g. 2G to 2.5G or 3G.

The capacity optimisation process ends up with a twofold solution. First, it would give solutions on how to use the existing capacity effectively; and secondly, on how to increase the capacity without/minimal disturbance to the network functionality, i.e. without disrupting the revenue inflow. Apart from this, the network would be more flexible and reliable.

How should capacity analysis be done? Engineers should collect the data that is required for the region they have to perform the analysis for. Apart from having the site information (site location, coordinates height, etc.), information collected should include capacity of the sites (present as well as forecasted), 2 Mbps routing plans and the media information (e.g. leased line, microwave radios, optical cables, etc.). The coverage plans would also help if new sites are being introduced.

Once this information is in place, the transmission planning engineers should calculate the existing capacity of each link and the forecasted capacity of each link. The next step is to identify the media, some base stations that have no LOS to another base

Table 3.2 Capacity analysis.

S. no.	Site ID	Connectivity before optimisation	Capacity (2 Mbps/site) (present)	Connectivity after optimisation	Capacity (2 Mbit/site) (After)	Capacity improvement (%)
1	82	82-50; 50-06	0.25	82-50; 50-06	0.50	100.00
2	83	83-82; 30	0.25	64-83; 64-06	2.00	700.00
3	87	70-87; 70-69; 25	0.33	70-87; 70-69; 69-06	0.80	140.00
4	2	61	1.00	61	1.00	0.00
5	35	35-82; 31	0.25	35-82; 50-06	0.50	100.00
6	36	4	1.00	36-06	4.00	300.00
7	73	73-76; 38	0.50	73-06	1.33	166.67
8	74	74-82; 50-82; 50-06	0.50	74-82; 50-82; 50-06	0.50	0.00
		Av. capacity per site (before)	0.11	Av. capacity per site (after)	0.28	160.41

station or BSC might be using leased lines/optical cable to get directly connected to the BSC. Such sites are usually small in number. The capacity utilisation of the rest of the sites should be assessed. The possibility of replacing the leased lines with microwave radios should be considered. When using the microwave radio, the sites can be connected to each other (if LOS exists) and the capacity increases many times. Leased lines are usually one PCM while the microwave links are usually 2E1 to 16E1 capacity. This not only increases the spare capacity of the network, but also increases the chances of creating topologies thereby increasing the protection level. One such analysis is shown in Table 3.2. There are eight sites that are analysed in a region. Each site is analysed and the topology change is proposed along with the change in the media from leased line to microwave radio. Though for some sites, e.g. site 74, there is no improvement in the capacity, the overall capacity of the system improves by over 160%. This example considers just a few sites, but when capacity optimisation takes place, minor changes in topology may lead to a tremendous improvement in the capacity.

If there are new sites that are getting added at the *hot spots* during/after the optimisation process, then the capacity increase in particular links/region should be considered apart from considering the impact of the new sites.

The results of the capacity plan analysis should be very well understood and its effect on the quality and cost should be analysed before making recommendation for changes in the master plan for the implementation of these recommendations.

3.5.2.3 Quality Analysis

Quality degradation of the network may be due to many reasons. These include the degradation in the microwave link performance due to propagation conditions, degradation in the network quality due to interference and synchronisation-related problems.

The most common reason for the degradation in quality of the microwave links is the varying and unpredictable propagation conditions. In the sections above, the conditions such as fading (flat, multipath and *k*-fading), layering, ducting, attenuation due to rain, etc. have been mentioned. Once the signal fades, and this phenomenon persists, then it becomes necessary to find the causes and solutions to rectify the problems. The quality of microwave links is measured in terms of error performance and availability analysis. Once the quality targets are checked and redefined, the process of data collection starts. The data that transmission-planning engineers collect generally includes:

- Pre-planning data and existing site information such as topology, hop lengths, planned link budget calculations, frequency plans, etc.
- Link performance data: This should contain the performance data at the time of link commissioning and during the period the problem is being studied. Also, data during different periods of the year and different times of the day would be of help. Mobile call statistics during the fading and non-fading phases would enable better understanding of the correlation between the call drop and fading, if any.

Analysis should begin with recalculation of the link budgets so as to reaffirm if there were errors in the initial design or not. The data of these links monitored at the time of commissioning and during the optimisation phase should be analysed against the backdrop of the design data in the master plans (made before the network launch). One such example of the link data calculated and monitored during the optimisation phase is shown in Figure 3.26. The unavailability figures of the four links in a region are calculated and the worst case is analysed by using the data recorder for a period of 24 hours.

Once it has been detected that the received level is below the desired level, path profile analysis should be done again in case the path lengths are more than a few kilometres. Also, the error performance statistics such as the ES (error seconds) and SES (severely error seconds) should be analysed. These statistics can be collected using the NMS, as shown in Table 3.3.

The results of the above exercise would lead to the problem existing in the microwave hop. The problem(s) may be due to phenomena such as multipath fading, ducting, k-fading, etc. and the possible solution to rectify the problems should be suggested (such as use of antenna diversity, etc.) and implemented subsequently.

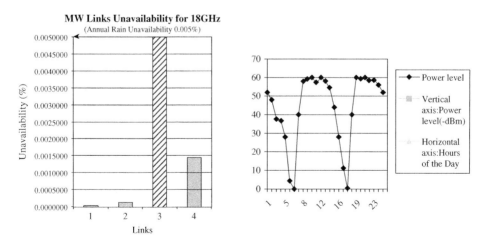

Figure 3.26 Unavailability graph for four links and power level observation of an MW link for a 24-hour period.

Table 3.3 Statistics from NMS.

	Avail_time	Degraded_min	Err_sec	Err_sec_severe
1/10/1997	86304	0	0	0
1/11/1997	86640	0	0	0
1/12/1997	83059	0	0	0
1/13/1997	83640	0	0	0
1/14/1997	63598	0	0	0
1/15/1997	63766	0	0	0

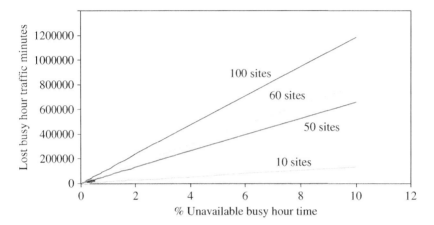

Figure 3.27 Lost traffic minutes versus percentage of unavailable time per year.

The general reason for dropping of calls in a mobile network is attributed to the radio network. In fact, in some cases, this may be due to problems related to interference and synchronisation. The traffic lost in the form of dropped calls can be sometime as high as a few thousand minutes for a small 100-site network as shown in Figure 3.27. This would result in huge revenue losses for the network operators.

With microwave transmission, the possibility becomes higher for interference being the cause of 'trouble'. The process starts with finding the cause of dropped power levels at a hop (or hops in a region). The interference problem analysis in the links or region may include the study of existing frequency plans or practical study by doing frequency scanning. The interference could be due to either an internal source or external source. An internal source of interference is easier to handle compared with an external source of interference. Once the interference has been identified, then solutions to the problems should be applied. These may include changing the frequencies, reducing the power level of the interferer, antenna discrimination, etc.

Synchronisation can be one of the reasons behind degraded network quality. This problem usually persists in mixed networks. The data collected would include mainly synchronisation plans (on whose basis the implementation was done) and NMS statistics on whether the slip frequency limit is exceeded or not. The primary source of error could be in the clock flow (i.e. MSC to BSC to BTS, etc.). In most cases, the accuracy of the PRC is unsatisfactory (against the given standards for the mobile networks) and needs replacement by an accurate clock such as mentioned earlier.

Typical outputs of the MW link budget calculations, frequency plans and interference are shown in Tables 3.4 and 3.5 and Figure 3.28, respectively.

3.6   GPRS Transmission Network Planning and Optimisation

The transmission network does not see any change from the transmission network planning in the GSM networks. The fundamental concepts remain the same for designing transmission networks for implementation of GPRS technology. However, one aspect that transmission planning engineers have to deal with is the packet control unit or PCU dimensioning. The PCU is located in the BSC as shown in Figure 3.29 and is responsible for the management of the GPRS (or packet) traffic.

The main aspect for the dimensioning of the PCU is its capacity with respect to:

- Maximum number of PDP contexts.
- Maximum number of TRXs.
- Maximum number of BTSs.
- Maximum number of PCM lines towards the A_{bis} and towards the G_b interface as well as the traffic on the G_b interface.
- Maximum number of location areas and routing areas.

The optimisation of the transmission network of GPRS is quite similar to that of GSM. There are two important aspects that might feature in a heavily loaded GPRS network. The first aspect is that as the GPRS is part of the network, i.e. the packet core may be added to the existing GSM network, there might be no changes required during the launch phase as there is less traffic. However, once the packet data users and thus PS traffic starts increasing, the capacity might need to be reviewed in the existing transmission network and there is a good possibility that the present configuration might need to be upgraded. The second aspect is the coding schemes in the GPRS network. Usually CS-1 and CS-2 are used in the GPRS network during the launch phase. If CS-3 and CS-4 are launched during optimisation or at a later phase, the A_{bis} timeslot allocation will be changed.

3.7   EDGE Transmission Network Planning

The transmission network planning process in EDGE networks is similar to that described earlier. However, due to increase in data rates, there is some new functionality added in the transmission network. EDGE transmission networks introduce a new

Table 3.4 Link data for the test network.

Link ID	Link type	Capacity	Length (km)	LOS status	Calculation method	pL value	C/K factor	Climate region/ rainfall rate	% Time	Band	Channel	Centre frequency (GHz)	Bandwidth (MHz)	Frequency designation	Latitude	Longitude	MTTR	Antenna size (m)	Antenna height (m)
Link 1	Default Microwave	16.0 Mb/s	1.46	OK	ITU-R P. 530-10	10.0	0.00042951	E	0.0100	38 GHz, 8×2	B51	39.02500	14.0000	High	60°21'58.76"N	024°38'5.03"E	6.000000	0.60	4.00
										38 GHz, 8×2	B51	37.76500	14.0000	Low	60°21'38.59"N	024°39'31.16"E	6.000000	0.60	6.00
Link 2	Default Microwave	16.0 Mb/s	1.54	OK	ITU-R P. 530-10	10.0	0.00043256	E	0.0100	38 GHz, 8×2	B52	39.03900	14.0000	High	60°21'38.59"N	024°39'31.16"E	6.000000	0.60	11.00
										38 GHz, 8×2	B52	37.77900	14.0000	Low	60°20'58.88"N	024°38'30.64"E	6.000000	0.60	13.00
Link 3	Default Microwave	16.0 Mb/s	2.26	OK	ITU-R P. 530-10	10.0	0.00042909	E	0.0100	38 GHz, 8×2	B51	39.02500	14.0000	High	60°20'5.75"N	024°40'11.93"E	6.000000	0.60	42.00
										38 GHz, 8×2	B51	37.76500	14.0000	Low	60°20'58.88"N	024°38'30.64"E	6.300000	0.60	44.00
Link 4	Default Microwave	16.0 Mb/s	2.68	OK	ITU-R P. 530-10	10.0	0.00042625	E	0.0100	38 GHz, 8×2	B52	39.03900	14.0000	High	60°21'12.28"N	024°42'4.14"E	6.300000	0.60	25.00
										38 GHz, 8×2	B52	37.77900	14.0000	Low	60°20'5.75"N	024°40'11.93"E	6.300000	0.60	25.00
Link 5	Default Microwave	16.0 Mb/s	2.48	OK	ITU-R P. 530-10	10.0	0.00043086	E	0.0100	38 GHz, 8×2	B52	39.03900	14.0000	High	60°21'38.59"N	024°39'31.16"E	6.000000	0.60	26.00
										38 GHz, 8×2	B52	37.77900	14.0000	Low	60°21'12.28"N	024°42'4.14"E	6.000000	0.60	27.00
Link 6	Default Microwave	16.0 Mb/s	2.18	OK	ITU-R P. 530-10	10.0	0.00043869	E	0.0100	38 GHz, 8×2	B52	39.03900	14.0000	High	60°21'12.28"N	024°42'4.14"E	6.000000	0.60	25.00
										38 GHz, 8×2	B52	37.77900	14.0000	Low	60°20'27.71"N	024°43'54.47"E	6.000000	0.60	25.00

Table 3.5 Detailed link performance data for the test network.

Link ID	Freespace loss (dB)	Atmospheric absorption (dB)	Obstruction loss (dB)	Total loss (dB)	Rx level (dBm)	Threshold value (dBm)	Threshold degradation (dB)	Composite fade margin (dBm)	Flat fade margin (dB)	Flat fade margin after interference (dB)	Req. FM against rain (dBm)	Interference margin (dB)	Dispersive fade margin (dB)	Flat outage (PnS) (%)	Selective outage (Ps) (%)	Red. of X polar discrimination (%)	Total worst month outage (Pt, w/o Div) (%)
Link 1	127.4243	0.1633	0.0000	138.5876	−35.1876	−85.0000	0.0000	48.8959	49.8124	49.8124	7.0421	0.0000	56.1024	0.0000000	0.0000000	0.0000000	0.0000000
	127.4243	0.1633	0.0000	138.5876	−35.1876	−85.0000	0.0000	48.8959	49.8124	49.8124	7.0421	0.0000	56.1024	0.0000000	0.0000000	0.0000000	0.0000000
Link 2	127.8877	0.1723	0.0000	129.0600	−25.6600	−85.0000	0.0000	54.4160	59.3400	59.3400	7.4068	0.0000	56.1024	0.0000000	0.0000000	0.0000000	0.0000000
	127.8877	0.1723	0.0000	129.0600	−25.6600	−85.0000	0.0000	54.4160	59.3400	59.3400	7.4068	0.0000	56.1024	0.0000000	0.0000000	0.0000000	0.0000000
Link 3	131.2242	0.2529	0.0000	132.4770	−29.0770	−85.0000	0.0000	53.0015	55.9230	55.9230	10.5880	0.0000	56.1024	0.0000001	0.0000002	0.0000000	0.0000000
	131.2242	0.2529	0.0000	132.4770	−29.0770	−85.0000	0.0000	53.0015	55.9230	55.9230	10.5880	0.0000	56.1024	0.0000001	0.0000002	0.0000000	0.0000000
Link 4	132.7115	0.3003	0.0000	134.0118	−30.6118	−85.0000	0.0000	52.1510	54.3882	54.3882	12.3770	0.0000	56.1024	0.0000001	0.0000004	0.0000000	0.0000000
	132.7115	0.3003	0.0000	134.0118	−30.6118	−85.0000	0.0000	52.1510	54.3882	54.3882	12.3770	0.0000	56.1024	0.0000001	0.0000004	0.0000000	0.0000000
Link 5	132.0359	0.2778	0.0000	133.3137	−29.9137	−85.0000	0.0000	52.5544	55.0863	55.0863	11.5339	0.0000	56.1024	0.0000001	0.0000002	0.0000000	0.0000000
	132.0359	0.2778	0.0000	133.3137	−29.9137	−85.0000	0.0000	52.5544	55.0863	55.0863	11.5339	0.0000	56.1024	0.0000001	0.0000002	0.0000000	0.0000000
Link 6	130.9188	0.2443	0.0000	132.1630	−28.7630	−85.0000	0.0000	53.1589	56.2370	56.2370	10.2530	0.0000	56.1024	0.0000000	0.0000001	0.0000000	0.0000000
	130.9188	0.2443	0.0000	132.1630	−28.7630	−85.0000	0.0000	53.1589	56.2370	56.2370	10.2530	0.0000	56.1024	0.0000000	0.0000001	0.0000000	0.0000000

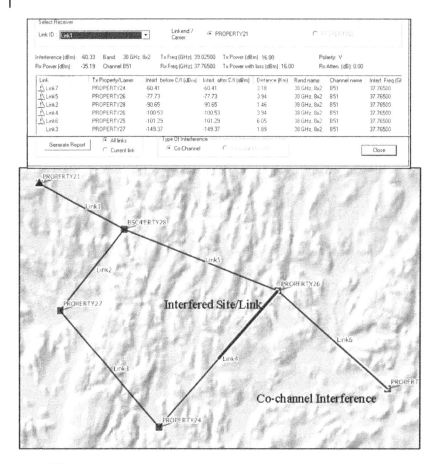

Figure 3.28 Interference analysis.

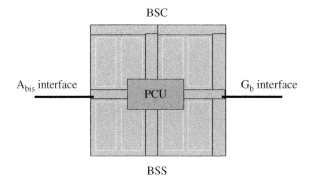

Figure 3.29 Packet core unit (PCU).

functionality known as 'dynamic A_{bis}.'[1] Thus, the following concepts assume importance in the design of an EDGE transmission network:

- dynamic A_{bis}
- dimensioning of dynamic A_{bis}
- dimensioning of PCU/BSC

Figure 3.30 A_{bis} interface.

We have already some aspects of related to dimensioning of the PCU/BSC. Here we focus on dynamic A_{bis} and its dimensioning.

The interface between the base station and the BSC is known as the A_{bis} interface as shown in Figure 3.30. The main difference between the A_{bis} of the GSM/GPRS networks and that of the EDGE networks is that in the former the nature of A_{bis} is 'static' while in the latter the nature of A_{bis} is 'dynamic'. As we have seen the earlier, in GSM/GPRS networks, the TRX channels are mapped onto the A_{bis} PCM timeslots. Each TCH uses two bits of PCM frame and these two bits together are known as PCM sub-timeslots. This is 'static' in nature because each TRX reserves its full capacity from the A_{bis} interface constantly even if there are no active users in the air-interface. This makes dimensioning of the A_{bis} interface a lot easier compared with that in EDGE networks.

3.7.1 Dynamic A_{bis} Functionality in EDGE Networks

The octagonal phase shift keying changes the data rates from as low as 8.8 to 59.2 kbps, i.e. from coding scheme MCS-1 to MCS-9. Although the voice signal is still carried in 16 kbps A_{bis} channels, for data this proves insufficient especially beyond coding schemes MCS-2. To carry more than 16 kbps traffic in the air-interface, the data traffic needs more than 16 kbps A_{bis} channels, probably 32, 48, 64, or 80 kbps. Now as this data traffic would not be there all the time, the concept of 'dynamic' Abis came into being. A group of these dynamic A_{bis} channels is known as a dynamic A_{bis} pool or DAP. Figure 3.31 shows an example of 'static' and 'dynamic' A_{bis}. DAP consists of a minimum of one timeslot and the maximum depends upon the system's capability within one PCM.

With EDGE TRXs, the BSC allocates A_{bis} capacity for data calls from the EGPRS dynamic pool (EDAP) when needed, i.e. when MCS-3 or higher is used. Standard GPRS (non-EGPRS) calls using CS-2, CS-3, and CS-4 can also use EDAP resources when allocated into EGPRS territory (EDGE TRXs).

3.7.2 Timeslot Allocation in Dynamic A_{bis}

Timeslot allocation in A_{bis} of an EDGE transmission network is almost the same as that of A_{bis} of the GSM/GPRS networks. Traffic channels on A_{bis} still occupy 16 kbps sub-timeslots for voice and data rates up to 16 kbps. However, for data rates more than 16 kbps, one master sub-timeslot of 16 kbps and up to four sub-timeslots from the DAP are required, as shown in Figure 3.32. The requirements for signalling channels for

---

1 Dynamic A_{bis} feature is vendor specific. In the equipment of some vendors, this feature may not be there. However, this feature clearly gives an edge to the EDGE transmission network as explained later in this chapter.

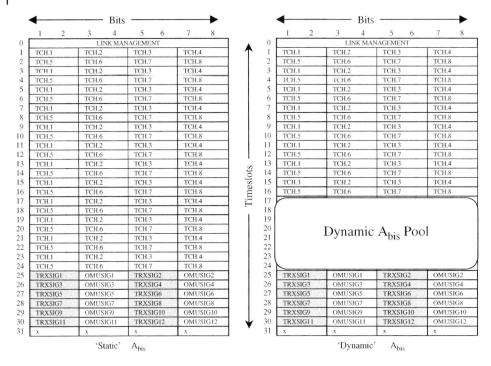

Figure 3.31 Static and dynamic A$_{bis}$.

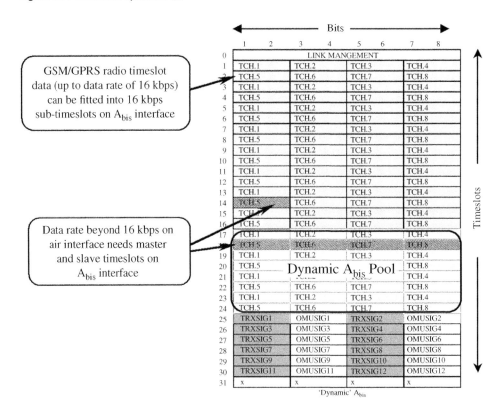

Figure 3.32 Timeslot allocation on dynamic A$_{bis}$.

TRXs and BCF are the same as that of GSM/GPRS networks. The pilot bits of synchronisation can be accommodated in the TS0 or TS31 (or any other timeslot).

3.7.3 Dimensioning of Dynamic A_{bis}

The dimensioning of the dynamic A_{bis} is the most important aspect of the transmission network planning process in EDGE networks. The dynamic nature of the A_{bis} and the allocation of a group of timeslots for the DAP changes the static capacity calculations of the GSM/GPRS networks. In the GSM/GPRS networks, the A_{bis} allocates the whole capacity for the TRX irrespective of usage; this makes the capacity calculations simpler, as the knowledge of number of TRXs and signalling rate would lead to the capacity required on the A_{bis} interface. However, with the increased data rates and dynamic nature of the DAP, the capacity calculations become slightly more complicated. However, before we look into the DAP dimensioning, it should be remembered that DAP is used only for the packet data for MCS greater than 2. Also, for each of the coding schemes, the number of required traffic channels will vary from none in MCS-1 (or CS-1) to four in MCS-8/9, as shown in Figure 3.33. However, designers should refer to the product information and the system capabilities before they design the pool. Also, the number of the A_{bis} channels will increase when signalling is added to it, which again depends upon the system being used in the network. The A_{bis} channels that are present in the pool are also known as 'slaves'. Based on the requirements of the mobile subscriber, the sub-timeslots from the pool are requested. The number of timeslots

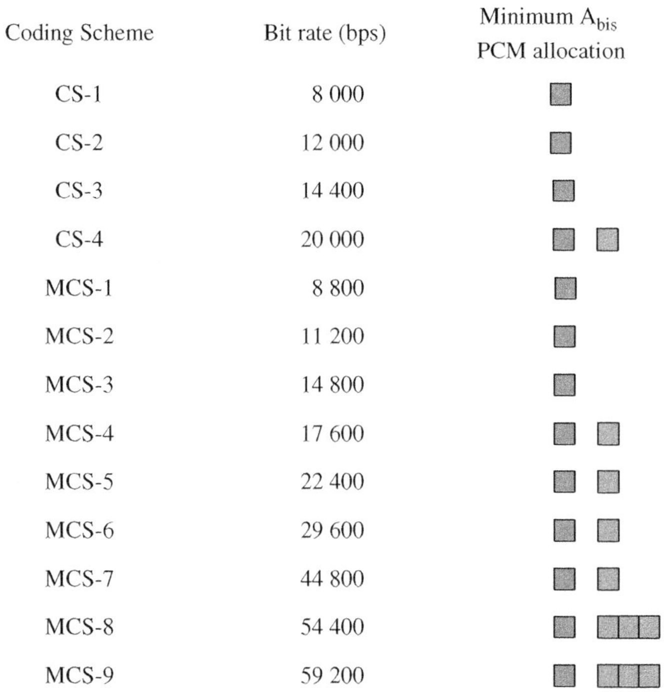

Coding Scheme	Bit rate (bps)	Minimum A_{bis} PCM allocation			
CS-1	8 000	▨			
CS-2	12 000	▨			
CS-3	14 400	▨			
CS-4	20 000	▨	▨		
MCS-1	8 800	▨			
MCS-2	11 200	▨			
MCS-3	14 800	▨			
MCS-4	17 600	▨	▨		
MCS-5	22 400	▨	▨		
MCS-6	29 600	▨	▨		
MCS-7	44 800	▨	▨		
MCS-8	54 400	▨	▨	▨	▨
MCS-9	59 200	▨	▨	▨	▨

Figure 3.33 Transmission requirements for EGPRS coding schemes on an A_{bis} interface.

allocated to the pool is usually in whole numbers, i.e. one, two, three, four or more timeslots, depending upon the DAP handling capacity of the system. Obviously, the number of timeslots allocated to DAP cannot be more than 30 (if TS0 and TS31 are 'reserved' for management purposes).

Dimensioning of dynamic A_{bis} means dimensioning of the pool. Once the number of timeslots required for DAP is known, the remaining process becomes similar to dimensioning/planning of the GSM/GPRS transmission network.

The main idea behind DAP dimensioning is to find the number of timeslots that can be assigned to a pool in such a manner that the A_{bis} does not becomes a limitation for the air-interface throughput. Thus, the main inputs for this would be the number of radio timeslots required for PS traffic (i.e. dedicated and default territory in EDGE radio network), capacity of radio timeslots, blocking that could take place on the A_{bis} interface, etc. Based on these inputs and the number of PCM timeslots available for the pool, the required number of PCM timeslots can be calculated. However, limitations and capability of the equipment needs to be taken into consideration. With this input, the blocking probability of DAP could be determined.

$$B(n,N,p) = \sum_{x=N+1}^{n} P(x) \tag{3.15}$$

where N is the number of timeslots available in the pool, p is the channel utilisation of the EDGE channels in the air-interface, n is the number of traffic channels used in the air-interface, and B is the blocking probability of the DAP.

The dimensioning would result in outputs such as the maximum throughput possible on the air-interface with A_{bis} acting as a bottleneck and the number of timeslots required to allow the maximum possible throughput on the air-interface to be carried on the A_{bis}.

3.7.4 Impact of Dynamic A_{bis} on Transmission Network Design

Assume that there is a base station site having a configuration of $3 + 3 + 3$. When operating in the GSM mode, using 16 kbps signalling, the required number of PCM timeslots per TRX is 2.25 (two for traffic and 0.25 for 16 kbps signalling). This means that the total number of timeslots required for these 9 TRXs is $9 \times 2.25 = 20.25$ or 21 TS (approx.). This would mean that of 32 timeslots, i.e. TS0 and TS31 along with these 21 timeslots, meaning 23 timeslots are utilised, nine timeslots available for possible future upgrade.

However, if the same site configuration is used for EDGE networks, then these nine timeslots could be used for the DAP. However, it should be seen whether these nine timeslots are sufficient or not. Now, consider that this EDGE network using MCS-9 coding scheme, i.e. each user that logs in the network for PS services requires four A_{bis} channels or one full additional timeslot apart from the master sub-timeslot on the PCM. This means at a given time only nine users can utilise the system at maximum rate of 59.2 kbps. As soon as the 10th subscriber logs into the network also requesting the MCS-9 scheme, the system starts sharing its resources. This means that though the access would be given to this 10th subscriber, not only will this subscriber's request be downgraded to lower coding schemes such as MCS-6 or MCS-7, but also the coding scheme of the existing subscribers logged into the network may be downgraded to MCS-6 or MCS-7, in order to accommodate this new subscriber. This would mean the throughput on the air-interface goes down due to the limitation on the A_{bis}. Thus, in this case, it would be advisable to have more timeslots in the pool. This would mean more PCMs would be required per site. This would affect the costs of the network. Thus, the number of timeslots planned for DAP should be a balance between the throughput and costs (i.e. additional PCMs required).

Example 6 *Dynamic A_{bis} dimensioning dimensioning for one one single single sitesite.Consider one single site of configuration 1 + 1 + 1. It should be able to deliver a data rate of 59.2 kbps (i.e. MCS-9) by using two radio timeslots. What is the number of timeslots that should be reserved for dynamic Abis so that the Abis data rate reduction factor is less than 1%? How many E1s are required for the BTS-BSC connection? The signalling is on the GSM layer.*

As one user is able to use two radio timeslots to get a data rate of 59.2 kbps, hence four users can use eight timeslots at a given time. This makes the number of data erlangs to be four.

EGPRS territory = 8
Data rate = 59.2 kbps
Data rates after C/I = 40 kbps
For a DAP = 12 TSL on A_{bis}.

Based on the binomial distribution formula stated, the results are as follows:

- *The sharing probability of the DAP is only 3.176%.*
- *The A_{bis} data rate reduction is less than 1%, which means that the A_{bis} will not be a bottleneck for the throughput of the air interface.*
- *The average throughput per RSL, limited by radio and A_{bis} is 19.88021.*
- *The A_{bis} data rate reduction is 0.60%.*

Now as 1 + 1 + 1, i.e. 3 TRXs would need only $3 \times 2.25 = 6.75$ timeslots and DAP will require 12 timeslots, hence the total number of timeslots needed would be 6.75 + 12 = 18.75. Apart from this TS0 and TS31 can be kept for management, etc. Thus, one E1 would be sufficient for connecting this site to the BSC.

The combined result of radio and transmission network dimensioning is shown in Table 3.6. Table 3.6 is the result of a dimensioning case, where the maximum throughput per user is 96 kbps. However, due to C/I limitations, the acceptable user throughput is reduced to 64 kbps per user. The dimensioning is done for coding scheme MCS-7. The EDAP size is nine timeslots.

3.7.5 EDGE Transmission Network Optimisation

The process of transmission network optimisation is quite similar to the transmission network optimisation in the GSM networks. However, there are two factors that need to be considered during the optimisation process: the dynamic nature of the A_{bis} and the variation of the air-interface throughput due to the bottleneck created by the A_{bis}. As there is a pool of timeslots reserved for packet data, it will always be the case that number of packet service users increase in the network. Though the pool itself would be

Table 3.6 RNP + TNP dimensioning output.

1	Data traffic in the BH	DT		96	kbps per cell
2	Acceptable average user throughput			64	kbps
3	RTSL per mobile	Nu		2	
4	Timeslot capacity	k		48	kbps
5	PS traffic intensity	Tps	DT / k	2	Data Erlangs
6	Number of RTSL per cell	Ns		4	
7	Utilisation		Tps / Ns	0.5	
8	Reduction factor f (Nu=2, Ns=4, utilisation)	RF	Calculated by RF	0.75	
9	Avg user throughput (without transmission limitations)	Trg	RF^*Nu^*k	72	
10	EDAP pool size			9	
11	A_{bis} data rate reduction factor	$A_{bis}RF$	Calculated by TRS	0.9962	
12	Avg user throughput (with A_{bis} transmission limitations)	E	$Trg^*A_{bis}RF$	71.728	

dynamic however, the number of timeslots that would be used for the pool would be static and once it is determined that the required timeslots (in the pool) need to be increased, some optimisation would be required. The case would be more severe if the number of PCM required per base station would need to be increased.

4

Core Network Planning and Optimisation

4.1 Basics of Core Network Planning

4.1.1 Scope of Core Network Planning

The core network in GSM is basically the circuit core. In GPRS, EDGE and UMTS, it has two components, the circuit core and the packet core, which are responsible for voice and data, respectively. In GSM, core or circuit core network planning is also known as switch planning, because this planning is mainly related to the mobile switching centre or MSC, also called 'switch' (Figure 4.1).

Core network planning in GSM consists of network elements such as MSC, VLR, HLR, AC, and EIR.

4.1.2 Elements of the Core Network

4.1.2.1 MSC/VLR

The MSC is the core elements of the network sub-system. It is responsible for the switching of subscriber calls and thus is also known as a 'switch'. It is also responsible for functions such as traffic management, paging, and collection of charging related information. Apart from this, the MSC also acts as an interface between the network and the networks of other operators and PSTN networks. Another important element that is hosted in an MSC is the VLR or visitor location register. The VLR contains the information on the subscribers that is handled by the MSC at a given moment. Owing to the nature of its functionality, the VLR participates in call processing and mobility management functions of the network such as the use of temporary mobile identification, IMSI attach/detach, etc., apart from location registration of the mobile subscriber. The VLR is also involved in security related functions such as IMEI (International Mobile Equipment Identification) checking.

4.1.2.2 HLR/EIR/AC

The home location register or HLR is responsible for keeping the subscriber data, i.e. all permanent information regarding subscribers and their equipment. It is also responsible for creating, modifying, deleting or managing this data. Apart from this, the HLR also participates in call processing functions such as routing, supporting the incoming call barring service/unconditional call forwarding service, etc. The HLR

Fundamentals of Network Planning and Optimisation 2G/3G/4G: Evolution to 5G,
Second Edition. Ajay R. Mishra.
© 2018 John Wiley & Sons Ltd. Published 2018 by John Wiley & Sons Ltd.

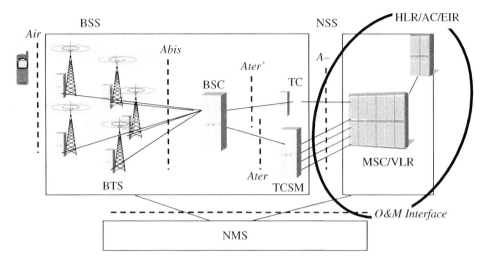

Figure 4.1 Scope of core network planning.

consists of an EIR and an AC. The EIR or equipment identification register contains databases for IMEI. The EIR is the place where mobiles that are missing or stolen are registered. Actually, the EIR contains three lists: white, grey and black. The white list contains IMEI for all authentic mobiles, the grey list has data for faulty mobiles, while the black list contains information regarding stolen/missing mobiles. Thus, when a call is being made, the number is crosschecked from the EIR and if it is found in the white list, the call goes ahead, but if the number is found in the black list, the call may be blocked. The AC or authentication centre is mainly responsible for security-related aspects. As the name suggests, it manages the security-related information for the authentication of the subscribers. This information is usually requested by the VLR and, in coordination with the EIR it prevents a non-authenticated card from accessing the network.

4.2   Core Network Planning Process

The core network planning process is slightly different from the radio or transmission network planning process. It contains two parts: switch network planning and signalling network planning. In the RNP or TNP, environmental factors and site selection processes assume much importance. However, these two are missing from the core network planning process, thereby reducing the process to three main steps before drawing the final core network plan. These steps are as shown in Figure 4.2:

- Network analysis
- Network dimensioning
- Detailed planning

The first two are explained separately for switch and signalling network planning.

Figure 4.2 Core network planning process.

4.2.1 Network Analysis

Core network planning becomes important due to the fact that the number of switches (or MSCs) in any network is far less than the number of sites or links. Some networks even have only one switch, which means that they cannot be changed on a 'daily basis' as that would have an impact on the performance of the whole network. Switch planners have to take into account the forecasts more seriously as switches are expected to cater for the present traffic and the expected traffic rise for a few years before their number is increased.

Network analysis needs data ranging from subscriber information to demographic information. The major dataset consists of the existing network data, the existing service plan, the subscriber base data, topographical data, and traffic data (both existing and forecast). Apart from these, some radio network planning data and/or transmission planning data might also be needed.

Existing network data information gives the core planning engineer information about the number of sites and the traffic it is expected to generate. Information such as existing and spare network capacity gives the network an idea about the existing traffic and maximum traffic that could be generated if the entire network were to be utilised to its fullest capacity. The total traffic that exists within a network usually consists of traffic that is generated within the network and traffic that was generated outside the network, e.g. calls being made to mobile subscribers from fixed lines. This traffic distribution is very helpful both during the planning phase and optimisation phase of the core network.

The next thing to find out is the services that need to be delivered to mobile subscribers. Apart from voice, this generally includes value added services such as SMS, MMS, internet, etc. Information on the types of service and the expected traffic it is expected to generate will be helpful during the planning phase.

Subscriber base data is one of the major inputs for switch planning. This includes the existing subscribers or subscribers expected at the time of network launch, the expected traffic that will be generated by these initial subscriber figures, forecast subscribers and forecast traffic. The forecast may be in phases that may depend upon expansion of the network or subscriber base. The information related to the different regions (urban, rural, hot spots, etc.) and the type of subscriber base (business users, residential users, etc.) would give a fair idea of the expected traffic. Other useful information comes from the numbering study in the network that includes the roaming numbers, IMSI numbers, signalling point codes (SPC), emergency numbers, allotted subscriber numbers, etc.

4.2.2 Network Dimensioning

All the information collected during the network analysis phase is required in the network dimensioning phase. So, what is the output of the dimensioning exercise? It is the number of nodes that are required for handling the subscribers (and traffic) efficiently

(a)

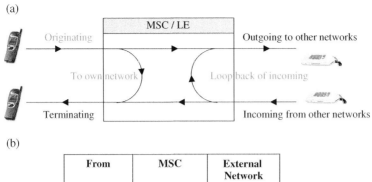

(b)

From	MSC	External Network
MSC	X	800
External Network	1600	X

Figure 4.3 (a) Traffic flow; (b) traffic generation (in Erlangs).

for a longer period of time. Here is the brief overview of the major outputs of dimensioning of the switch network:

- Expected traffic generated in the network.
- Number of switches required handling the subscribers and traffic (both present and forecast).
- The most efficient location of the switches in the network.
- How the switches will be connected to each other, i.e. transmission plan for the switches.
- Most efficient way to route the traffic.

The traffic calculation is perhaps the most important aspect of switch planning. It has to be as accurate as possible and all the factors discussed in the previous section contribute to the accuracy of this calculation. One such example is shown in Figure 4.3a and b. When there is more than one switch or more than one external network (i.e. PSTN, other operator, etc.), the generation table become bigger and more complicated as traffic flow from each of the switches to/from the other switches and external network need to be taken into account. Of all the traffic existing in a cellular network, some percentage of the traffic is being generated and terminated in the same network, while some traffic is terminating in other networks and vice versa. A simple example follows. The usual case would be that of a multi-switch. In such cases, the traffic scenario (also known as the traffic matrix) becomes more complicated as the traffic locations where traffic is originating and terminating become more. Thus, inputs like the number of subscribers calling from mobile to mobile, mobile to fixed, and fixed to mobile are required.

Example 1   Basics of traffic calculation

Consider a simple core network with one switch. The traffic present is due to its own subscribers and traffic coming in from external sources (or local exchange). This is a scenario that can be represented by Figure 4.3a. Before the dimensioning starts, a few parameters need to be understood and defined.

Subscribers originating (SO): This is the traffic amount that is originated by subscribers of the network. Generally, this input comes from the network (or network operator). Typical values can be 65% for switch and/or 50% for the LE.

Subscribers terminating (ST): This is the traffic that is being terminated in the mobile network. This value can be calculated as ST = 100% – SO.

Own network (terminating) traffic: This is the traffic that is originated in the network and getting terminated there. It is usually a product of subscribers in the own network and SO.

Loop back of incoming traffic: This is the traffic originating from the external networks and routed back to them.

Outgoing and incoming traffic: This is the traffic going outside the mobile network to an external network is outgoing traffic while traffic coming from external networks is incoming traffic. These can be calculated as follows:

$$Incoming\ traffic = (ST - own\ network\ traffic)/(1 - loop\ back\ of\ incoming\ traffic)$$

$$Outgoing\ traffic = SO - own\ network\ traffic + loop\ back\ of\ incoming\ traffic$$

Voice mail system traffic/interactive voice response (IVR) traffic: This is traffic related to voice mail. It is of two types: forwarded and listen. When the traffic is diverted to voice mail when the subscriber is not available, it is known as VMS forwarded traffic, while traffic generated to listen to voice message is the VMS listen traffic. IVR is the traffic generated through the interactive voice responses, i.e. when a recorded voice responds rather than a subscriber, e.g. press 1 for Hindi, press 2 for English, etc.

Inter-switch traffic: This is calculated on the basis of the incoming and outgoing traffic. Generally tools are used for this.

Another important parameter related to subscriber behaviour is the 'calling and moving interest'. Calling interest indicates the distance to which calls are made (i.e. long distance or short distance) as shown in Figure 4.4. This figure is useful for inter-switch modelling when calculating the number of subscriber-originated calls. Moving interest is another parameter that defines the subscriber behaviour of moving within the network. The external calls would then be diverted via the short path using the HLR-call enquiry feature.

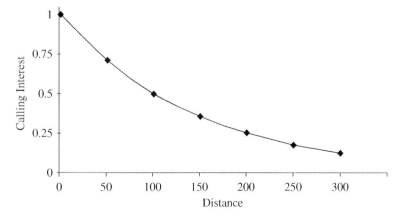

Figure 4.4 Calling interest.

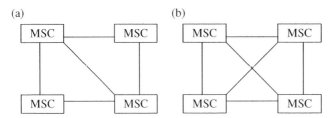

Figure 4.5 Interswitch connection.

Once the number of subscribers has been determined, the number of switches can be calculated. The subscriber number directly affects the VLR or the visitor location register. Thus:

$$\text{Number of switches} = \text{Number of subscribers}/\text{VLR (or HLR) capacity}$$

Once the number of switches has been determined, the next step is to locate them. If there is only one switch, the decision is quite easy. It is generally located near the mobile headquarters or at a place with easy accessibility. In this case, the expected traffic is also low and there are no routing plans required as all the traffic is going only to one switch. However, the scenario changes when there is more than one switch. The main idea is to keep the switch location in an area of high subscriber density. This would save transmission costs.

After the number of switches and their location is decided, the traffic route needs to be defined. There are different ways to route the traffic. It may be done in two ways: partly mesh or completely mesh. Assume that there are four switches. The traffic can be routed as shown as in Figure 4.5a, where all the MSCs are connected in cyclic fashion with some protection afforded by a diagonal connection. However, as shown in Figure 4.5b, all the MSC are inter-connected, i.e. traffic from each flows into the rest of the MSCs. Though the former technique is simple, it results in a higher amount of transit traffic as compared with the latter where there are more connections but the transit traffic is less. Obviously, the traffic in Figure 4.5b is more protected. This is a simple case. If the number of switches increases, then the mesh technique becomes more complicated and routing of all of the traffic to all the switches become a difficult task. In such cases, another dimension is added to routing, i.e. transit switches are added.

If a transit switch were added to the network, it would look as shown in Figure 4.6. All the traffic, generated both internally and externally, will be routed through the transit switch. As the transit switch in this case has assumed importance, it is often advised to have a redundant transit switch and the traffic through each of the switches is routed through both the transmit switches. If the cellular network is spread over a very large region and has many switches, then the number of transit switches would increase. Although the MSCs are connected to the nearest transit switches, the network topology between the transits switches is themselves is fully meshed.

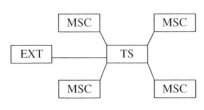

Figure 4.6 Switch network with transit switch.

Input				
	MSC1	MSC2	MSC3	MSC4
X	0	0	200	100
Y	0	0	0	100
subscribers	50000	50000	50000	50000
tr/subs (mErl)	27	27	27	27
total traffic	1350	1350	1350	1350
subs originating	60%	60%	60%	70%
subs terminating	40%	40%	40%	30%
to own network	20%	20%	20%	20%
to own NW of total	12%	12%	12%	14%
to own NW of total	162	162	162	189
loopback	5%	5%	5%	5%
initial outg ISW	110	110	93	136
initial outg ISW%	8%	8%	7%	10%

call int	215
move int	50

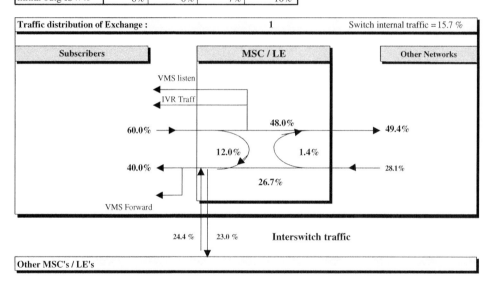

Figure 4.7 Example of traffic dimensioning.

Next the routing plan is devised. The routing plan is responsible for the flow of traffic between the switches. One single route is not overloaded with traffic but the traffic is evenly divided between various routes in the best possible way. The routing plan gives the direct path that is taken by the traffic and in the case of failure of the primary route, the secondary path taken by traffic. The secondary path choice is dependent upon the traffic it is carrying. Thus, even in the initial phase of the network, spare capacity planning is required so as to route the calls through the secondary paths when the primary are blocked.

One example of traffic dimensioning is shown in Figure 4.7. There are four MSCs and an external network. The number of subscribers handled by each MSC is 50 000. With the traffic generated by each subscriber being 27 mErl, the total traffic generated by each MSC is 1350 Erl. The percentage of calls originating within their own switch is 60% of which 40% terminate in the same network. Calling interest and moving interest are 215 and 50, respectively.

4.3   Basics of Signalling

4.3.1   Signalling Points

Once switch network analysis and dimensioning is done, the next step is the analysis and dimensioning of the signalling network. As mentioned in Chapter 1, the signalling in the NSS network is SS7. Signalling can be transferred on the 64 kbps timeslots in PCM. Any point that is capable of sending or receiving the signalling is known as a SP or signalling point. This can be a STP or signalling transfer point, or signalling end point or SEP. Signalling traffic passes through the STP and reaches its intended location or SEP. The SP and SEP are the same. The SPs are connected to each other via PCM links. A SP can also act as a SCP or signalling control point. Through this point, access to advanced services can be made such as to freephone numbers.

Signalling networks can be of different types. Signalling network indicators give the information on the type of signalling network used. In GSM networks, as SS7 is used for signalling, four different types of network are possible, namely NA0, NA1, IN0 and IN1. The former two are for national networks while the latter codes are used for international networks. A SP can support one or more than one network address. It can also support all four networks, but in that case it would have four different SPCs, one for each network.

4.3.2   Signalling Links

A signalling link is a logical connection between two SPs. SPs are connected using PCM links (also known as PCM circuits). The purpose of the signalling link is to carry messages of higher layers. Thus, the signalling link has the ability to define the start and end of the frame structure, locate the frame for initial alignment, maintain the signalling link, detect errors, etc. Usually two signalling links are used between two SPs. This is done for protection purposes. These two signalling links are sent on two physically separate PCM links. These signalling links constitute a signalling link (SL) set as shown in Figure 4.8. The maximum number of signalling links that can constitute a signalling link set is also limited, e.g. 16. These signalling links are usually assigned priorities. If there is a failure, the highest priority link carries the traffic of the failed signalling link.

For protection purposes, there are different routes designated to carry the signalling from one destination to another. These destination points are known as destination point codes or DPCs. There is also a limit to the maximum number of routes that can be assigned between two DPCs. However, there should be a minimum of two routes for protection purposes.

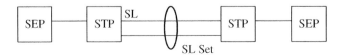

Figure 4.8 SEP, STP, SL and SL set.

4.3.3 Signalling Network Dimensioning

Signalling is transferred on 64 kbps timeslots on the PCM. High capacity links use more capacity, i.e. more than one timeslot. But, as traffic increases, a delay will take place in delivering the messages. A acceptable delay used in signalling dimensioning is usually 0.2 Erl. For performance, 0.4 Erl is used and for satellite connections a value of 0.6 Erl is used. Thus, one of the main aspects to be remembered during dimensioning is to not exceed the traffic load recommended. The process of dimensioning of the signalling network is as follows:

- dimension the end-to-end traffic (based on capacity assessment)
- route the traffic
- transmission required

The end-to-end traffic calculation is perhaps the most tedious as it is the whole basis of signalling dimensioning and planning. The main parameter inputs required for this include the call related parameters, i.e. related to successful calls. Short calls are essentially considered in core network planning because they generate large signalling traffic. They also include parameters related to the ISDN user part and telephony user part, i.e. TUP/ISUP. Another aspect to be considered is the traffic generated by messages, i.e. MAP or mobile application part parameters. Routing related parameters such as signalling link utilisation, signalling link set, etc. form another set of parameters to be considered for dimensioning. TUP/ISUP are call-related signalling while the MAP is non-circuit-related signalling. Then there is signalling related to the intelligent network.

4.4 Intelligent Network (IN)

The intelligent network concept in the core network permits services to be introduced in a network in a more cost-effective way. Also, this allows these services to be managed and controlled more effectively. How is this done? This is done by implanting these services in a common/standard database instead of implanting them in each network element. This implementation makes the network more 'intelligent'. Pre-paid SIM cards, originating call screening, reverse charging, freephone (1–800-numbers), etc. are some examples of services provided by an IN network. For planning such a network, inputs such as the number of initial and forecast subscribers, traffic behaviour, network topology, etc. are used. However, when doing the switch and signalling planning, IN network requirements should also be taken into account. Once the inputs are in place, the IN network analysis is done for setting the performance objective and connection requirements based on the traffic estimates. This is done for each of the IN module/services and then finally the total traffic demand from all the IN services offered by the IN platform is calculated. Based on these inputs, INAP or IN application protocol requirements and the subsequent IN platform (on which IN services run) is determined. Then some of the switches are upgraded by addition of IN capable platforms and traffic from the rest is routed through these switches. Impact on the core network would be in terms of capacity used by these IN service apart from the fact that IN services would generate their own signalling, thus increasing the total signalling in the core network. One of the most important element in the IN networks is the IP or intelligent peripheral. This is basically a stand-alone processor that provides the additional services for which IN is implemented in a network. IP functions include IVR, DTMF (dual tone multi-frequency) translation, speech recognition, providing access to the signalling networks, etc.

Based on the above factors, the number of signalling links can be calculated. Usually, network-planning tools are used for the purpose *(see Appendix A)*. The output of the signalling network dimensioning is the required number of signalling links. Once this is done, the number of ET ports can be calculated for the same.

4.5   Failure Analysis and Protection

As mentioned earlier in this chapter, the number of switches is fewer than the number of base stations or base station controllers, and in some cases there is only one switch in the network. A small failure in the core network may lead to a large part of the radio network becoming non-functional leading to huge revenue losses. Thus, protection of the network becomes an inseparable and important part of core network planning. For designing (or assigning) protection, it is important to know the main failures that may take place in a network. These failures are:

- site failure
- equipment/node failure

A site rarely fails. If it does, it is usually due to conditions that are beyond human control, e.g. floods, earthquakes, fires, etc. Though these natural calamities cannot be prevented or their effects predicted, some steps should be taken to minimise their effects. These may include choosing a site at higher altitude in regions where floods are a common (or annual) phenomenon and avoiding buildings or locations that do not have proper fire detection and prevention equipment. Proper cooling arrangements can avoid over-heating (and subsequent damage) of network elements.

These above reasons may lead to equipment or node (MSC/HLR, etc.) failures. Apart from the site failures, small failures such as power problems or transmission problems may lead to temporary failure of the equipment/node. One MSC or one HLR handles about 0.5 million and one million subscribers, respectively (these values change depending upon the equipment manufacturer and individual equipment capacity). Thus, if one MSC or HLR fails completely, almost 0.5 million or one million subscribers, respectively, are 'lost', resulting in huge revenue losses to the operator. Using redundant MSC/HLR is one way to reduce the lost traffic due to equipment/node failures. The situation in networks having one switch is quite severe in cases of failure. However, in networks that have more than one switch, the failure can be recovered from. One of the methods is to move the base stations from a failed MSC to another 'live' or redundant MSC. It should be noted that when such a solution is being implemented, the second 'live' MSC should have existing capacity to handle the traffic of the failed MSC. In traditional networks, the HLR data was transferred using DAT tapes. However, these days, real-time update of the subscriber data takes place between the 'live' and redundant HLR. In some cases, all of the equipment may not fail, but only slight damage/malfunctioning may lead to link or route failures. Typical examples of this may include a damaged or disconnected PCM cable or ET port failure. There may be cases when route failure takes place. In such cases, the existing routes may experience higher load leading to higher delays. Thus, for these cases, it is always recommended to define more than one signalling link in each link set, and/or have a redundant signalling route. During dimensioning, each of the links should be over-dimensioned, so that if there is a failure, the link should be able to carry extra traffic. Thus, there has to be a trade-off between the resources and failure protection in core networks.

4.5.1 Output of Network Dimensioning

As mentioned above, network planning tools are used for dimensioning and planning of core networks. Some results of the dimensioning exercise are explained below.

- Traffic calculations: Based on inputs such as the number of subscribers, calling interest, moving interest, forecast number of subscribers, etc., a traffic matrix is generated.
- Number of switches, their capacities and their location: The number of switches required is one of the main outputs of dimensioning. This is based upon subscriber capacity, geographical conditions, type of equipment available (in terms of capacity), type of routing done, etc.
- Transmission connections: This refers to the inter-connection between the switches and the actual capacities required for these transmission links. This transmission can be optical fibre cables or microwave links (for dimensioning and designing the latter, refer to Chapter 2).
- Signalling plans: The various aspects of signalling such as routing, protection and related synchronisation should be planned during the network analysis and dimensioning phase.

The above-mentioned plans constitute the major outputs of the dimensioning phase. Some typical examples of switch and signalling network dimensioning plans are shown in Figures 4.9 and 4.10, respectively.

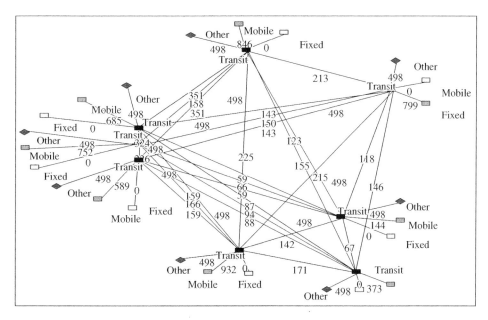

Figure 4.9 Typical outputs of switch network dimensioning.

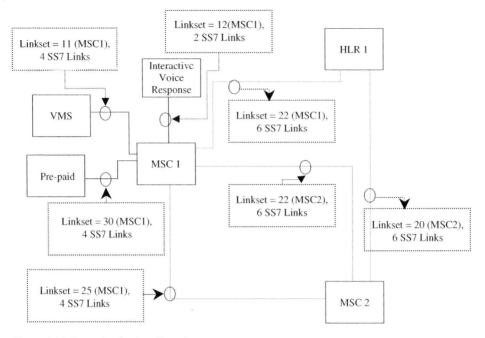

Figure 4.10 Example of a signalling plan.

Figure 4.9 shows a switch network plan. This is an example of a big network. Thus, instead of interconnecting the switches directly, transit switches are used. Each region is generating traffic from mobile, fixed and other sources. The traffic is getting routed through the transit switches. The numbers shown in the figure indicate the traffic in millierlangs (mErl).

Figure 4.10 shows a typical signalling plan with the signalling links and signalling link sets (SLS), along with the routing plans. This is a simplified example though in practice with more complicated networks involving a higher number of switches, IN, etc. signalling plans would become more complicated.

4.6   Detailed Planning

Detailed planning of a core network generally consists of signalling plans, routing plans, numbering and charging plans, DCN settings, synchronisation plans, etc. The input information for creating these detailed plans comes from the network analysis already described. The major outputs of the detailed core network plans are described below.

4.6.1   Routing Plans

Routing type and minor information is decided in the analysis and dimensioning phase. However, in detailed planning routing is fixed, names and naming convention are adopted and created for destinations and sub-destinations, circuit groups, etc. This is explained in detail in the next section.

4.6.2 Signalling Plans

Signalling points, SEP, STP, etc. are defined. Signalling link numbers and signalling link sets are also finalised. Signalling routes are also defined, keeping in mind the protection aspects.

4.6.3 Numbering and Charging

The numbering group used by each switch is finalised. This is usually done by taking into account aspects such as geographical locations in order to make it easier. Simultaneously, charging zones and their cases are defined, along methods of collection and charging record transfer. The main categories of the numbering plans include IMSI (International Mobile Subscriber Identity), MSISDN (Mobile Subscriber ISDN number), value-added services number (e.g. in a virtual private network), roaming numbers, handover numbers, test numbers, emergency numbers, etc.

4.6.4 Synchronisation Plans

Synchronisation is defined at switch level along with the priority settings. The principles of synchronisation were defined in Chapter 2.

The source data for each of these aspects is generated for each switch. An example of a detailed plan for SLS is shown in Figure 4.11.

4.7 Core Network Optimisation

4.7.1 Basics of the Optimisation Process

Core network optimisation focuses on both switch and signalling network optimisation. The process starts with the defining of key performance indicator definitions along with data collection for the process. This is then followed by analysis of the data collected. Performance analysis is done for both the switch and signalling aspects of the network. In some cases, the transmission network (for the core) and other aspects such as IN may be included and their effects on the core network might also be included. Analysis of the network and data might bring to light some bottlenecks and lead to suggestions for a network upgrade, or parameter tuningc. These corrections are then made/suggested in the final core network optimisation plan for implementation (Figure 4.12).

4.7.1.1 Key Performance Indicators

Any information that is relevant to the quality of service of the network can be considered to be part of the key performance indicators. These may include parameters related to traffic related parameters, signalling performance related parameters, and measurements related to HLR, VLR and other network elements in the core network. Data can be collected by using test equipment, using counters on NMS, or by any other measurements/reports from the network.

Signalling Link Sets {SLS}

SIGNALLING NETWORK TYPE	SIGNALLING LINK SET NUMBER	SIGNALLING LINK SET NAME	SIGNALLING POINT CODE	SIGNALLING POINT NAME	SIGNALLING LINK NUMBER	SIGNALLING LINK CODE	SIGNALLING LINK PRIORITY	EXTERNAL PCM-TSL	CONTROLLING UNIT		REMARKS
									TYPE	NUMBER	
NA0, NA1 IN0, IN1	NUMERIC 0...299	IDENTIFIER 5 CHARACT.	HEX/DEC 0..3FFF/ 0....16383	IDENTIFIER 5 CHARACT.	NUMERIC 0...254	NUMERIC 0...15	NUMERIC 0...15	NUM-NUM 64...511–1...31	CCSU	0...	INFO
NA1	0		100H	DUS01	0	0	0	130-16	CCSU	0	
NA1	1		150H	FRA01	1	0	0	132-16	CCSU	1	

Figure 4.11 Detailed plan for SLS.

Figure 4.12 Core network optimisation process.

4.7.2 Data Collection and Analysis

4.7.2.1 Data Collection: Switching

In Chapters 2 and 3, we have seen that data related to the existing network topology is required. Similarly in core network optimisation, data related to the existing network topology should be collected and analysed. This mainly consists of information related to the existing network elements (in the core network) and the interconnections between them (as this will be the basis of routing analysis and optimisation).

Optimisation in the core network is dependent mainly on traffic measurement inputs, which is also the most critical input for the whole process. The data generally contains information such as cell measurements, traffic category measurements, measurements related to the busy hour such as traffic load on a cell/network in the busy hour, incoming traffic, outgoing traffic, etc. This data helps in knowing the condition of the system and giving further recommendations to improve the existing and anticipated traffic related problems. The network in general should have a nomenclature consistency for the efficient network performance. The inputs for this generally consist of naming network elements (in the core network), naming destinations and sub-destinations, naming routes, etc. Another input required is the configuration parameters of the core network elements mainly MSC/VLR and HLR. From MSC/VLR, this consists of parameters related to charging, roaming, authentication, ciphering, etc., while from HLR parameters related to transferring subscriber data, basic and supplementary services, etc. are required.

Unlike in radio and transmission network optimisation, where tools other than the NMS are also used for data collection, in the core network optimisation process, NMS plays perhaps a more important role. However, the type of measurements that need to be done should be created in the switching platform. These measurements include the network elements and/or part of the system that is connected to the switch. A typical example is a control group measurement that covers all the control units in the core network, as shown in Figure 4.13. When taking measurements, the schedule and the duration of the measurements need to be specified. Planning engineers should take into consideration the system's capability and requirements for such measurements. Every equipment manufacturer has its own specification such as the number of measurements available, number of measurements that can run at a given time, limitation on the duration of measurements, delivered outputs (in ASCII/word/text files, etc.).

4.7.2.2 Data Collection: Signalling

The existing signalling plan is the basis for the signalling optimisation process. Information such as signalling link, signalling link sets, capacity of the signalling links, signalling routes, signalling route sets, signalling network topology, signalling load and signalling load sharing, etc. is important for this process. Some of the information, e.g.

```
SWITCHING PLATFORM      LONDON          29-0-1999  02:02:29

TRAFFIC MEASUREMENT REPORT

SOURCE: STU-1        OUTPUT INTERVAL: 45 MIN

UNIT              CALLS      ACCEP      ANSW     SFAIL IFAIL  EFAIL     ERLANGS

LSU-0                 0          0         0         0     0      0        0.0
LSU-1                 0          0         0         0     0      0        0.0
CCSU-0                0          0         0         0     0      0        0.0
CCSU-1              121        121       121         0     0      0        1.4
CCSU-2              633        633       633         0     0      0       12.6
BSU-0                 0          0         0         0     0      0        0.0
BSU-1              1890       1890      1890         0     0      0       21.4
BSU-2              3139       3139      3139         0     0      0       26.9
IWCU-0                0          0         0         0     0      0        0.0
IWCU-1                0          0         0         0     0      0        0.0

END OF REPORT
```

Figure 4.13 Example of control group measurements.

```
METERS OF LAST PERIOD:  01:45:00 - 02:15:00  (30 MIN)

        3.1              3.2              3.3            3.4            3.5
LINK    3.6        L1    L2         L3   TOT
&       3.7        L1    L2         L3          TOT
BIT     3.10       L1    L2         L3          TOT
RATE    3.11       L1    L2         L3          TOT
=====   ==========  ==========  ==========  ==========  ==========
    1   0000008071  0000000000  0000000170  0000014343  0000000173
  64K   0000000000  0000000000  0000000000  0000000000
        0000000000  0000000000  0000000000  0000000000
        0000000000  0000000000  0000000000  0000000000
        0000000000  0000000000  0000000000  0000000000
```

Figure 4.14 Statistics showing signalling link utilisation.

signalling topology, exists already, however, for information such as existing load, some measurements need to be performed. Again, the NMS can be used to collect information related to the performance of the signalling links. This can be done with the help of statistical counters available for monitoring the performances of the signalling network. One such example of signalling link utilisation is shown in Figure 4.14. If possible, measurements and statistics from all the signalling elements in the core network should be fetched. These measurements and statistics are analysed and compared with the existing plans and any changes needed are suggested.

4.7.2.3 Data Analysis: Switching and Signalling

Once the data is collected, it needs to be analysed. Based on the analysis, suggestions for improvements and optimisation must be given. Traffic measurement reports are the inputs for analysis. Usually traffic generated is divided into individual categories as seen in the planning phase which includes traffic originating and terminating in the network, traffic originating and terminating in an external network, internal and transit traffic, etc. This kind of division of traffic distribution would make the analysis much easier. Traffic and signalling analysis would result in information such as:

- Traffic handled by the switches/exchanges.
- Exact amount of traffic under each traffic class.
- Subscriber calling related measurements (subscribers/calls/successful call attempts/ traffic intensity).
- Traffic load in the switching exchanges and their availability (leading to congestion figures).
- Configuration of the signalling network.
- Load on the signalling network.

The analysis is usually done using network planning tools (refer to Appendix A). Improvement suggestions based on the above analysis is the next step in the optimisation process leading to the final optimisation plans.

4.7.3 Core Network Optimization Plan

Suggestions to improve the core network quality after the analysis constitutes the optimisation plan. This usually contains proposals for both the switch and signalling parts:

- Switch network optimisation plan
  - If the congestion problem is witnessed, extra PCM connections should be suggested at the location where congestion is experienced. If the congestion is severe and a new switch is required, then it should be proposed. This would mean that planning engineers should produce a whole of network topology with details such as the location of the new switch, traffic routed through it, etc.
  - Inter-switch connection, traffic routing between the MSC and transit switches should be modified in locations/regions where the transit switch is carrying excessive traffic. This would mean devising new routing plans.
  - When the networks are rolled out, mismatch in the naming conventions may happen. One of the outputs of the optimisation process is to clear the 'naming mess' and naming convention should be done in a manner that is similar in the whole network.
- Signalling network optimisation plan
  - The signalling link number should be optimised with an increase or decrease done wherever required.
  - The signalling links and SLS should be distributed uniformly across the network. If it is not uniform, new signalling links and link sets should be proposed.
  - Usually load sharing is not equal. New signalling plans/network topology should be created.
  - Proposals should be made for redundancy of the signalling control units.

4.8 GPRS Core Network Planning

Due to the introduction of the packet data, core network planning gets subdivided into two major parts: circuit core and packet core. Circuit core network planning remains more or less the same as discussed in Chapter 3. Here we discuss the packet core network planning.

4.8.1 Packet Core Network Planning

The most important aspect of packet core network planning is the dimensioning of the three interfaces: G_b, G_n, and G_i. However, Gb interface planning is of most importance.

4.8.1.1 G_b Interface

The interface between the BSS and SGSN, which allows the exchange of both the data and signalling information, is called the G_b interface, as shown in Figure 4.15. The G_b interface is more dynamic than the A interface. It not only allows multiple users to share its resources, but also reallocates the resources once the data transfer is stopped, as compared with the A interface where the physical resources are dedicated to the user irrespective of usage.

The protocol stack of G_b interface is shown in Figure 4.16. There are three layers on top of the physical layer. The physical layer serves the upper layers and transfers the data and signalling information from one end to another along with the overheads that are generated by each of the layers. The interconnection between the BSS and the SGSN can be done by using any of the physical media or by implementing the frame relay network. The advantage of using the frame relay network is that the interfaces at the two ends can be different. What is a frame relay network? A frame relay network is

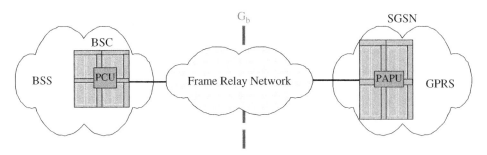

Figure 4.15 G_b interface.

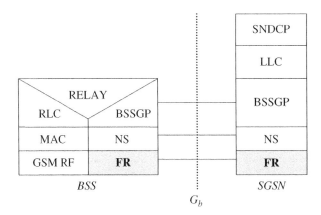

Figure 4.16 Protocol stack of a G_b interface.

basically a packet switch network that transfers the (packet) data in packets of variable lengths. These packets are called as frames. The network service or NS layer controls the frame relay layer.

There are two hardware units associated with the G_b interface or the frame relay network: the packet control unit or PCU and the packet processing unit or PAPU. The PCU is located on the BSS side while the PAPU is located on the SGSN side. For the G_b interface dimensioning there are three major aspects: BSC (or PCU) dimensioning, SGSN dimensioning, and the frame relay planning. We have already seen the inputs that are required for the PCU dimensioning. For the BSC dimensioning, from a packet core network planning perspective, the only input required is the BSC's PCU handling capacity.

4.8.1.2 SGSN Dimensioning
SGSN dimensioning basically results in the number of SGSNs that may be required in a GPRS network. Major inputs are SGSNs own capacity in terms of handling the amount of data, subscribers, and the number of PCMs.

The data handling or the processing capacity would determine the number of SGSNs required in a given network based on the amount of traffic. Calculations should not be done based on the traffic generated by the total number of BSCs because at a given moment, all the BSCs would not be fully loaded. Similarly, not all the subscribers would not using the GPRS at the same time. So, the reduction factors (i.e. what percentage of users will be using the GPRS network at the same time) should be planned carefully. Apart from this every SGSN (based on the vendor) has its own capacity for handling the number of frame relay links.

4.8.1.3 Frame Relay Planning
There are many ways to transfer data from the BSC to the SGSN. This may include techniques such as sending voice and data traffic separately or multiplexing both the voice and data traffic. In the initial phases of network launch, the PS data might be negligible in terms of the amount of data handled by each PCU. Thus, the traffic from some PCUs is then multiplexed and transmitted to the SGSN. The method of transmission can be similar to that from BTS to BSC, or BSC to MSC, or it can be also sent through the frame relay method. If the frame relay network is not present, then the ATM network could also be used to transfer the frame relay. Even dedicated PCM links (also known as frame relay links) could be used to send the packet data from the BSC to the SGSN. The main result of the frame relay dimensioning is to find the number of timeslots that are needed to send the packet data traffic from the BSC to the SGSN.

4.8.1.4 G_n Interface
The interface connecting all the GPRS networks elements (e.g. SGSN-SGSN, SGSN-GGSN, etc.) is the G_n interface. The interface dimensioning is based on information that includes the amount of data flow during the peak hours, number of subscribers, number of G_b interfaces, information on the IP network, etc. The core planning engineers classify the GPRS network based on this information. Generally, the GPRS network could be one of a few types. Either it can be a GSM network in which the GPRS network elements are introduced to launch the packet data services or it can be an advanced network that is focusing on issues like coverage and quality. The main

difference between the two kinds of GPRS networks is the GPRS network elements that would be used. Thus, in an advanced network, the number of SGSNs or GGSNs would increase thereby increasing the number of G_b or Gn interfaces. An Ethernet LAN connection is the GPRS IP backbone. All the network elements inside the GPRS packet core network such as NMS, firewalls, DNS, SGSN, and GGSN, etc. are connected through this switch. Thus, it becomes necessary to not only use a good quality switch but also keep a redundant switch.

4.8.1.5 G$_i$ Interface

The core network planning engineers have also to look into the interface between the GPRS network and the external network, i.e. interface between the GGSN and the external network. As different kinds of data will be flowing through this network, Gi is not a standard interface but acts as a reference point. As the Gi interface interacts with the external networks, security is also an important issue here. Firewalls are used between the GGSNs and the external networks that protect the GPRS network. The firewall protects both the subscribers and the QoS of the network (by not letting it become overloaded). The connection GGSN and the external network are done through virtual routers that are also known as access points. A gateway tunnelling protocol or GTP in the GPRS network is responsible for the routing of the information either way, i.e. from the GPRS network to the external network and from the external network to within the GPRS network. Another important aspect that is taken care of by the core planning engineers is the IP addressing of the network elements. Some fundamentals of IP addressing and routing are given below.

4.8.2 Basics of IP Addressing

Data exchange within two network elements is a three-step process: addressing, routing and multiplexing. Every network element in the Internet should have a unique IP address making it possible for data to reach the right host. The routers route the data to the correct network while multiplexing makes it possible for the data to reach the correct software within the host.

There are three classes of IP address: A, B, and C. The number of networks in each class can be computed as:

Class A: 128
Class B: 64 * 256 = 16 128
Class C: 32 * $(256)^2$ = 2 097 152

The number of addresses in each class can computed as:

Class A: $(256)^3$ = 16 777 216
Class B: $(256)^2$ = 65 536
Class C: 256

Thus, class A has the least number of networks and the largest number of addresses, while class C has the largest number of networks and the least number of addresses. How can one identify an address as to which class it belongs to? The addresses are of 32 bit lengths. If the address starts with bit 0, then it is a class A address. Class B addresses start with bits 1,0 and class C addresses start with 1,1,0.

Class A:

First bit 0: acts as class identifier.
Next seven bits: network identifier.
Last 24 bits: host identifier.

Class B:

First two bits 0,1: acts as class identifier.
Next 14 bits: network identifier.
Last 16 bits: host identifier.

Class C:

First three bits 1,1,0: acts as class identifier.
Next 21 bits: network identifier.
Last eight bits: host identifier.

How are IP addresses written? The IP addresses are written as four decimal numbers separated by dots, with each of the four numbers ranging from 0 to 255. If the value of the first byte is less than 128, it represents class A addresses, 128–191 represents class B addresses, while 191–223 represents class C addresses. Addresses above the value of 223 are reserved addresses.

Another aspect to know of when doing IP planning is that of 'sub-network' or 'subnet'. With the standard addressing scheme, a single administrator is responsible for managing host addresses for the entire network. As the networks are growing with high speed, local changes may become impossible, thereby requiring a new IP address for the 'new' network. By using the concept of 'subnet-working', the administrator can delegate address assignment to smaller networks within the entire network. By subnetting it is possible to divide the whole network into smaller networks in such a way that each of these smaller networks has its own unique address. This address is still considered to be a standard IP address. The combination of the subnet number and the host number is known as the local address. The subnet addressing is done in such a way that it is transparent to remote networks.

4.8.3   IP Routing

Once the addresses are defined, the next step is to route the traffic. As the networks are complicated structures, proper routing would enable traffic to reach the host. Routing involves transportation of the packet data to the right host and though the right or optimal path. To route the traffic, there are many parameters that determine the selection of the path, e.g. path length. There are algorithms that are used to do the calculations and find an optimal path for the traffic to be routed. These algorithms generally need information such as the source address, destination address and the next hop address. The core gateways have all the information related to the networks, which is shared between them to find the optimal route to process the data. This is done through the GGP or gateway-to-gateway protocol. For routing the traffic in a complex network, a routing table is used. This table consists of all the information that is required to route the data to the required destination within the network or to the local gateway if the destination address belongs to an external network. The routing table consists of information such as the source and destination address, flags network interface name, etc.

Packet Core Optimisation of the GPRS network focuses on four issues: PCU/BSC and SGSN capacity; PDP functioning, and overall network security.

Generally, the SGSN capacity is not an issue. However, if the number of port requirements for the G_b interface increases, then it is possible that a new SGSN may be required. However, based on the criteria and limitations discussed in Chapter 3, there is a possibility that the number of PCUs needs to be increased.

There may be problems related to the PDP functionalities. This may be due to incompatibilities between different network elements or signalling such as HLR and SGSN or signalling between SGSN and GGSN (GTP signalling). Another problem can be congestion on the G_b interface due to sudden increase in the data traffic. The G_b interface might need to be readjusted for such a scenario. Security is another issue in a data network. Due to variation in the type of data that is being exchanged through the GGSN, the rules of security filtration may need to reviewed.

4.9 EDGE Core Network Planning

Both the CS and PS core network planning remain similar to those described earlier in the chapter. The major change in the packet core section would be enhanced capacities of the BSC and SGSN that would take place to cater to enhance data rates on the radio network. Dimensioning of the packet core network mainly involves the PCU, the G_b interface and the SGSN dimensioning:

- PCU dimensioning:
  - Number of TRXs that can be supported by one PCU.
  - Amount of (data) traffic that can be handled by one PCU.
  - Number of traffic channels that can be handled by one PCU.
- G_b interface dimensioning
  - Minimum number of G_b interfaces required between the PCU and SGSN.
  - Uplink and downlink traffic calculations for the frame relay links.
- SGSN dimensioning
  - Number of subscribers.
  - Processing capacity of the SGSN.
  - G_b interface capacity.

Part II: 3G

UMTS Network Planning and Optimisation

5

3G Radio Network Planning and Optimisation

Unlike in the GSM networks where speech traffic was dominant, the 3G universal terrestrial mobile system or UMTS networks have a predominance of data traffic. The rate at which this data traffic can move would be significantly higher than that offered by the GSM/GPRS/EDGE network. For this reason, the 3G UMTS networks are fundamentally different from the existing GSM systems.

5.1 Basics of Radio Network Planning

We have seen some basics of 3G network based on WCDMA technology in Chapter 1. As in the GSM network, radio network planning takes an important role in the network planning process owing to the proximity to the subscribers.

5.1.1 Scope of Radio Network Planning

As mentioned in Chapter 1, the radio and transmission networks have been combined to form the RAN or radio access network. For ease of study, we look here at the scope of the radio network planning engineer's work. The focus area is quite similar to that of the GSM/GPRS network, as shown in Figure 5.1, i.e. planning capacity and coverage for an upcoming 3G network having a desired quality. Before we go into the details of radio network planning and optimisation in 3G networks, let us try to understand a few concepts that will help in the planning process.

5.1.2 3G System Requirements

As the 3G networks came into being for a different purpose, the requirements also changed from the existing GSM networks. The major changes are:

- Maximum user bit rates up to 384 kbps.
- Efficient handover between different operators and technologies (e.g. GSM and UMTS).
- Ability to deliver requested bandwidth.
- Ability to deliver different services (both CS and PS) with the required quality of service (QoS).

Fundamentals of Network Planning and Optimisation 2G/3G/4G: Evolution to 5G, Second Edition. Ajay R. Mishra.
© 2018 John Wiley & Sons Ltd. Published 2018 by John Wiley & Sons Ltd.

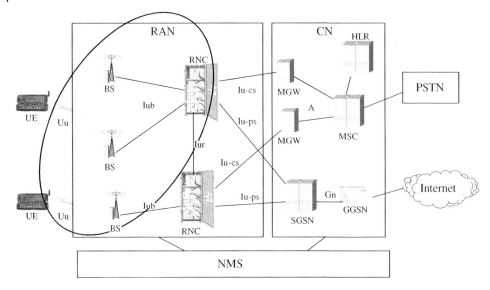

Figure 5.1 Scope of radio networking planning in a 3G system (WCDMA).

5.1.3 WCDMA Radio Fundamentals

Before we go into radio network planning, let us try to understand some principles that will help in designing a radio network that would give a desired performance. WCDMA or wide-band code division multiple access technology has emerged to be the most preferred and adopted technology for the 3G air-interface. As air-interface technology has changed, there are also some major differences between the WCDMA and GSM air-interfaces:

- The WCDMA system supports higher bit rates, so a large bandwidth of 5 MHz is used as compared with 200 kHz in GSM.
- Packet data scheduling in WCDMA is load based while in GSM/GPRS it is timeslot based.
- Theoretically, only one frequency channel is used in WCDMA while GSM uses many frequency channels.
- Limited bandwidth of 5 MHz is sufficient for radio network design. Multipath diversity is possible with the rake receivers while in GSM techniques like frequency hopping are used for (frequency) diversity.
- Quality control in WCDMA is done using RRM algorithms while in GSM it was done by implementing various techniques such as frequency planning.
- Users/cells/channels are separated by codes instead of time or frequency

5.1.4 Service Classes in UMTS

In a 3G network, mobile equipment will be able to establish and maintain multiple connections simultaneously. The network will also allow efficient cooperation between applications with diverse quality of service requirements, as well as adaptive applications that will function within a wide range of QoS settings. From the user perspective,

3G networks are the service networks, i.e. will be able to give high quality for a variety of services. This means that all the sections of the network RAN and CN will be trying to achieve these quality standards (from the user perspective and defined by the ITU). The quality can be defined by two main parameters:

- guaranteed and maximum bit rate (kbps) possible
- permissible delays (ms)

Both single-media and multi-media services will be handled in the 3G networks. Based on the QoS criteria, multi-media services have been further classified as:

- conversational
- streaming
- interactive
- background

Conversational class, as the name suggests, is for applications like speech. It is the most delay-sensitive of the four classes. A typical example of this class is video telephony, voice over IP (VoIP). In this class the delay is based on the human perception of the application, hence has strict requirements for QoS.

The *streaming class* refers to the traffic flow, which is steady and continuous. It is server-to-user type. The most common example in this class is the Internet. In 3G networks, the Internet would be faster than the present day, i.e. the user will be able to see the data before it is completely downloaded. There are two components of this class: messaging and retrieval. A typical example is downloading of streaming videos, e.g. news.

Web browsing is a typical example of the *interactive class.* In this case the user requests data from a remote entity, e.g. server. Location-based services in a 3G network is an example of this class. A user will be able to access information like bus and train timetables, flight schedules, restaurants, and any local data that might be useful.

Short messages, file transfers, etc. come into the *background class.* Nearly all the traffic that does not fall into the first three categories fall into this category, e.g. emails. This class of service has the least stringent QoS requirement of all the four classes.

5.1.5 Element in a WCDMA Radio Network

5.1.5.1 User Equipment (UE)

The mobile terminal is known as user equipment. This basic principle remains the same as explained in Chapter 2, but with the addition of the capability to handle data. There whole of the UE can be subdivided into three parts: USIM, ME and TE, as shown in Figure 5.2. The SIM card in the UE, also known as the USIM, i.e. UMTS SIM, contain information such as authentication information and associated algorithms, encryptions and subscriber-related information. Unlike the USIM, which is a user-dependent part of the UE, the mobile equipment or ME is user-independent. The elements within the ME make it possible for the UE to behave as a voice or data terminal (both RT and NRT) according to the call requirements. While the terminal equipment or TE, as the name suggests, is responsible for the termination of the entire control and user plane bearer with the help of the ME.

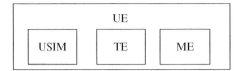

Figure 5.2 Simplified block diagram of user equipment.

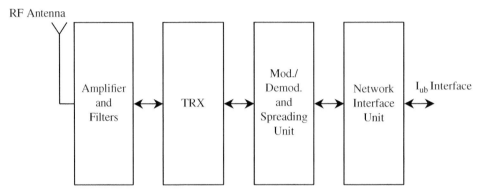

Figure 5.3 Simplified block diagram of a WCDMA base station.

5.1.5.2 Base Station (BS)

The base station is also known as node B in WCDMA radio networks. It is more complex than the base stations of the GSM networks. The functions of the bases station in a WCDMA radio network include handover channel management, base band conversion (TX and RX), channel encoding and decoding, interfacing to the other network elements, etc. A simplified version of it is shown in Figure 5.3.

5.1.5.2.1 Amplifiers and Filters

This unit consists of signal amplifiers and antenna filters. The amplifiers are used to amplify the signal coming from the TRX and going towards the RF antenna, i.e. downlink signal, while the filters are used to filter the required frequencies coming in from the RF antenna (i.e. uplink signal) and amplify them for further processing before sending them to the RX part of the TRX.

5.1.5.2.2 Transceiver

The TRX is capable of the transmitting and receiving signals, i.e. capable of handling uplink and downlink traffic. It consists of one transmitter and one or more receiver.

5.1.5.2.3 Modulation/Demodulation and Spreading Unit

This unit is responsible for modulating the signal in the downlink direction and demodulating in the uplink direction. It is responsible for summing and multiplexing the signals and also processing the signals. This unit contains the digital signal processors that are responsible for coding and decoding the signals.

5.1.5.2.4 Network Interface Unit

This unit acts as an interface between the base station and the transmission network or any other network element, e.g. co-sited cross-connect equipment.

5.1.5.3   Radio Network Controller (RNC)

The radio network controller or the RNC is similar to the BSC in the GSM/GPRS networks. However, it is rather more complicated and has more interfaces to handle than a BSC. The RNC performs radio resource management and mobility management (RRM and MM) functions such as the handovers, admission control, power control, load control, etc. The RNC plays a dual role in WCDMA radio networks, which should be understood from a network planning perspective. A given RNC can be known as an SRNC (serving RNC) or a DRNC (drifting RNC). From one mobile, if the RNC terminates both the data and related signalling then it is known as an SRNC. If the cell that is used by this UE is controlled by any other RNC other than the SRNC, then it is known as a DRNC.

More details on the structure of RNC are given in Chapter 8 where access transmission in UMTS networks is discussed.

5.2   Radio Interface Protocol Architecture

Based on the OSI reference model, Figure 5.4 shows the first three layers of the WCDMA radio interface protocols. These three layers are needed for the functioning (set-up, release, configuration) of the radio network bearer services.

Layer 1 is the physical layer, i.e. the actual medium of transfer. Planning engineers should note that this layer is not just a 'physical medium' but should also be able to perform some functions to qualify as a physical layer. The main functions of layer 1 include RF processing, modulation/demodulation of the physical channels, multiplexing/demultiplexing of the physical channels, error detections and corrections, rate matching, power control, synchronisation, etc.

Layer 2 is the link layer. It is required because of the need to allocate minimum resources for a constantly changing data rate. It has two main sublayers: RLC and MAC. There are two other layers known as the Packet Data Convergence Protocol (PDCP) and Broadcast-Multicast Control (BMC) layers. They exist only in the user plane. The MAC layer is responsible for mapping the logical channels to the transport channels. It provides the data transfer services on the logical channels. As it is an interface between L1 and L3, it also provides functions like multiplexing and demultiplexing of the packet

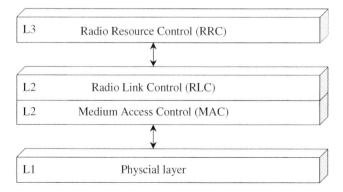

Figure 5.4 Basic radio interface protocol architecture.

data units and to/from the physical layer. The MAC layer is also responsible for the measurement related to traffic volume on the logical channels and further reporting to the L3 or RRC. Functions like the segmentation and reassembly of the variable length packet data into smaller payload units is done by the radio link control (RLC) layer, which is a sublayer of layer 2. Another important function of this sublayer is error correction by retransmission in an acknowledged data transfer mode. Other functions include controlling the rate of information flow, concatenation, ciphering, preservation of the higher order PDUs, etc. There are three modes of configuring an RLC by an RRC: transparent (no protocol overhead added), unacknowledged (no retransmission protocol is used, i.e. data delivery not guaranteed), and acknowledged (retransmission protocol is used and data delivery is guaranteed). PDCP and BMC protocols exist only in the user place. PDCP is only for the packet data with its major function being compression of the PDUs at the transmitting end and decompression at the receiving end in all three modes of operation, i.e. transparent, unacknowledged, and acknowledged. BMC functions only in the transparent and unacknowledged modes, providing the broadcast/multicast scheduling and transmission to the user data.

Layer 3 also contains sublayers but the radio resource control (RRC) sublayer is the one that interacts with layer 2. It handles the control plane signalling between the UE and network in connected mode. It is also responsible for the bearer functions like establishment, release, maintenance, reconfiguration in the user plane and of the radio resources in the control plane. The functions of RRC include RRM and MM, as well as functions like power control, ciphering, routing (of PDUs), paging, etc.

5.2.1 Protocol Structure for Universal Terrestrial Radio Access Network (UTRAN)

The protocol structure for UTRAN is based on the above-mentioned model and can be used further for studying the protocol structure for different interfaces in detail. As shown in Figure 5.5, there are two main layers: the radio network and the transport network. There are also two planes: the user plane and the control plane. The visible part of the network is the radio layer, while the transport layer elements (or equipment/technology) can vary without making any changes to the radio layer characteristics. Transportation of all the user specific data CS or PS is done through the user plane while the transport plane is responsible for all the signalling activities in the network. The control plane protocol present in the radio network layer is the known as the application protocol and includes protocols such as RANAP, RNSAP, and NBAP. The transport network layer has another control plane called the transport network control plane that is responsible for all the signalling within it. The protocol here is known as ALCAP. The transport layer has its own user plane also for the data bearers in the transport layer.

RANAP or the Radio Access Network Application Protocol is the signalling protocol defined between the RAN and CN. The main function of RANAP is to control the resources on the I_u interface, providing both control and dedicated control services.

RNSAP or the Radio Network Subsystem Application Protocol provides the signalling on the I_{ur} interface. As the I_{ur} interface is present between the two RNCs (one is SRNC and the other is DRNC), it is managed by these two RNCs. RNSAP is responsible for the management of the bearer signalling on the I_{ur} interface apart from the transport and traffic management functions.

Radio Network

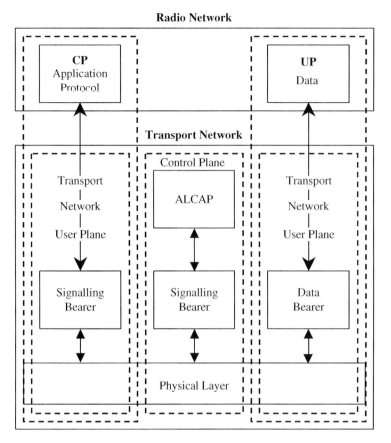

Figure 5.5 General protocol structure for UTRAN.

NBAP or Node B Application Protocol is the one maintaining the control plane signalling on the I_{ub} interface. This is protocol is responsible for the communication between the node B (WCDMA base station) and UE.

ALCAP or the Access Link Control Application Part is required for setting up of the data bearer and signalling in the transport network control plane. The presence of ALCAP is necessary as the user plane and the control plane can be separated and independent from each other. ALCAP may or may not exist depending upon the type of data bearer.

5.2.2 Channel Configuration in WCDMA Radio Network

As mentioned above, the MAC layer is responsible for mapping the logical channels on to the transport channels. It provides the data transfer services on the logical channels. As usual, logical channels are composed of control and traffic channels. These are again subdivided into common and dedicated channels. These channels along with their functions are mentioned in Table 5.1.

Table 5.1 Logical channels in a WCDMA radio network.

Channel	Abbreviation	Function/application
Broadcast common control channel (DL)	BCCH	Transmits the system control information
Common control channel (UL/DL)	CCCH	Used (usually by the UE) for transmitting control-related information between network and UE
Common traffic channel (DL)	CTCH	Used to transmit dedicated user information to a group of UEs
Dedicated control channel (UL/DL)	DCCH	Dedicated channel for control-related information between the UE and network
Dedicated traffic channel (UL/DL)	DTCH	Similar to DCCH except that it is used for user information
Paging control channel (DL)	PCCH	Used to page information to the UE

5.3 Spreading Phenomenon

The spreading phenomenon is the basis of the (W)CDMA technique. The word 'spreading' refers to the 'spreading of bandwidth' of the actual information. This means that the actual information is transmitted at a larger bandwidth than the original bandwidth. Through this process, the modulated signal becomes tolerant towards the narrow band interfering signal. This solves the problem faced by the TDMA and FDMA systems, which limit the simultaneous number of users. By using the spread spectrum technique, efficient use of the spectrum will allow multiple users to use the same frequency band. Using a very common technique known as direct sequencing, the information can be spread over a frequency spectrum. The most common techniques used for this purpose are:

- DS-WCDMA_FDD (direct sequence WCDMA frequency division duplex)
- DS-WCDMA_TDD (direct sequence WCDMA time division duplex)
- MC-CDMA (multi-carrier code division duplex)

In the DS-WCDMA_FDD method, the information is spread over the frequency spectrum, while in the DS-WCDMA_TDD method the frequency band is located on both sides of the WCDMA_FDD signal. In MC-CDMA, the whole of the frequency band is used with a number of carriers instead of one. Though both of the techniques FDD and TDD would be implemented, in the initial phase of a 3G system, only the FDD mode would be used because it is more suitable for outdoor coverage purposes. The frequencies for the DS-WCDMA_FDD are:

$$Downlink: 2110 - 2170 \, MHz$$
$$Uplink: 1920 - 1980 \, MHz$$

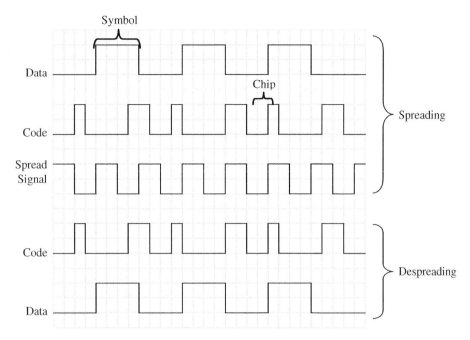

Figure 5.6 Spreading and despreading.

As seen from the frequencies, the bandwidth is 60 MHz in either direction while the separation between the uplink and downlink is 190 MHz.

The phenomenon of spreading (and despreading) is shown in Figure 5.6. The signal is 'spread' by modulating the original signal (referred to as data in the figure). Usually, BPSK or binary phase shift keying modulated signal is used as the original signal. This means that when 'spreading' the signal modulation is performed a second time on the original signal. This original data signal (obtained after BPSK modulation) is then modulated by multiplying with a sequence of bits also known as the wide-band spreading signal, thus converting the original BPSK narrowband signal into a spread signal of wider bandwidth. Each bit of the original signal, also known as a 'symbol', is multiplied by a sequence of bits called 'chips'. These chips are a part of the signal that is used to multiply the original data, called the 'code' signal. The resulting data or the spread signal is a product of the data and code and has a wider bandwidth.

This signal is then transmitted across to the receiver. At the receiving end, the received signal is then again multiplied by the same code to obtain the original data. The original data stream can be obtained only if both the codes (spreading and despreading) are similar. In the radio network, there are many subscribers that would be transmitting and receiving signals at the same time. These spreading codes achieve the separation between the signals. Here we can define the term 'spreading factor'. It can be defined as the number of chips used by one symbol. The spreading factor is also sometimes known as 'processing gain'. Processing gain can also be defined as the ratio between the chip rate and the bearer bit rate. The processing gains, together with the wideband nature, suggest a frequency reuse of 1 between different cells/sectors. Thus, we can arrive at some important information:

- One WCDMA carrier is 5 MHz. With the guard band taking 1.16 MHz of bandwidth, the effective bandwidth is 3.86 MHz.
- One bit of the base-band signal is known as a 'symbol'.
- One bit of the code signal is known as a 'chip'. The chip rate is defined to be 3.84 Mcps (million chips per second). This means that one chip is 0.26 µs.

Now these spreading codes contain two more codes: scrambling and channelisation. Actually, *spreading code = scrambling code × channelisation code.*

5.3.1 Scrambling Code

Scrambling codes are used on top of the spreading codes, thus not changing the bandwidth of the signal but only making it different for the purposes of separation. These codes are used to separate the users and cells (and/or base station) in the uplink and downlink directions, respectively. In the downlink direction, it is possible to generate about $2^{18} - 1 = 262\,148$ scrambling codes. However, all these codes cannot be used. These codes are divided into 512 groups and each of these groups has one primary and 15 secondary scrambling codes. Based on these principles, it is possible to have 8191 codes. For the uplink direction, there are 2^{24} scrambling codes. These uplink codes are subdivided into long and short codes.

5.3.2 Channelisation Code

Channelisation codes are used in the downlink direction for the separation of the users and channels within the same cell. In the uplink direction, they are used for the separation of data and control channels from the same user equipment. These codes are based on a technique known as OVSF or the orthogonal variable spreading factor. The OVSF technique maintains the orthogonality between the different spreading codes while allowing the spreading factor to be changed. These codes increase the transmission bandwidth.

5.3.2.1 Rate Matching

This is a process by which the number of bits that are to be transmitted are matched to the number of bits that are available on a single frame. If the number of bits transmitted is lower than the maximum, then the transmission may get interrupted. Thus, processes such as repetition and puncturing are used to make the two numbers (transmitted and maximum) equal. In the uplink direction, repetition is the preferred way of rate matching while in the downlink direction, puncturing is the preferred way. Uplink rate matching is a dynamic process that varies with the processing of every frame. This process is important in 3G networks as it is one way in which the desired QoS can be achieved by fine-tuning the bit rates.

5.4 Multipath Propagation

Multipath propagation is an interesting phenomenon in the WCDMA radio networks. Multipath propagation takes when the signal takes multiple paths and reaches the receiver. This means that the signal would reach the receiver at different points in time, i.e. with different amplitudes and phases. This may lead to a phenomenon known as fast fading, especially when the signals are arriving with a difference of half a wavelength, which means deep fades are taking place in the received signal. RAKE receivers are used in WCDMA radio networks, which is a more efficient way to receive the multipath signals. A RAKE receiver contains many receivers and is able to allocate them based on the timing of the arriving signal, code generators and amplitude/phase detecting equipment. These receivers, also known as RAKE fingers, are able to track the fast changing amplitudes and phases originating from the fast fading process and remove them. Typically, RAKE is able to handle four fingers. The number of RAKE fingers used in the base station and user equipment is usually different.

5.5 Radio Network Planning Process

The radio network planning process for WCDMA is nearly the same as that for GSM networks. The main aspects include preplanning/dimensioning, fieldwork of site surveys, etc., and subsequent modification of the parameters based on the fieldwork, before the final plans go to the commissioning phase. These aspects continue till the sites go on air. One major difference is that the QoS targets get more stringent in these WCDMA radio networks as compared with the GSM radio networks. With each kind of services demanding a different QoS, the planning of the network becomes an even more daunting task.

5.5.1 Pre-Planning Phase

The most important work in the pre-planning phase is to dimensioning of the network based on the inputs and assumptions for getting a desired coverage and capacity (Figure 5.7).

Coverage of the WCDMA radio network is dependent upon a few conditions such as the expected area to be covered, type of area to be covered, network configurations (and system capabilities) and propagation conditions. Defining the base station location is an important part of this process. The number of base stations required to achieve the coverage and quality associated with it is also dependent upon the capacity of the base stations. A link budget calculation and propagation models form an important part of the coverage predictions.

Capacity planning in the initial phase is a major challenge in the WCDMA radio networks. With so many kinds of applications having a varying quality and capacity requirements, the capacity along with the quality may prove to be a bottleneck, especially for the real time data service applications. The frequency availability, subscriber base and growth, and type of services required would be major inputs for the dimensioning exercise and the main results would include the number of sectors and the transceivers that would be required for these base stations.

As mentioned in Chapter 2, coverage, capacity and quality go hand-in-hand. Network configuration plays a big role in achieving the desired standards for these services. There are four different types of service classes and numerous different kinds of services in each class, with each service having a different quality requirement. Enough coverage, sufficient capacity and desired throughput would be key in the network configuration design to achieve the desired quality standards.

For proper dimensioning of the WCDMA radio networks, some phenomena affecting the coverage, capacity and quality need to be understood.

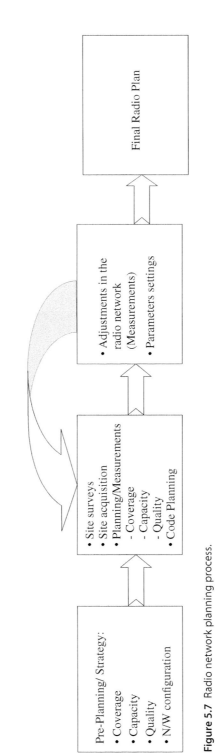

Figure 5.7 Radio network planning process.

5.5.2 Structure and Performance of the Physical Layer

In a WCDMA radio network, the structure of the physical layer directly affects the performance of the network, so some understanding of this is necessary to dimension the network. We would look first at the transport and physical channels, some phenomena associated with the physical layer (e.g. power control and handovers) and the measurements associated with it. We have already seen the logical channels in Table 5.1. Here we briefly look at the transport and physical channels.

5.5.2.1 Transport Channels

Transport channels, like the logical channels, are of two types: dedicated and common. These channels are given in Table 5.2 for the WCDMA_FDD mode of operation. There is only one dedicated channel and the remaining six are common channels. A dedicated channel or DCH is one that supports the soft-handover phenomenon discussed later in the chapter. Both dedicated and some common channels support fast power control.

5.5.2.2 Physical Channels

There are two kinds of physical channels: dedicated and common. The transport channels are mapped on to the physical channels. These physical channels are a layered structure of radio frames and timeslots carrying information related to the physical layers. These physical channels are identified by a specific carrier frequency, codes (channelisation/scrambling), timings, etc.

Table 5.2 Transport channels in a WCDMA radio network.

Channel	Abbreviation	Function/application
Broadcast channel (DL)	BCH	Transmits system- and cell-specific information
Common packet channel (UL)	CPCH	Transports packet-based user data in uplink direction
Dedicated channel (UL/DL)	DCH	Transmits user or control information in either direction
Downlink shared channel (DL)	DSCH	Shared by several UEs; carries both user and control information
Forward access channel (DL)	FACH	Transports control information to the UE
Paging channel (DL)	PCH	Used to page information to the UE
Random access channel (UL)	RACH	Received from complete cell and contains control information from the UE

Table 5.3 Physical channels in a WCDMA radio network.

Channel	Abbreviation	Function/application
Dedicated physical control channel (UL/DL)	DPCCH	Dedicated higher link information such as user data and signalling is carried on this layer
Dedicated physical data channel (UL/DL)	DPDCH	Transmits dedicated physical layer control information
Physical random access channel (UL)	PRACH	Transmits data part of RACH and layer 1 control informations
Physical common packet channel (UL)	PCPCH	Carries CPCH transport channel
Physical downlink shared channel (DL)	PDSCH	Carries DSCH transport channel
Primary common control physical channel (DL)	PCCPCH	Carries BCH transport channel and contains only data
Secondary common control physical channel (DL)	SCCPCH	Carries FACH and PCH transport channels

There are two dedicated channels and five common channels, see Table 5.3. The physical channels mentioned in Table 5.3 are the ones on which the transport channels are mapped, although there are a few more common channels, such as the synchronisation channel (SCH), common pilot channel (CPICH), paging indication channel (PICH), etc.

5.5.3 Uplink and Downlink Modulation

In the WCDMA radio networks, modulation characteristics are different in the uplink and downlink directions. Both the uplink and downlink directions have a chip rate of 3.84 Mcps and use quadrature phase shift keying or QPSK modulation. In the uplink direction, dedicated channels are multiplexed using complex coding schemes (also known as I/Q coding), while in the downlink direction they are time multiplexed with respect to time. If time-based multiplexing is used in the uplink, then during the DTX, interference that occurs will be audible, which is not the case in the downlink because transmission is continuous in the case of common channels.

5.5.4 Uplink and Downlink Spreading

Both uplink and downlink spreading are based on channelisation codes. The dedicated channel spreading factor variation is on a frame-by-frame basis in the uplink, while in the downlink it is not except for the DSCH (downlink shared channel) that may use the variable spreading factor on a frame-by-frame basis. Uplink scrambling uses complex sequences of spreading codes. In the uplink there are two spreading codes, long and short, while in the downlink, only long spreading codes are used. Long codes have a frame length of 10 ms, which means that a chip rate of 3.84 Mcps will result in 38 400 chips, while short codes have a length of 256 chips. As the number of spreading codes is high, a network planning tool should be used. Usually, only one scrambling code is used per sector in the downlink direction, as orthogonality between the channels (and users) needs to be maintained. Bad code planning would lead to less orthogonality and thus more interference resulting in less coverage and capacity.

5.5.5 Code Planning

Owing to the spreading phenomenon, scrambling and channelisation code assignment becomes necessary in these networks. Uplink code allocation for both the scrambling and channelisation is done by the system. As there are 64 code groups used in downlink direction, having 8 codes each, i.e. a total of $64 \times 8 = 512$ codes, each cell that the user equipment has connectivity to, should be assigned a different code. This leads to a re-use factor of 64, which is quite high, making the cell search process less complicated. The high re-use factor also means that the code planning can be done manually.

5.5.6 Power Control in WCDMA Radio Networks

Fast power control tied with a high level of accuracy is an essential feature of WCDMA radio networks. As the frequency re-use factor is 1, fast and accurate power control becomes even more essential. The reason is that in absence of a power control feature, the mobile that is nearer to the base station can easily 'overshout' the other mobiles that are handled by the cell, thereby producing a blocking effect. Thus, a closed loop power control feature is used. In the uplink direction, the base station makes measurements of the power received from different mobiles in terms of SIR (signal to interference ratio). It then compares these with the target SIR. If the $SIR_{measured} > SIR_{target}$, the base station will request the mobiles to reduce their transmitted power. The same phenomenon happens in the downlink. However, in this case, the targets are the mobiles that are located on the cell edges, i.e. to provide them with power and reduce external cell interference. Another aspect that is used in these networks is 'slow power control'.[1] This is also known as the 'outer loop power control'. This phenomenon is used for controlling the SIR_{target} in a base station. The process is based on the needs of individual single links and is responsible for maintaining the quality targets in the base station and the network. As the open outer loop power control process is based on the propagation aspects in the downlink, and is not sufficient as the uplink and downlink frequencies are quite different from each other, a closed loop power control process is used. The BTS receives the signal from the user equipment and, based on its measurements, directs the mobile to increase or decrease the power level. These

---

1 The slow power control can be easily confused with a fast power control algorithm that is slowed down by only executing a few consecutive commands.

measurements are based on the received signal level per bit to the interference signal level.

In the uplink direction, after the higher layers have set the uplink power, the control procedure for uplink starts. The power in the uplink direction is controlled (or set) by the network. In the fast power control procedure, it controls (or adjusts) the uplink power, i.e. power transmitted by the user equipment. This control is based on the SIR_{target}. Another power control mode is the compressed mode.[2] The compressed mode is used in both the uplink and downlink directions, which is used to provide fast power control. In this mode, larger steps are used so as to recover the power level for SIR to move closer to the target SIR. Power control processes have a direct impact on the coverage and capacity of the network. Air-interface capacity in the downlink is more critical and is dependent upon the transmitted interference. If the required transmitted power can be reduced to a bare minimum, the capacity can be increased. However, in the uplink, both the transmitted and received power would increase the interference to adjacent cells and users of the same cell, respectively. The higher the isolation levels between the cells, the higher the capacity. Decreasing the power levels to the minimum required can increase this isolation.

5.5.7 Handover in WCDMA Radio Networks

WCDMA networks experience a tremendous increase in the number of handovers taking place in their radio networks. Here we will consider different types of handovers that take place in WCDMA radio networks.

Handover is a phenomenon that takes place when the mobile subscriber is moving around. This is done so as to give the best coverage and quality to the mobile subscriber. The following types of handover exist in WCDMA radio networks:

- intra-mode handover
- inter-mode handover

Apart from this, there are two categories of handovers: soft and hard.

Handover was a phenomenon existing in GSM networks as well. In GSM networks, as every cell had a different frequency, the handover was called as hard handover. In WCDMA radio networks as the frequency re-use factor is 1 only, a different kind of handover exists which is described as soft.

Intra-mode handover involves soft, softer, and hard handovers. The inter-mode handover is the one that takes place where the FDD and TDD modes exist simultaneously. Here we will concentrate only on intra-mode handovers. The intra-mode handover procedure is based on the measurements performed on the common pilot channel. When the mobile is in a position between two cells, i.e. in a region of overlapping coverage, then it is connected to two different sectors of two base stations (case 1, Figure 5.8) simultaneously. In this case, signal is coming to the mobile from two directions, however, due to Rake receivers, the mobile antenna receives a combined signal. Another

---

2 Compressed mode or slotted mode is needed to make inter-system or inter-frequency handovers. The mobile/BTS do not receive power control commands during compressed frames; this is why larger steps need to be used. This is a special case that is used during inter-system measurements.

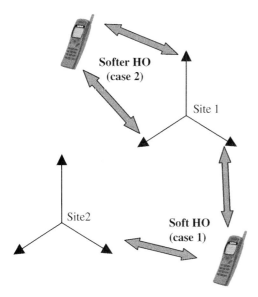

Figure 5.8 Handover in a WCDMA radio network.

type of handover is called the softer handover. In this case, the mobile is connected to two different sectors of the same base station (case 2, Figure 5.8).

Again, the mobile receives both signals due to the presence of Rake receivers. In both cases, as the mobile is using two different paths (interfaces) for communicating with the base station, there are two different codes used so that the mobile can recognise the different paths. There is not much difference in the downlink direction, however there is a big difference in the uplink direction for the two handovers. In the case of the soft handover, the received signal is routed to the RNC from both the base stations so that the best RNC can take a decision on keeping the signal which is giving a better quality. As this decision-making process takes time (about 10–80 ms), double air-interface capacity is utilised by a single mobile for one call. Hence, radio engineers have to take into account this process when doing capacity planning for the radio network. About 30% of the links witness the soft handover phenomenon against 10% for softer handover. The measurements generally include received signal code power (signal power received on one code), signal power received within the channel bandwidth, and E_c/N_o (ratio of the received signal power to the received signal power indicator).

5.5.8 Coverage Planning

As mentioned above, coverage is an issue that is dependent upon the area to be covered, terrain type and the propagation conditions. Coverage should include all the regions such as dense urban, urban, rural, etc. As in GSM networks, all these factors are responsible for the distance that can be covered by a given cell, also called a cell range. Unlike in GSM, capacity within a cell has a deeper impact on the coverage especially when the network is densely populated. A densely populated network would have more interference, having a negative impact on its quality. The calculation for a cell range can be done using link budget calculations. A link budget calculation in a GSM radio network

Table 5.4 Equipment and network parameters.

Equipment parameters	Network parameters
BTS receiver noise figure	Soft handover gain
BTS noise power	E_b/N_o
Cable loss + connector loss	Body loss
TX antenna gain	Processing gain
RX antenna gain	Interference margin
Thermal noise density	Power control headroom
Isotropic power	Peak EIRP

is relatively simple. However, in a WCDMA radio network it becomes more complicated as the number of parameters affecting it directly increases. Let us consider these parameters before we look at an example of a link budget calculation. Some of the parameters are listed in Table 5.4. Most of these parameters are obvious but some of them are new or have more impact in a WCDMA network.

5.5.8.1 E_b/N_o

E_b/N_o is the ratio of the received bit energy to the thermal noise. E_b or received energy per bit is the modulating signal power and the bit rate. N_o is the noise power density divided by bandwidth. Link budget calculations are basically done to calculate the E_b/N_o ratio and the interference signal density.

5.5.8.2 Soft Handover Gain

The soft handover phenomenon gives an additional gain against the fast fading that takes place in the network. Due to the soft handover phenomenon, the mobile connectivity is to a base station that gives a better signal quality. Thus, due to macro diversity combining, the soft handover gain has a positive impact on the base station.

5.5.8.3 Power Control Headroom

This is commonly called the fast fading margin. It is the fade margin needed to maintain the closed loop power control in action.

5.5.8.4 Loading Effect

Interference from neighbouring cells has an impact on the cell's performance. This is also known as cell loading, and the parameter to describe this is called the loading factor. Owing to this loading, the degradation in the link budget takes place, also known as interference degradation. If the loading factor is α, then the interference degradation margin can be calculated as:

$$L = 10\log(1-\alpha) \tag{5.1}$$

Theoretically, the parameter α may vary from 0 to 100%, but practically it is in the range of 40–50%. The principle of the load factor is similar for uplink and downlink power budgets, however the parameters are different. In the downlink power budget, an orthogonality factor needs to be considered as well. As we saw at the beginning of this chapter, mobile subscribers are assigned codes and these codes are orthogonal to each other. Orthogonality of the codes is maintained only in direct propagation conditions, which is usually not the case, as there will always be some degree of multipath propagation taking place. This would cause a delay in the signal and the mobile would treat some part of the incoming signal as an interference source. Perfect orthogonality would mean unity, while practically it may vary anywhere between 0.4 and 1.0.

5.5.8.5 Propagation Model

The propagation models that have been described in Chapter 2, such as Okumara–Hata and Walfish–Ikegami, are valid for WCDMA radio networks as well. As in GSM networks, the correction factor procedure needs to be followed for these networks too.

5.5.8.6 Link Budget

An example of an uplink and downlink power budget is shown in the Table 5.5. The fundamental principle behind the link budget remains the same as described in Chapter 2. However, in WCDMA radio networks, the link budget calculations need to be done individually for voice and various data rates, e.g. 64, 144, and 384 kbps, for a load of 50% in this example and for different applications.

5.5.9   Capacity Planning

As compared with GSM radio networks, capacity analysis finds itself more than ever dependent upon the coverage. The uplink and downlink coverages are quite different in the WCDMA radio network. The downlink coverage decreases with an increase in the number of mobile subscribers and their transmission rates. To achieve a service rate of 144 kbps, more sites would be required than for achieving a service rate of 12.2 kbps. Moreover, as the number of users increases, the cell coverage area decreases. As downloading is more prevalent than uploading, the downlink capacity will be a critical factor for the range of area covered by the cell. Once these two factors have been dealt with, cell configurations can be decided upon in terms of sectors and carriers (TRXs).

Once coverage and capacity plans are made and the process reiterated many times by using the radio network planning tools (refer to Appendix A), the radio network planning process moves forward to the site survey and site acquisition phase. This process is explained in Part I of this book. However, at this point is helpful to also understand the concept of adaptive multi-rate or AMR. An understanding of AMR helps in choosing the right parameters to analyse and optimise the quality of the network.

5.5.10   AMR

Unlike in GSM where speech codecs are fixed, e.g. FR (full rate) and HR (half rate), and their channel protection is also fixed rate, in WCDMA radio networks, AMR is used thereby making it possible to adapt speech and channel coding rates according to the quality of the radio channel. This improves the error protection and channel quality. The codec basically has one single integrated speech codec with eight source rates. This is controlled by the RRM functions of RAN. These eight source rates are

Table 5.5 Link budget calculation in a WCDMA radio network.

Link budgets		Voice		LCD		UDD		UDD		UDD	
Data rate (kbps)		12.2	12.2	64	64	64	64	144	144	384	384
Load		50%	50%	50%	50%	50%	50%	50%	50%	50%	50%
		Uplink	Downlink	Uplink	Downlink	Uplink	Downlink	Uplink	Downlink	Uplink	Downlink
		Node B	UE	Node B	UE	Node B	UE	Node B	UE	Node B	UE
RECEIVING END											
Thermal noise density	dBm Hz^{-1}	-174	-174	-174	-174	-174	-174	-174	-174	-174	-174
BTS receiver noise figure	dB	3.00	8.00	3.00	8.00	3.00	8.00	3.00	8.00	3.00	8.00
BTS receiver noise Density	dBm Hz^{-1}	-171.00	-166.00	-171.00	-166.00	-171.00	-166.00	-171.00	-166.00	-171.00	-166.00
BTS noise power (NoW)	dBm	-105.16	-100.16	-105.16	-100.16	-105.16	-100.16	-105.16	-100.16	-105.16	-100.16
Required Eb/No	dB	4.00	6.50	2.00	5.50	2.00	5.50	1.50	5.00	1.00	4.50
Soft handover MDC gain	dB	0.00	1.20	0.00	1.20	0.00	1.20	0.00	1.20	0.00	1.20
Processing gain	dB	24.98	24.98	17.78	17.78	17.78	17.78	14.26	14.26	10.00	10.00
Interference margin (NR)	dB	3.01	3.01	3.01	3.01	3.01	3.01	3.01	3.01	3.01	3.01
Required BTS Ec/Io (q)	dB	-17.97	-16.67	-12.77	-10.47	-12.77	-10.47	-9.75	-7.45	-5.99	-3.69
Required signal power (S)	dBm	-123.13	-116.83	-117.93	-110.63	-117.93	-110.63	-114.91	-107.61	-111.15	-103.85
Cable loss	dB	2.50	2.50	2.50	2.50	2.50	2.50	2.50	2.50	2.50	2.50
Body loss	dB	0.00	5.00	0.00	0.00	0.00	0.00	0.00	0.00	0.00	0.00
Antenna gain RX	dBi	18.00	0.00	18.00	0.00	18.00	0.00	18.00	0.00	18.00	0.00
Soft handover gain	dB	2.00	2.00	2.00	2.00	2.00	2.00	2.00	2.00	2.00	2.00
Power control headroom	dB	3.00	0.00	3.00	0.00	3.00	0.00	3.00	0.00	3.00	0.00
Sensitivity	dBm	-137.63	-111.33	-132.43	-110.13	-132.43	-110.13	-129.41	-107.11	-125.65	-103.35

TRANSMITTING END		UE	Node B	UE	Node B	UE	Node B	UE	Node B	UE	Node B
Power per connection	dBm	21.00	27.30	21.00	28.30	21.00	28.30	26.00	33.30	26.00	33.30
Maximum power per connection	dBm	21.00	40.00	21.00	40.00	21.00	40.00	26.00	40.00	26.00	40.00
Cable loss	dB	0.00	3.00	0.00	3.00	0.00	3.00	0.00	3.00	0.0C	3.00
Body loss	dB	5.00	0.00	0.00	0.00	0.00	0.00	0.00	0.00	0.0C	0.00
Antenna gain TX	dBi	0.00	18.00	0.00	18.00	0.00	18.00	0.00	18.00	0.0C	18.00
Peak EIRP	dBm	16.00	42.30	21.00	43.30	21.00	43.30	26.00	48.30	26.00	48.30
Maximum isotropic path loss	dB	153.63	166.33	153.43	165.13	153.43	165.13	155.41	162.11	151.65	158.35
Isotropic path loss to the cell border		153.63	153.63		153.43		153.43		155.41		151.65

12.2, 10.20, 7.95, 7.40, 6.70, 5.90, 5.15 and 4.75 kbps. The process of AMR selection is based on the channel quality measurements for which both the user equipment and BTS are involved. Once it is confirmed that the quality of the signal is bad, the number of speech codec bits is reduced thereby increasing the number of bits that can be used for error protection and correction. The speech frames in the AMR coder are of 20 ms, so if the sampling rate is 8000 samples s^{-1}, 160 samples are processed. Thus, the bit rate can be changed at each of these 160 samples through in-band signalling or through DCH. This process is known as link adaptation (as seen in Chapter 6). LA is a characteristic of AMR through which the codec of the link radio link can vary depending upon the radio link conditions. Thus, by using AMR, speech, coverage, quality, and performance can be improved.

5.6 Detailed Planning

Once the pre-planning phase is over, and the site search has begun, the detailed planning process begins. The process includes definition of the practical aspects of the site such as the antenna locations, antenna heights, etc., but also the study of link performances and the interference analysis necessary. Along with this, the all-important process of defining the parameter settings takes place. The detailed planning process is also sometime referred to as the pre-launch optimisation. Radio network planning tools play an important role in this process. In addition, some data collected from the drive tests is also utilised.

5.6.1 Coverage and Capacity

The main aspect of detailed planning is more accurate coverage and capacity plans. This starts with performance analysis of the radio links. The power budgets are analysed by taking some data from the actual drive tests results. The factors that usually cause a deviation of the practical link budget (i.e. actually received power) from the theoretical link budget calculations are:

- propagation model
- link budget parameters
- propagation conditions

The above three factors are inter-dependent. As in the GSM networks, tuning of the propagation model is a necessity. The model tuning should be done for (near to) actual propagation conditions. A network in a big region would mean that the analysis should be done for each region separately as the propagation conditions (apart from topography) would change from one region to another. This would lead to changes in the propagation model parameters (propagation models are similar to the ones used in GSM radio network, e.g. Okumara–-Hata and Walfish–Ikegami) such as clutter type corrections, diffraction, topographical corrections, etc. One big change in a WCDMA radio network's link performance analysis is the fact that data starts playing an important role. The type of data and its individual delay requirements makes it impossible to define one set of parameters for the whole network. Each of the data rates would have individual coverage probability requirements. This analysis would lead to results that

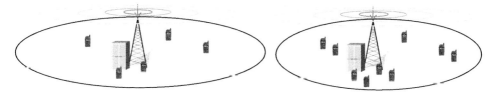

Figure 5.9 Cell breathing (decrease in coverage area with increase in mobile subscribers).

would give quite accurate antenna heights, antenna tilts, bearing, location of the sites, etc. Both the uplink and downlink transmitted power need to be found, and then the number of mobile subscribers can be predicted more accurately.

An accurate prediction of the number of subscribers would lead to a more accurate interference analysis compared with the results in the pre-planning phase. The frequency re-use factor is 1 in a WCDMA radio network, but code planning needs to be done accurately. As there are 512 sets of codes, it is recommended to use the automatic code allocation technique rather than manual allocation as the lattter may lead to errors, thereby degrading the quality.

One concept that is important to understand at this point is that of *cell breathing.* The coverage area of the cell depends upon the number of subscribers and their data rates. If the number of subscribers using the cell increases, the area covered by the cell would decrease. As the load decreases, the area increases. This phenomenon is called cell breathing, as shown in Figure 5.9.

Another phenomenon that directly affects the capacity and quality is *soft capacity.* Erlang B tables can be used for the capacity calculation based on hardware limitations. However, in WCDMA radio networks, as the capacity is interference (on the air-interface) limited, it is called soft capacity. The less the interference from neighbouring cells, the more the capacity increases. The concept of soft capacity is important for the data users. More capacity would lead to a better connection, i.e. lower blocking for the real time data users. Soft capacity can be obtained in one cell only if the adjacent cells have a lower loading (or fewer subscribers in them). Uplink soft capacity is based on the total interference at the base station, which includes own-cell and other-cell interference.

5.6.2 Radio Resource Management

For parameter planning in a WCDMA radio network, RRM and MM should be understood. RRM and MM, as in GSM networks, consists of concepts such as radio resource control, admission control, power control, handover control, etc. RRM is important from the perspective of air-interface resource utilisation, offering optimum coverage, capacity and, above all, guaranteeing the QoS.

As seen in Figure 5.4, layer 3 or the radio resource control layer has two main sublayers beneath it which also constitute layer 2: RLC and MAC. MAC forms the I_{ub} interface and channels the information coming in from the RLC layer. RLC produces information related to RRC signalling, CS and PS data. RRC has two main states: idle and connected. In the idle state, the subscriber is not connected to the network. In the connected state, the user equipment is performing data transfer activity (voice, CS, or PS). For the user equipment to change state from idle to connected, the call admission function comes into action.

The admission control (AC) resource management function is responsible for the control of the number of subscribers logged on to the network. It may admit or deny access to a new subscriber based on the resulting effect the new subscriber would cause to the existing users on the network. These results are based on the algorithms that are executed in the RNC. These algorithms are also called AC algorithms or CAC (call admission control) algorithms. AC is responsible for preventing congestion in the network and thereby maintaining the quality of the connected subscribers. Although AC algorithms are executed separately for uplink and downlink, the requesting subscriber can be admitted only after gaining clearance from the uplink and downlink algorithms.

AC algorithms can be power based or throughput based. In the power-based calculations, the uplink AC takes a decision based on whether the new subscriber mobile will increase the interference level over the planned load target and degrade the quality of the network. Each new user's entry affects the interference levels of the cell and adjacent cell. Thus, interference level calculations are based on these cell measurements and the receiver noise.

In the throughout-based calculations, both uplink and downlink AC take a decision on whether or not a new subscriber is admitted based on the load factor calculations. The load factor has a direct effect on the interference margin (IM) as seen from:

$$IM = 10 \times \log_{10}\left(\frac{1}{1-L}\right) \tag{5.2}$$

where L is the load factor from Eq. (5.1).

Interference does not form an automatic part of the calculation, but does affect it indirectly. An enhanced functionality in these networks is congestion control. This feature monitors, detects and corrects the situations when the network congestion is degrading service quality. Overloading can be controlled by using functionalities such as fast power control (both uplink and downlink), handing over the new subscribers (or the ones trying to congest the network) to another carrier, or even by reducing the throughput of the packet data (a feature handled by packet scheduling). Mobiles that move within cell range cause different amounts of interference, so overload can occur when the mobile(s) moves towards the cell edge.

The power control function becomes all the more important in the WCDMA radio networks operating in the FDD mode. The reason behind this is that mobiles in WCDMA radio networks transmit continuously, unlike in GSM where every subscriber transmits in different timeslots. Also, the WCDMA network uses only one frequency, i.e. frequency re-use factor is 1. Each subscriber has an individual code and appears as a noise source to other subscribers. Hence, any inaccuracy in power control will lead to increase in interference, which directly affects the number of subscribers getting admission in the network. This makes it necessary for the power control to be accurate and fast. Thus, the phenomena of open-loop power control and closed-loop power control take place.

The power control feature is responsible for handover management in the WCDMA radio networks. There are many types of handovers possible. There are two main types of handover procedures: soft handover and hard handover. In soft handover, a mobile subscriber is always connected to two base stations simultaneously while in hard handover, the existing radio link is released before a new one is connected. The decision for handover however lies in the RNC. The algorithms based in the RNC make a decision based on the measurements received by the mobile stations.

Another important function in RRM is the allocation of codes. Owing to the single frequency, the separation of subscribers is based on these codes. Allocation of codes is based on information such as the network configuration and RRM features such as admission control (including packet scheduling as well). The allocated codes have to be orthogonal to each other. However, due to some practical conditions such as delays, etc., the codes are not perfectly orthogonal. Thus, interference may occur. Code allocation is done by RNC in two steps: channelisation and spreading. The coding is done in such a way that the scrambling codes are different from each other having low cross-correlation properties with OVSF channelisation codes being responsible for maintaining the orthogonality between the subscribers.

5.7 High Speed Packet Access

High speed packet access or HSPA is a 3.5G technology which increases the data rate by adding transport channels and also improves latency and increases throughput. Increase in data is done through additional downloading transport channels known as high speed downlink shared channels (HS-DSCHs). HSPA is based on 3G release 5 solutions. The major advancements of HSPA are the downlink direction known as High Speed Downlink Packet Access (HSDPA) and the uplink direction known as High Speed Uplink Packet Access (HSUPA).

Specifications	HSDPA	HSUPA
Full form	High Speed Downlink Packet Access	High Speed Uplink Packet Access
3GPP Standard	Release 5	Release 6
Direction	Downlink	Uplink
Data rate	1.8–10 Mbps	Up to 4 Mbps
Modulation schemes	QPSK and 16QAM	QPSK
Transmission time interval	2 ms	2 ms, 10 ms
Physical channel	E-AGCH (Enhanced Absolute Grant Channel) and E-RGCH (Enhanced Relative Grant Channel)	E-DPCCH (Enhanced Dedicated Physical Control Channel) and E-DPDCH (Enhanced Dedicated Physical Data Channel)

5.7.1 HSDPA

HSDPA is part of release 5 of 3GPP. The main features of HSDPA include high order modulations, short transmission time intervals, Hybrid Automatic Repeat reQuest (HARQ) retransmission protocol, high speed channels, fast packet scheduling and fast link adaption. The transmission time interval in HSDPA reduced to 2 ms which increases

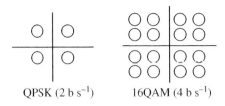

QPSK (2 b s^{-1})   16QAM (4 b s^{-1})

Figure 5.10 Modulations in HSPA.

the rate of tracking channel variations; QPSK and 16QAM are the two modulation schemes which make more efficient use of the bandwidth. HSDPA is a soft combining technique which improves system performance and is used to request the retransmission of data (Figure 5.10).

5.7.1.1 WCDMA Network to HSDPA Network

A new bearer HS-DSCH allows several users to be multiplexed and hence provides effective utilisation of resources. A fixed spreading factor of 16 allows a maximum of 15 parallel codes – both for user and signalling traffic. A transmission time interval of 2 ms increases the rate of tracking, the variations and allowing the system to adapt faster. Adaptive modulation and coding (AMC) is used to compensate for the variations in radio transmission conditions. A new MAC-hs entity is added which handles new features: the HARQ (hybrid automatic repeat request) retransmission mechanism is associated directly with the base station rather than the RNC.

5.7.1.2 HSDPA Protocol Structure

The MAC layer is split into two, i.e. MAC-d and MAC-hs. MAC-d is for the dedicated channel which is located at the RNC and is used for mapping between transport and logical channels. MAC-d includes multiplexing/demultiplexing upper layer PDUs, identifying user equipment in common transport channels. MAC-hs is fast scheduling done at the base station which is responsible for packet scheduling, link adaption and error correction.

5.7.1.3 User Equipment

The characteristics of the HSDPA handsets are more complex than the 3G handsets as they include Equalizer and Diversity, 16QAM modulation, Hybrid ARQ, increased and faster buffer memory, and faster turbo decoder.

5.7.1.4 HSDPA Channel

The packet data is transferred in the downlink direction through the DCH (data control channel), the FACH (forward access channel) and the DSCH (downlink shared channel). DSCH is replaced by high speed DSCH. FACH is used for signalling when no activity takes place in the user equipment. DCH can be used for any data (CS/PS) which has a fixed spreading factor. Some other channels have been added in HSDPA, namely, HS-DSCH, HS-SCCH, and HS-DPCCH.

5.7.1.5 HS-DSCH (High Speed Downlink Shared Channel)

HS-DSCH is a transport channel, which is mapped on to HS-PDSCH (High Speed Physical Downlink Shared Channel). Synchronisation of transmission and scheduling is complex. Soft handover and power control is not supported. It can use multi-codes with spreading factor 16 which allows carrying over 2 ms transmission time interval or TTI slots. For better signal quality, it can use 16QAM and QPSK.

5.7.1.6 HS-SCCH (High Speed Shared Control Channel)
HS-SCCH uses QPSK modulation and a fixed spreading factor of 128 which allows 40 bits per slot to be carried. When HSDPA is operated on time multiplexing, HS-SCCH is configured. Data rates available for users are dependent upon the type of terminal, power allocation and environment.

5.7.1.7 HS-DPCCH (High Speed Dedicated Physical Control Channel)
HS-DPCCH carries the Ack/Nack (repetition) and CQI (Channel Quality Information) which informs the base station about the data rates and also carries uplink feedback information. This information is used for link adaptation and physical layer retransmission purposes. It uses BPSK modulation and a spreading factor of 256; the channel has a three slot (2 ms) structure. The first is used for HARQ while the remaining two slots are used for CQI information. The HARQ channel provides decoded information.

5.7.1.8 Radio Resource Management
The physical layer operations in a HSDPA network from data perspective is synchronous from the UE side while it is asynchronous from the network side. Each date user is evaluated by the base station every 2 s. After a data UE is identified, HS-DSCH parameters are identified and start getting transmitted by the base station. UE decodes the HS-SCCH, first part 1 and then the rest. After decoding part 2, the UE determines data in the ARQ process and combines data in the soft buffer. After decoding, Ack/NAck is sent in uplink for the needed L1 procedures. The call ends with the Ack/NAck field.

5.7.1.9 Adaptive Modulation and Coding Scheme
An important feature is link adaptation which is done by changing the modulations and number of codes. With the absence of a power control feature in HS-PDSCH, the impact of LA is so high that it leads to cell capacity being lost and fast scheduling not working. The UE uses the CQI signalled through HS-DPCCH which tells the network the highest data rate. Based on this information, the scheduler chooses a coding and modulation format for the next TTI. This is based on an inner loop algorithm. The major parameters are the CQI and DPCH power measurement reports.

5.7.1.10 Power Control
HS-SCCH should be power controlled at each TTI which uses inputs such as CQI reports and the downlink DPCCH power. The power allocated to HS-SCCH should be sufficient as HS-DSCH can be decoded if HS-SCCH is correctly received. With a power control phenomenon, higher power is needed at the cell edge and less near the base station.

5.7.1.11 HARQ (Hybrid Automatic Repeat reQuest)
In a HSDPA system, HARQ adds tremendous robustness. There are two techniques: incremental redundancy (IR) known as non-identical retransmission as it uses different rate matching between retransmissions; and chase combing (CC) known as identical

retransmission as the same rate matching is used between different retransmissions. HARQ is a multiple stop and wait mode, instead of just stop and wait which would reduce the efficiency of the system.

When the base station receives an ACK from a UE, then everything is fine. When the base station receives a NACK from a UE, then the packet was received but not detected properly and retransmission (IR) should take place from the base station. In cases where the base station does not receive any Ack or Nack, but rather a DTX, it should retransmit using another self-decodable rate matching scheme.

5.7.1.12 Fast Packet Scheduling

This algorithm is not specified in 3GPP. It defines how to share the available resources with the available UEs. However, the capability of the UE also affects the scheduling process. There are a few different types of algorithm. Round Robin is one of the most popular and least complex as it allocated the resources with equal probability without taking into account the quality conditions. Proportional fair is one of the more complex and uses quality information before resource allocation by the user. The maximum C/I scheduler gives the resources to a very small number of users and at the end of the cell the user might never get scheduled. The minimum bit rate scheduler provides more advanced QoS differentiation.

5.7.1.13 Code Multiplexing

When several UEs are present in the same TTI, then data is sent to them by using a different set of codes known as code multiplexing. This is used to optimise the performance of the data users. If there are 10 codes used in the HS-PDSCH, 5 codes could be used to send data for UE1 and the remaining 5 codes could be used for UE2 in the same TTI. An example that requires code multiplexing is VoIP over HSDPA. Implementation of both time and code multiplexing is expensive.

5.7.2 HSUPA (High Speed Uplink Packer Access)

HSUPA (High Speed Uplink Packet Access) is complementary technology in the uplink direction to downlink HSDPA. The 3GPP standard is release 6 known as E-DCH (Enhanced Uplink Dedicated Channel) that includes investigations related to shorter TTI for faster link adaptation, HARQ for transmissions that are more effective, faster scheduling and high order modulations. This technology works with release 99 with many fundamental features such as RRM and random access; cell selection remains the same. The uplink transport channel, namely E-DCH, is similar to that of the downlink though this is a dedicated channel and not a shared one. Higher order modulations are not used in HSUPA.

5.7.2.1 Protocol Structure

A new aspect of protocol structure in HSUPA consists of a layer, namely a MAC layer called MAC-e, added in the base station, RNC and UE and its main task is to make sure that packets that were sent to the upper layers in a sequence are sent to the base station. This is because scheduling is controlled and combining functionality is at the base station. The MAC-e layer performs reordering after the data is received by the base station.

5.7.2.2 HSUPA Channels

HSUPA consists of E-DCHs (Enhanced Dedicated Channels) and its main features include supporting faster base station scheduling, shorter TTI and faster HARQ. This channel is similar to HSDPA's HS-DSCH but is different from the perspective of soft handover, no adaptive modulation, faster power control and a variable spreading factor.

E-DPDCH (E-DCH dedicated Physical Data Channel): Data is carried in this and the data bits are transmitted from the mobile to the base station using BPSK modulation and faster power control. It supports fast HARQ, fast base station scheduling and 2 ms TTI and OSVF.

E-DPCCH (E-DCH Dedicated Physical Control Channel): This exists to support the E-DPDCH by providing the information needed to decode the data channel transmission along with the information related to channel estimation and power control. This contains three segments: the E-DCH Transport Format Combination Indicator (E-TFCI) indicating the format of transportation; the retransmission sequence number which gives the HARQ sequence number; and a single bit known as the 'happy bit' which indicates if the mobile is content with the data rate or not.

E-AGCH (E-DCH Absolute Grant Channel): This is a downlink physical channel that tells the mobile the maximum transmission power which can be used for data transmission. The five-bit information tells the power level that E-DPDCH can use.

E-RGCH (E-DCH Relative Grant Channel): This is a downlink physical channel, BPSK modulated and transmitting-cell dependent used for adjusting the uplink data rate by scheduling relative transmission power.

E-HICH (E-DCH HARQ Indicator Channel): This is a downlink channel and used to acknowledge (positive/negative) uplink packet transmission. Positive acknowledgment is used when the base station receives TTI correctly and negative acknowledgement is used when TTI is received incorrectly. The structure is similar to that of E-AGCH.

5.7.2.3 HSUPA Radio Resource Management

5.7.2.3.1 HARQ

HARQ in HSUPA is fully synchronous. IR and CC are present in HSUPA with IR making retransmissions more effective and the buffer is maintained by the base station. All the timing sequences are defined with respect to 2 ms and 10 ms and tell which HARQ process is being used. The number of processes is not required to be configured.

5.7.2.3.2 Scheduling

HSUPA scheduling is moved to the base station and its resources are distributed across all users. This approach is called a shared approach. The scheduler in the HSUPA system does the task of allocating resources of existing users when admitting new users in the network. The scheduling has better spectral efficiency (L1 HARQ) and it is faster (base station based scheduling). Three physical channels are used, i.e. E-DPCCH in uplink while E-AGCH and E-RGCH in the downlink direction. There are two scheduling methods: long-term grants and short-term grants. Long-terms grants are issued to many mobiles that can send data simultaneously and is done in the code domain. Short term grants allow multiplexing mobiles in the time domain.

5.7.2.3.3 Soft Handover

In release 99, scheduling and HARQ are not required to operate in many base stations (to reduce the near–far problem) while in HSUPA, as such a problem does not exist, it is needed in a maximum of four cells. Scheduling is impacted by the uplink soft handover function. The handover control in RNC decides on the cells that are in the active set and the cell that is a serving HSUPA cell.

5.8 WCDMA Radio Network Optimisation

The fundamental process of WCDMA radio network optimisation is quite similar to that for GSM radio network optimization. However, the presence of data, both real time and non-real time, with four classes, and each class having its own types of application, with each application demanding a different quality aspect, makes this process of optimisation much more complicated and critical (for the success of the network). The process of optimisation starts at the beginning of the network planning, i.e. when pre-planning starts. It is slightly different to the network planning process; it is more oriented towards optimising the radio network, rather than the launch of the network. The optimising process that is done before the network launch focuses more on the performance optimisation by configuration changes, e.g. improving coverage by antenna height adjustments or by changing antenna tilts, etc.

The optimisation process starts with the existing data. The present network data and the master plan (made during the initial phase of network planning) are recommended input data for this process. Present network data is needed to verify if there have been any deviations from the original plans during the implementation phase. The master plans will show the targets (both short term and long term) for coverage, capacity and quality for which network design was done, apart from the configuration of the network (Figure 5.11).

5.8.1 Key Performance Indicators

Deciding on the key performance indicators or KPIs is the most important part of the optimisation process as the methodology of optimisation is decided. Owing to the complexity of the networks, there is a large number of parameters in the WCDMA networks. However, few of these parameters are chosen; the ones that are have a significant impact on the radio network. These are the parameters whose values/performance is studied in the whole process and then their values/performance during the optimisation process is improved upon. The standard values are usually derived from some standard formulas. As the radio network varies in performance from terrain to terrain or region to region, these formulas and values may change. However, these key chosen parameters are the ones that are responsible directly for the coverage, capacity and quality of the network. The main KPIs include the call success/failure rate, dropped call rate, (soft) handover success rate, average throughput on uplink/downlink, and average throughput on various channels such as RACH, FACH, and PCH, etc. When defining the KPIs, it is important to know whether or not tools are available that could measure the performance of the chosen parameters.

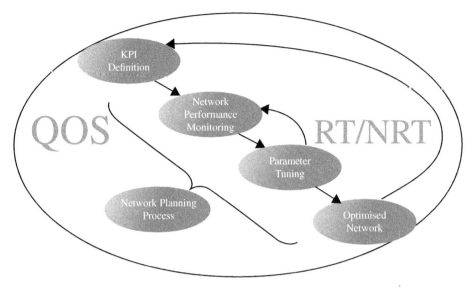

Figure 5.11 WCDMA radio network optimisation.

5.8.2   Network Performance Monitoring

The network performance can be monitored through two ways similar to the ones described in Chapter 2: drive tests and the network management system or NMS. However, the major difference between the monitoring processes of two networks (GSM and WCDMA) is the application that is being monitored. In WCDMA radio networks, real time data transfer delay would perhaps be the most critical aspect to be monitored (Table 5.6). Even in real time data, each application would be monitored/ measured against the backdrop of its own QoS requirements. In the WCDMA radio network, this process and especially the tools involved, i.e. drive test tool, and the NMS are much more complicated. For the NMS measurements, as in GSM radio networks, most of the measurements are performed on the RNC, based on the KPIs on which these measurements are run. The statistics collected from the drive test tool and the NMS are then utilised in conjunction with some post-processing tools that help in providing different kinds of outputs (or reports).

The quality of the network is usually viewed from the perspective of mobile subscribers. This is the reason why drive tests are usually important. The quality can be assessed right from the time when the first site goes live. The NMS usually comes into place when the network is near launch or has launched. Moreover, the NMS usually is able to provide some statistics when there is a certain number of subscribers. Drive tests help in assessing the coverage and the quality of the networks, apart from the verification of the bit rates. For this purpose, some field measurement tools are used. On the other hand, NMSs should have the capability to handle the various aspects of the WDCDMA radio network, e.g. various service classes, individual applications and their individual quality requirements. They should not only be able to handle open interfaces but also a multi-technology and multi-vendor environment.

Table 5.6 Drive test result summary for video requirements.

	Conditions					Required performance			
						Call set-up failure rate		Call set-up delay, 95th percentile	
Date	DPHC_E_b/N_o (downlink) (dB)	DPHC_E_b/N_o (uplink) (dB)	Loading	BLER (downlink)	Dropped call rate	Mobile originated call	Mobile terminated call	Mobile originated call	Mobile terminated call
R1.5.2	>6.5	>4	<70%	<1%	<3.5%	<2.5%	<2.5%	<5 s	<5 s
Drive test results summary				0.79%	1.78%	2.19%	1.34%	95.11%	86.93%
Sample size				11 997	354	549	571	634	505
Call set-up failure rate					95th percentile		Delay		
2.19%								6.34 s (Alerting)	

5.8.3 Coverage, Capacity and Quality Enhancements

5.8.3.1 Coverage Enhancements

Coverage is directly related to the link performance. Thus, if the link performance is enhanced, coverage of the network increases. An increase in the coverage would increase the average TX power of a base station in the downlink direction. If the system capacity were downlink limited, then an increase in coverage would lead to a decrease in capacity. However, if this system is uplink limited, then the capacity is not affected. Thus, link performance increase is directly related to the increase in coverage. There are many ways to improve the coverage. We have already seen the main parameters that affect the link performance. Parameters such as link BER, BLER (block error rate) E_b/N_o, power control headroom, etc., directly affect the power budget and hence the coverage. Uplink coverage can be improved by decreasing the interference margin or by reducing the base station noise figure, or even by increasing the antenna gain. However, the processing gain and E_b/N_o are the two major values that affect the coverage.

A reduction in values such as E_b/N_o increases the coverage of the network. This is because for a lower E_b/N_o, less power is required for the same performance and so more area can be covered. The performance of E_b/N_o is dependent upon a number of factors including the bit rates, channel accuracy, SIR algorithms, etc. Uplink coverage becomes an issue at higher bit rates as compared with the GSM radio networks as these networks (GSM) offer only low bit rate services. Thus, (accurate) traffic distribution will play a major role in coverage improvement. If the bit rate of the uplink direction can be reduced, the coverage can be improved (as the transmit power required would be less). This is possible only for the NRT data (which are less delay critical) or for voice, which has lower bit rates. However, E_b/N_o cannot be lowered below the requirements of the requested service.

One more way to improve the E_b/N_o ratio is to increase the multipath diversity. The two signals coming to the two antennas instead of one can be combined coherently while the receiver noise can be combined non-coherently. Not only would this technique provide better gain, but it would also give protection against fast fading, etc. This technique is also common in the GSM system. Concepts like that of antenna tilts are also used in the WCDMA radio network to improve the coverage area (a typical example shown in Figure 5.12).

5.8.3.2 Capacity Enhancements

As mentioned before, in WCDMA networks, capacity and coverage are heavily dependent. The higher the uplink coverage, the lower is the uplink capacity and vice versa. This is because lower capacity would mean a lower number of mobile subscribers, which means less interference. Moreover, the uplink power budget is used to calculate the cell range, which is further used to calculate the downlink power budget.

The load factor along with the link budget calculations could be used to study the capacity in the network. Load factor is used for capacity analysis for both the uplink and downlink directions. Load factor is dependent upon E_b/N_o, processing gain, interference, activity factor, etc. The orthogonality factor and soft handover are a couple of other factors associated with the load factor in the downlink direction.

The best way to improve the capacity is always to increase the number of cells/carriers. The increased number of sectors increases the capacity of the network and the ratio of

Site ID	Sector	Scrambling Code	Actual tilt angle		Proposed tilt angle	
			E tilt	M tilt	E tilt	M tilt
9205	A	172	6	0	4	0
9024	B	84	3	0	4	0
	C	92	5	0	6	0
9133	B	202	4	5	6	5

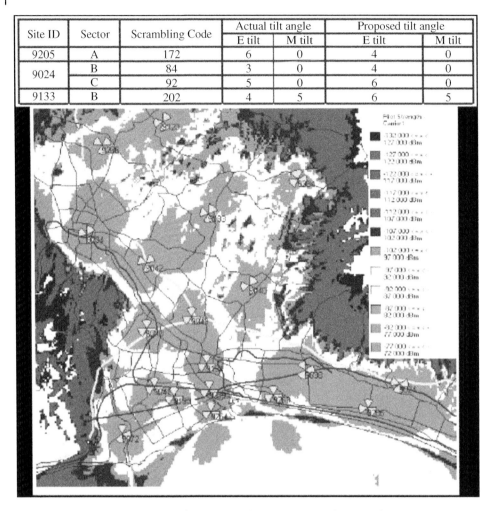

Figure 5.12 Typical coverage plans after antenna tilts in a WCDMA radio network.

increase in capacity is directly proportional to the increase in the percentage of carriers. Another factor is the orthogonal codes that should be ideally orthogonal although due to multipath, some orthogonality is partly lost (thereby increasing interference). Multipath diversity, as stated in the section above, improves the coverage but also reduces the orthogonality. Multipath diversity is more important at the cell edges as it improves the performance.

Another way to improve the capacity is by transmit diversity. If multipath diversity is less, then the downlink transmit diversity increases the capacity to quite an extent. Lower bit rates would also increase the capacity. This is possible by using the adaptive mean rate (AMR) codes. AMR is the speech codec scheme that is used in UMTS.

Increasing the number of sectors can also increase the capacity. A two-sector site will offer a higher capacity than an omni site. This increases the amount of area covered along with the quality. Of course, this upgrade would mean additional costs for antenna changes, etc.

Table 5.7 PS service requirements and drive test results.

Downlink data rate (kbps)	RAN	Conditions				Required performance				
		Down link DPCH_E_b/N_o (dB)	Uplink 64 kbps DPCH_E_b/N_o (dB)	Airlink capacity (kbps per carrier/sector)	BLER (uplink and downlink)	Data session drop rate 1 drop/N h	Data session activation failure rate	Session activation delay time	Average peak throughput (uplink and downlink)
64	R1.5.2	>5.0	>3.5	800	<10%	N>15 h	<5%	<5 s	>50 kbps
128	R1.5.2	>5.0	>3.5	800	<10%	N>15 h	<5%	<5 s	>101 kbps
384	R1.5.2	>5.0	>3.5	800	<10%	N>15 h	<5%	<5 s	>303 kbps
Drive test results summary					6.34%	No drop/15 h	1.41%	89.40%	57.06 kbps (UL) 341.36 kbps (DL)
Sample size					12 320		594	538	14 711 (UL) 1194 (DL)

5.8.3.3 Quality

Though end-to-end QoS would be an area to focus on in 3G networks, here we focus on the delay on the air-interface, which will have a direct impact on the quality. As mentioned in the earlier chapters, QoS would be application-dependent but an immediate concern would be to reduce the delay at the air-interface for the PS services. Unlike in GSM where voice quality was the only 'big' concern, in WCDMA attention turns to PS service requirements and performance. Delay may or may not take place at the air-interface (it may be due to transmission or the core networks), yet the first step for end-to-end quality is the performance of the application at the air-interface.

One such example is shown in Table 5.7. It specifies the conditions and required performance for the PS services. PS service performance requirement is based on parameters such as BLER, DCR, application failure rate, delay and the throughput for 64, 128 and 384 kbps data rates. Drive test results give the monitored results for these five parameters.

5.8.4 Parameter Tuning

Once the KPIs and related parameters have been identified, the process of tuning these parameters begins. The tuning is done in order to enhance coverage, capacity and quality of these networks. As mentioned before, the WCDMA networks have huge number of parameters and some of the key ones are chosen, measured/analysed and optimised. These parameters can be subdivided into various groups based on the functions they affect the most. These groups may include parameters affecting handover control, packet scheduling, call admission control, power control, etc.

5.8.4.1 Handover Optimisation

Soft handover gain is one of the parameters in the link budget calculations. Soft handover gives some protection against both slow and fast fading. With respect to slow fading, due to uncorrelation between the base stations, the mobile is able to select a better base station (based on the measurements analysed in the RNC) while with fast fading, due to the effect of macro diversity combining, the required E_b/N_o is reduced. The soft handovers also induce overhead in capacity calculations, as at a given time the mobile is connected to more than one cell, thereby increasing the capacity requirements. Thus, both the overheads and the gain should be optimised. The idea behind optimising the overheads is to save the downlink capacity. A typical value for the soft handover overhead is between 30% and 40%. Soft handover gain on the other hand can be estimated by using parameters such as DCR, CSR (call success rate), transmit and receive powers. For the capacity and coverage optimisation (leading to better network quality), it is necessary to optimise the handover control feature in these networks. One important parameter to mention at this stage is the transmitting power of CPICH. This parameter affects the coverage and it should be set as low as possible. The optimum value of this parameter would determine the coverage and capacity, i.e. should a new user be permitted in the cell. This parameter also affects packet scheduling. If the value of CPICH were not optimum, then it would lead to either the network being underutilised or even huge interference (if the number of user is more than required), thus degrading the quality of the network. These parameters also affect the call success rate and dropped call rates. As in GSM networks, these two factors directly determine the network quality.

5.8.4.2   Packet Scheduling Optimisation

Packet scheduling is one of the most important aspects when controlling the congestion in the network. This handles the non-real time packet data deciding the timings of the packet initiation and the rate at which it should be delivered. As seen at the beginning of the chapter, there are four traffic classes defined. Each of these classes has different applications to take care of. The NRT packet data is 'bursty' in nature, containing one or more data calls. Packet scheduling is carried out for both uplink and downlink for non-real time bearers. The packet can be scheduled by using the time division or code division techniques or both. Packet scheduling and load control (inclusive of admission control) work in tandem. A higher load would lead to higher interference, which would mean less calls being admitted in the network. This would affect the bit rates assigned to the NRT packet data. Thus, for packet scheduling, i.e. less delay and higher bit rates for NRT data, load control would be an important parameter to analyse and optimise. The transmitted power in the downlink and interference power in the uplink would be important parameters for optimisation. When the thresholds of these two parameters are crossed, preventive measures to control the load are initiated. From the perspective of packet scheduling, this is important because assigning of the higher bit rates or lower bit rates is dependent upon the load control and AC.

5.8.4.3   Power Control

Efficient and fast power control is the key to success in WCDMA technology. As seen previously, power control is based on the SIR. The power control feature has a direct effect on the coverage area. Another aspect related to power control is interference that may lead to capacity limitation. Both uplink and downlink power control are necessary, with downlink being more critical. Power control of the common channels is necessary and the most important ones that should be analysed and controlled are the CPICH, AICH, PICH, and CCPCH. However, the process is more of control rather than to optimise.

5.8.4.4   Admission Control

The admission control function is directly related to the load control process. The process is critical for both the RT and NRT traffic generators/users. As the AC function is power based and throughput based, parameters that are related to both these features are important for capacity and coverage optimisation. The most important parameters would then be the transmitted power and received power, both in uplink and downlink, orthogonality, and the throughput in both uplink and downlink directions.

6

3G Transmission Network Planning and Optimisation

6.1 Basics of 3G (WCDMA) Transmission Network Planning

6.1.1 Scope of Transmission Network Planning

The scope of transmission network planning includes the region constituting the interfaces between the BS and RNC, between the RNC and CN, and between RNC and RNC, i.e. I_{ub}, I_u (CS and PS) and I_{ur}, respectively (Figure 6.1).

6.1.2 Elements in 3G Transmission Networks

6.1.2.1 Base Station (Node B)

Base stations play an important role in 3G transmission networks. Their role changes significantly in 3G as compared with GSM transmission networks. In GSM transmission networks, if the number of transceivers or TRXs was knows, the A_{bis} capacity calculation could be done. However, with data coming to play an important role, the components of the base station also change, thereby changing the dimensioning and functioning of the network. A simplified block diagram is shown in Figure 6.2. A brief overview of the functions of these blocks is given below.

6.1.2.1.1 Antenna Filter and Power Amplifier (AF and PA)

The filter isolates the transmitted and received signals. The amplifier amplifies the received signals.

6.1.2.1.2 Transceivers (TRXs)

TRXs contain one transmitter and two receivers. A TRX consists of two frequency bands that are separated by a frequency of 190 MHz. The major differences form the GSM TRX include accurate power control, a different frame structure, and channelisation is different as seen in Chapter 5.

6.1.2.1.3 Summing and Multiplexing Units (S and M)

These sum the signals from the various other summing and multiplexing units and/or from signal processing units.

Fundamentals of Network Planning and Optimisation 2G/3G/4G: Evolution to 5G,
Second Edition. Ajay R. Mishra.
© 2018 John Wiley & Sons Ltd. Published 2018 by John Wiley & Sons Ltd.

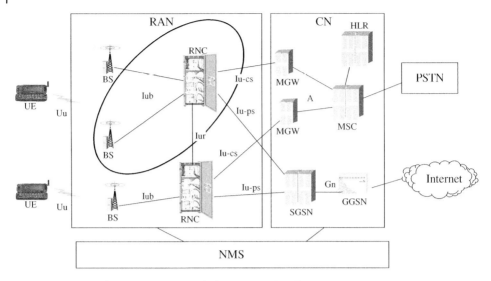

Figure 6.1 Scope of transmission network planning in 3G (WCDMA).

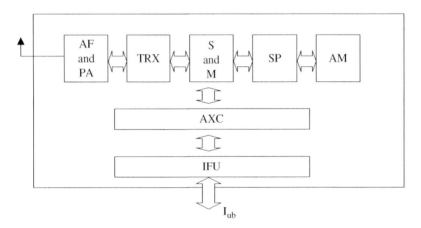

Figure 6.2 Block diagram of 3G base station.

6.1.2.1.4 Signal Processor (SP)

The major function of this unit is to perform coding, decoding, and code channel processing (both in receiving and transmitting directions).

6.1.2.1.5 Application Manager (AM)

This unit is responsible for operation and management functions and carrier control. From a transmission network planning perspective, the AM takes the same place as a TRX in GSM transmission network planning. The number of AMs is dependent upon the traffic that is carried by the base station and is in no way dependent upon the number of cells or number of TRXs.

6.1.2.1.6 ATM Cross Connect (AXC)

This is responsible for cross-connection at the ATM level. It also acts as an interface between the AM and the interface units. AXC can be a standalone unit.

6.1.2.1.7 Interface Unit (IFU)

This is the interface between the base station and I_{ub} interface. This may vary depending upon the type of transmission, i.e. E1 /T1/JT1 or PDH/SDH, etc.

6.1.2.2 Radio Network Controller (RNC)

As mentioned in previous chapters, the RNC in WCDMA 3G networks is equivalent to the BSC of radio networks. Apart from similar functions, there is one main addition in the RNC, which is an interface, i.e. I_{ur} (interface between two RNCs). Though the RNC is a very complicated network element, a simplified block diagram is shown in Figure 6.3.

6.1.2.2.1 Interface Unit (IFU)

There are mainly two kinds of IFU: PDH and SDH. On either of the interfaces, i.e. I_{ub}/I_{ur}/I_{u}, a PDH or SDH interface can be configured.

6.1.2.2.2 Switching Unit

The asynchronous transfer mode or ATM forms the backbone of WCDMA networks. In the RNC, the functionalities of the switching unit include providing the required support for the ATM traffic, ATM cell switching, AAL2 switching, and multiplexing of traffic.

6.1.2.2.3 Control Unit

Radio resource management or RRM is considered to be the most important function of the RNC. All three network planning entities – radio, transmission, and core – are affected by it. The control unit is responsible for RRM functions such as handovers, admission control, power control, and load control. It is also responsible for the control mechanisms for packet scheduling and location-based services.

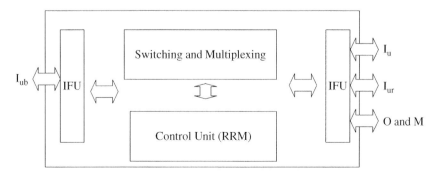

Figure 6.3 Block diagram of radio network controller (RNC).

6.2   Transmission Network Planning Process

Fundamentally, the transmission planning process is the same as in GSM transmission networks apart from an important addition: the inclusion of ATM technology. The process as usual consists of nominal planning and detailed planning. However, in the 3G networks, there are two areas which have major changes in the transmission planning process compared with the 2G GSM transmission networks: in nominal planning it is dimensioning and in detailed planning it is (ATM) parameter planning. All other areas are quite similar to those of GSM transmission network planning. The process is shown in Figure 6.4.

Note, according to 3GPP Rel'99, ATM is the main technology in the transmission network. However, IP is expected to replace ATM as the main technology.

6.2.1   Nominal Planning

The constituents of nominal planning remain the same as in GSM transmission planning, i.e. protection, link budget calculations, topology, equipment dimensioning, etc. However, the main change takes place in dimensioning of the transmission network. Owing to the presence of traffic voice, CS data and PS data, with the data consisting of both RT and NRT traffic, dimensioning becomes much more complicated and important compared with the GSM transmission network. However, before we venture into dimensioning, let us consider some aspects that affect it.

6.3   Asynchronous Transfer Mode

B-ISDN or broadband integrated services digital network is an integrated services network that provides digital connections capable of supporting rates greater than the primary rate (i.e. >2 Mbps) between user-network interfaces. B-ISDN was developed to

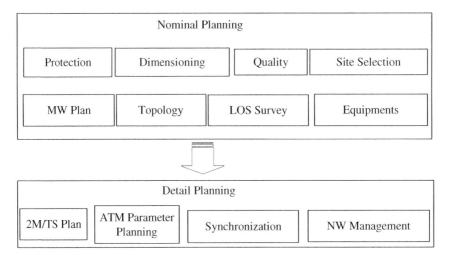

Figure 6.4   Transmission network planning process.

support business and residential customers, with constant and variable data rates, data, voice and still and moving pictures, transmission and multi-media applications combining several service components. The basic idea behind the development of this technology was to provide a platform for development of all future technologies. Owing to the variable nature of B-ISDN, ATM was selected as a switching and multiplexing technique as it is capable of giving a desired quality of service or QoS for different applications.

Information transfer takes place in small packets, also known as cells. One ATM cell is 53 bytes long and is divided into two parts: header and payload. The header is 5 bytes long while the payload is 48 bytes long. These cells are multiplexed to form virtual channels, which are subsequently multiplexed to form virtual paths, as shown in Figure 6.5. Transmission media carry these virtual paths, which may vary from one to numerous (depending upon the capacity of the transmission media).

6.3.1   Cell Structure

The cell structure of the ATM depends upon the type of interface. There are two types of interface. The interface between the ATM terminal and ATM switch is known as the user-to-network interface or UNI and the interface between two ATM switches is known as the network-to-network interface or NNI. Both types of cell have a 5-byte header and 48 bytes of payload, as shown in Figure 6.6. The difference lies in the structure of the header. The various components of the header are described in the following.

6.3.1.1   Header Error Control (HEC)

HEC is for error detection and one-bit error correction of the five-byte header. It is eight bits long and its sequence is calculated and set in the HEC field at the transmission side while the detected errors are corrected at the receiving side.

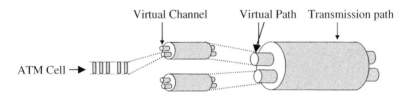

Figure 6.5 Multiplexing of ATM cell.

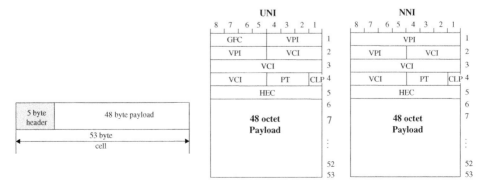

Figure 6.6 ATM cell structure.

6.3.1.2 Virtual Path and Virtual Channel Identifiers

As mentioned before, the ATM cell is multiplexed in virtual channels or VCs and virtual paths or VPs, respectively. For the cell to be able to identify its VC and VP, identifiers are needed. There are known as virtual channel identifiers or VCIs and virtual path identifiers or VPIs, respectively. The cell routing is based on the values of the VCI and VPI. The VPI has 8 bits for UNI and 12 bits for NNI.

- Virtual channel: A sequence of ATM cells belonging to a particular type of service or destination.
- Virtual path: Number of VCs sharing one single link, which bundles different VCs into one VP. Also, to simplify routing and switching of cells belonging to a particular destination and particular type of service, VCs are combined into a single VP.
- Virtual channel identifier: To identify a particular type of VC, the header of the ATM cell has a VCI.
- Virtual path identifier: To identify a particular type of VP, the header of the ATM cell has a VPI.

6.3.1.3 Payload Type (PT)

The payload type basically indicates the kind of information a cell is carrying, i.e. user information or management information. If the payload is carrying user information it may have information about congestion in the network while if it contains management information, it may contain information to distinguish different types of cells such as whether it is an associated cell or a resource management cell.

6.3.1.4 Cell Loss Priority (CLP)

CLP is a one-bit information field that gives the priority for cell loss, basically used for traffic management. There are two priorities for ATM cell loss: 0 and 1. Cells with CLP = 1 are discarded before cells having CLP = 0, in periods such as congestion.

6.3.1.5 Generic Flow Control (GFC)

This is a feature of a UNI cell header which basically helps the user to control the traffic flow to obtain a desired QoS. It is a four-bit space in the UNI cell.

6.3.2 ATM Protocol Layers

ATM protocol layers (shown in Figure 6.7) are based on the B-ISDN protocol reference model consisting of three planes: user, control, and management. The user plane handles the user-related information such as flow control, while the control plane handles the control-related information such as call control.

6.3.2.1 Physical Layer

Generally it is said that the physical layer is the physical (transmission) medium. However, this is only a part of the physical layer. In the ATM protocol structure, the physical layer consists of two sublayers: physical medium (PM) and transmission convergence (TC). The PM is the lowest sublayer and its functions are dependent on the PM. The main functions usually include bit alignment, line coding and electrical/optical conversion. TC lies above the PM and its main functions include maintaining the

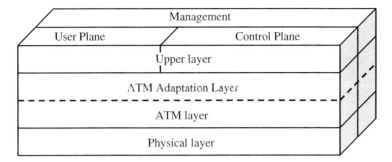

Figure 6.7 ATM protocol layer.

cell rate of the ATM cells. It does this by adding/extracting cells to adapt the rate of ATM cells. Other functions of TC include cell delineation, frame generation, and frame recovery.

6.3.2.2 ATM Layer

This is the layer in which VPs and VCs are established and is responsible for providing the connection-oriented service, i.e. establishes connections before transmission. The features of this layer are independent of the physical layer below it. It is responsible for generic flow control (only for UNI) and generation (and extractions) of cell headers. Another important function includes cell VPI/VCI translations at the switching/cross-connect nodes. It is in this layer that cells from individual VPs and VCs are multiplexed into one resulting composite cell stream.

6.3.2.3 ATM Adaptation Layer

As the name suggests, the ATM adaptation layer or AAL 'adapts' the applications in a way to be understood by the ATM layers (or network) below. It also has two sublayers: segmentation and reassembly (SAR) layer and convergence (CL). The SAR layer takes the cells from upper layers and sends them to the ATM layer. Another function of this layer is 'reassembly'. It reassembles the cells it receives from the ATM layer and passes them to the CL layer. The CL layer is an interface between the application and the SAR. It splits the cells for transmission towards the SAR and on receiving the information from SAR, it reconstructs the cells into the original message. CL is a service-dependent sublayer making AAL also a service-dependent layer. There are several AAL classes to support the different service classes. Connections falling under a service class have similar QoS requirements. The network can handle each service class separately by providing separate virtual paths. The following service classes are defined by the ATM forum:

- Constant bit rate (CBR): It requires a static bandwidth that is available for the connection lifetime. Traffic is fixed and synchronous. A typical example is real-time voice or video services.
- Real-time variable bit rate (RT-VBR): This is for services that require variable bit rate during the time of use. When the application is real time, service falls under this category. A typical example is compressed video.
- Non-real-time variable bit rate (NRT-VBR): Another form of VBR, it is for non-delay critical application. A typical example is email.

- Available bit rate (ABR): Bursty traffic with a known bandwidth requirement falls under this category. Basically, the traffic would follow whatever bandwidth the network can give to this. A typical example is web browsing.
- Unspecified bit rate (UBR): Applications not having strict requirements for delay and delay variations fall under this category. Also known as best effort traffic, there is no QoS guaranteed for this service class. Typical examples are emails and FTPs.

Different types of AALs can handle different types of traffic. There are five AALs present from AAL1 to AAL5. A brief description is given below.

6.3.2.3.1 ATM Adaptation Layer 1 (AAL1)

This is used for the traffic that requires a constant bit rate and a permanent connection during its execution, e.g. voice. Due to the kind of application that this supports, the data should be transferred with minimum delay and high quality. Usually when data transfers from one layer to another some 'overheads' are added to the data stream. These overheads are basically some bits that are added to make the data stream 'layer' compatible. The larger the overhead (i.e. the higher the number of bits added), the longer the delay. However, in this case overheads are kept to a minimum so as to reduce the delay.

6.3.2.3.2 ATM Adaptation Layer 2 (AAL2)

This is for the variable bit rate information that requires a strict relationship between the transmission and reception clocks. It provides bandwidth efficient transmission for short, variable-length packets. This is able to multiplex the short packets from the multiple users to one ATM connection. AAL2 is designed for applications such as compressed video, i.e. the RT-VBR type of traffic class. This layer can also do error checking and sequencing of the data. Sometimes, when data is flowing, i.e. payload is empty, this layer also does the padding in the payload so as to maintain the real-time aspect of the traffic.

6.3.2.3.3 ATM Adaptation Layers 3 and 4 (AAL3/4)

Both these layers are used for data transfer functions. Another feature of these layers is multiplexing, i.e. allowing same virtual channel to carry traffic from multiple sessions. This allows effective usage of the VC in applications such as X.25.

6.3.2.3.4 ATM Adaptation Layer 5 (AAL 5)

This is similar to layers 3 and 4 but is more efficient. It can be used for both the message and stream modes of data transfer.

6.3.3 Multiplexing and Switching in the ATM

As ATM cells transverse through the network, VCI and VPI values may change depending upon the routing and addressing needs of the network and ATM elements. VCI and VPI have local significance only to each interface. The cells may enter a particular ATM switch with certain VPI/VCI values and leave with different VPI/VCI values. VPI/VCI identifies the next ATM pair of segments that the cells have to travel to in order to reach the final destination. Depending on the level of the switching performance at these cells, the cells can be switched on either the VC or VP level. If a cell is switched on the VC level, it is switched from one VP to another. If the switching is done on the VP level, all the VCs are included in a particular VP through the transmission path.

6.4 Dimensioning

We saw in Chapter 5 how radio link budget changes with data rates. As the data rates increase, coverage area decreases. Similarly, data rates have a big impact on the transmission network planning right from the dimensioning phase. Once the number of base stations has been decided, all the RAN interfaces except the U_u interface, i.e. capacity required for I_{ub}, I_{ur} and I_u (both CS and PS) interfaces, should be dimensioned. In addition to the interfaces, the number of RNC required should be dimensioned. For dimensioning the interfaces, some knowledge about protocol stacks for CS data and PS data is required.

Figure 6.8a is a protocol stack for the user plane CS traffic. It basically shows the movement of the traffic from the mobile station to the MGW or media gateway. Figure 6.8b shows a similar movement of the traffic for PS data from a mobile station to the GGSN. Some hardware units in the equipment perform the functioning of these layers. One such example is node B or the base station. SP units perform the layer 1 functions while FP/AAL2/ATM layers are present in the application manager (as shown in Figure 6.2).

An important aspect from a dimensioning perspective is that when the traffic/signal moves from one layer to another, it adds some overhead bits. These bits increase the capacity of the signal and thereby increase the actual capacity requirements of the interface. Thus, when doing dimensioning, each overhead has to be carefully calculated.

Assume that a mobile subscriber is trying to establish a connection with the network. As the call is voice (CS), it will follow the protocols shown in Figure 6.8a. The overheads that would be attached to the call's capacity calculations would include voice activity factor (VAF), soft handover (SHO), protocol overhead, and signalling overhead.

6.4.1 VAF

Once a mobile is connected to the network, the subscriber will either speak or listen, i.e. it would be either in active mode or silent mode. In either mode, some bits would be required for the functioning of that mode. Obviously, overheads during the active mode are more than in the silent mode. The type of voice connection can be defined in terms of AMR (adaptive multi-rate) codec modes. There are eight AMR codec rates, from AMR-0 to AMR-7 with the rates ranging from 12.2 to 4.75 kbps, respectively. Usually dimensioning for voice is done for AMR-0 (or 12.2 kbps codec rate).

6.4.2 SHO

As seen in previous chapters, there are many types of HOs in the WCDMA radio network which are very much advantageous to the mobile subscriber as they ensure continuous connectivity. However, as soft handovers take more capacity, the capacity

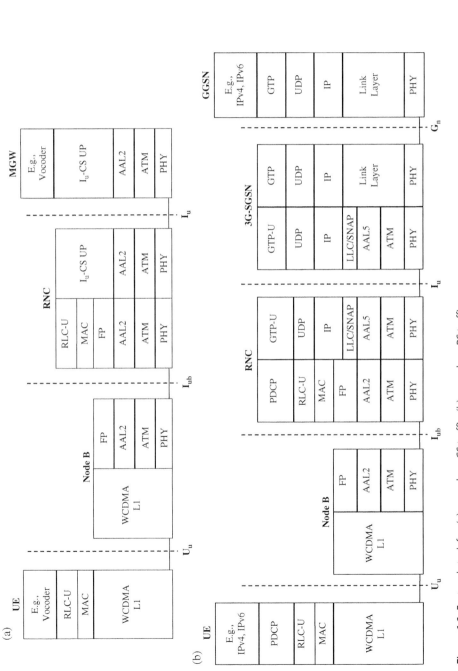

Figure 6.8 Protocol stack for: (a) user plane CS traffic; (b) user plane PS traffic.

required on the transmission interfaces also increases. In some case the capacity requirements may go up to 150% due to SHO.

6.4.3 Protocol Overhead

As the signal moves from one layer to another, some bits are added at each of these layers. This includes the RLC, FP, and AAL2, etc., layers. The overhead will vary with the type of call, i.e. AMR codec.

6.4.4 Signalling Overhead

For each call, signalling is required. Thus, for each interface for which capacity is being calculated, signalling would be required. This signalling also needs capacity to associate itself with the data (CS/PS) when a call is being made from one subscriber to another. Thus, this capacity also needs to be added.

Now, if the user is making a data (PS) call, then some of the above factors will change depending upon the type of call, i.e. data rate, being made. Each different type of data rate will have a different type of overhead. For a data call, however, VAF is not considered as there will be no voice activity. However, overheads for retransmission and buffering are required. The amount of capacity required for buffering the user data is usually similar to that required for transmission.

6.4.5 Signalling on Transmission Interfaces

We have seen the general protocol structure of UTRAN in earlier chapters. Based on it, the following types of signalling are required at the various interfaces.

6.4.5.1 ATM Adaptation Layer 2 Signalling
AAL2 signalling is required for establishing, maintaining, and terminating the AAL2 connections. It also performs functions such as error checking or carrying signalling for lower ATM layers.

6.4.5.2 NBAP (on I_{ub})
NBAP is of two types: common (C-NBAP) and dedicated (D-NBAP). C-NBAP is responsible for the handling of channels like RACH/FACH, cell configurations, fault management, etc. D-NBAP is required for the handling of dedicated/shared channels, air-interface fault management, configuration of radio links, etc.

6.4.5.3 RNSAP (on I_{ur})
RNSAP or the Radio Network Subsystem Application Part is a protocol responsible for functions such as radio link management, physical channel reconfiguration, measurements on dedicated resources, compressed mode control, power drifting correction, error indications, UL/DL signalling transfers, etc.

6.4.5.4 RANAP (on I_u)
RANAP or the Radio Access Network Application Part is a protocol responsible for paging, flow control, processor/CCCH overload at the UTRAN, cipher mode control, location reporting, resource check, direct transfer, trace invocation, error handling, HO procedures, etc.

6.4.6 Other factors

When a call is transferred from the base station to the MSC (for voice) or towards the SGSN (for data), overheads are added. As the RRM functions are located in the RNC, there is no SHO taking place on the $I_{u\text{-}cs}$ interface. Thus, no SHO factor is taken into account. However, factors such as blocking probability and signalling should be taken into account. However, if the call is a data call, some additional overheads are added especially related to the GTP protocol apart from the overheads added due to UDP, IP, etc. (refer Figure 6.8b) and those related to signalling.

Another focus area is RNC dimensioning. It is dependent upon the capacity of the RNC to handle:

- interface traffic, e.g. Iub/Iur/Iu
- number of TRXs
- Number of BTS

Example 1   Transmission network dimensioning
Consider an upcoming network for which transmission network dimensioning is to be done. The inputs required/given are as follows:

- *Dimensioning is to be done for Phase 1 only.*
- *There are 10 000 subscribers each in the dense urban, urban, suburban, and rural type of areas.*
- *Each region has 50 base stations and each site has a configuration of 1 + 1 + 1.*
- *SHO = 40%.*
- *VAF = 67%.*
- *Total CS voice: 20 mErl/subscriber.*
- *Total CS data (64 kbps) = 1.50 mErl/subscriber.*
- *Total PS data (64 kbps) = 0.012 kbps/subscriber.*
- *Total PS data (128 kbps) = 0.272 kbps/subscriber.*
- *Total PS data(384 kbps) = 0.06 kbps/subscriber.*

Total I_{ub} traffic = CS (voice + data) traffic + PS traffic + signalling + overhead + O&M.
Assuming that the ATM overheads are 30%, signalling OH in both RAN and CN is 10%, and packet data OH (GTP, IP, etc.) is 20%, the results are shown in Table 6.1.

6.5 Microwave Link Planning

Microwave link planning is similar to that for GSM transmission planning. However, in 3G networks, ATM over microwave links seems to generate concern over the quality of the links. In GSM transmission, a BER of 10^{-3} is considered to be good enough to meet the desired quality aspects; the GSM transmission links carry more voice traffic than data. However, in 3G networks, with the traffic scenario changing, i.e. data traffic increasing compared with voice, new standards are required for the transmission systems. Moreover, networks in 3G are ATM based. The main difference in microwave planning between GSM and 3G is the BER threshold consideration. As the quality requirement for data is more stringent, a BER threshold of 10^{-6} is considered during link planning. More standards such as ITU-T G.828, ITU-T I.356 an ITU-T I.357 are recommended to be used as ATM requires high quality transmission. The recommended threshold values are 10^{-3} and 10^{-6} for PDH and SDH, respectively.

Table 6.1 Results of transmission interface dimensioning.

Area	No of subs	BS type	TRX	Total sites (BS)	Total TRX	CS voice (kbps)	CS data (kbps)	PS data (kbps)	Total traffic/site (kbps)	Total Iub per site (kbps)	Total traffic to RNC (Mbps)
Phase 1											
D Urban	10000	1+1+1	3	50	150	178.3	349.4	0.0	527.70	1.2	58.0
Urban	10000	1+1+1	3	50	150	178.3	349.4	474.0	1001.70	1.2	58.0
Suburban	10000	1+1+1	3	50	150	178.3	349.4	474.0	1001.70	1.2	58.0
Rural	10000	1+1+1	3	50	150	178.3	349.4	474.0	1001.70	1.2	58.0
				200	600						

Area	No of subs	BS type	#Sites (BS)	CS Mbps	Iu to MSC CS Mbps RNC → MSC	Iu to SGSN PS Mbps RNC → SGSN	Iur Mbps RNC → RNC
Phase 1							
D Urban	10000	1+1+1	50	4.94	5.44	12.89	0.27
Urban	10000	1+1+1	50	4.94	5.44	12.89	0.27
Suburban	10000	1+1+1	50	4.94	5.44	12.89	0.27
Rural	10000	1+1+1	50	4.94	5.44	12.89	0.27
	40000		200		21.74	51.57	1.10

Another aspect to be considered is the ATM performance itself. ATM performance is measured using parameters such as the cell loss ratio or CLR. To keep this value to a minimum, availability targets are usually kept larger than 99.99%. However, this value may change depending upon the operator's requirements.

6.5.1 Topology in Transmission Networks

Transmission Networks in GSM saw all kinds of topologies: star, chain, loop, etc. However, in 3G networks, this will not be the case, at least in the beginning. There are two main reasons for this: capacity requirements and delay. Traffic in these networks consists of voice traffic, CS data, PS data and common channels such as RACH, FACH, etc. If there are about seven users generating voice traffic, and three users generating CS-64 and PS-64, then the total traffic generated by the base station (configuration 1 + 1 + 1) would be more than 50 Erl, of which common channels alone constitute more than 30 Erl, as shown in Figure 6.9. This increases the required capacity to almost four E1s, while in GSM transmission network, a configuration of 1 + 1 + 1 would need less than one E1 of capacity. Thus, longer chains/loops are not preferred in these networks.

Delay is another reason for making star the preferred topology in 3G transmission networks. For real-time traffic, especially data, delay will be the basis of quality. Longer chains would mean more time required for the traffic to reach the RNC/SGSN, which will degrade the quality of the call. Thus, to reduce the delays, star topology is used.

6.6 Detailed Planning

As seen in Figure 6.4, detailed planning has four major aspects: 2 Mbit plans, parameter planning (ATM), synchronisation, and network management system or NMS planning. The principles of 2 Mbit planning and NMS planning are the same as described in Chapter 3. The details of parameter planning are given in the following.

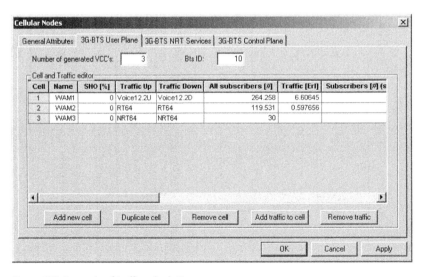

Figure 6.9 Example of traffic calculations.

6.6.1　Parameter Planning

Due to the involvement of ATM, parameter planning has now become an essential part of transmission network planning. Due to this, transmission planning in 3G networks has become a more complicated and challenging task as compared with transmission planning in GSM networks.

We have seen the basics of ATM in the sections above. Now there are two main aspects to be focused on: ATM configuration parameters and ATM performance (related) parameters. However, before we venture into parameter planning for the services provided by an ATM network, let us understand the traffic management on the ATM itself.

6.6.2　Traffic Management on the ATM

Apart from virtual channels and virtual paths, there are also permanent virtual connections or PVCs and switched virtual connections or SVCs. A PVC is set up on a permanent basis. PVCs are set up by the NMS or element managers while SVCs on the other hand require signalling protocol (UNI/NNI) to set up the connection. ATM signalling is used to set up the virtual connections, allowing the switching and end station to communicate, and to exchange various management and QoS information for these connections. Signalling in ATM is always considered to be outband as there is always a dedicated pair of VCI/VPI values to send the signalling messages to an end station from a station requesting a virtual connection. How does this process take place? When one ATM device wants to set up a connection with another ATM device, it sends a signalling request directly to the connected ATM switch. The signalling request packets contain the end destination point address and any desired QoS parameter for the virtual connection. The signalling packet is reassembled by the switches and is examined to see if it has the resources, capability along with the desired QoS parameters to set up such a connection. It then sets up a VC on the input links and subsequently forwards the request out of the interface as specified in its switching table to the next ATM switch for further analysis till it reaches the destination. 'V' connections are then set up along the path going to the end destination as every switch in the route examines the signalling packet and forwards it to the next switch. If the switch is not able to provide the desired QoS, the request is rejected and a message is then sent back to the point of origin. If the end point can support the desired QoS, it responds with accept messages as well as VCI/VPI values that the originator should use for the cells destined to the end point. Upon acceptance of the connections, traffic management functions ensure that each of the connections adheres to the traffic contract as agreed. This involves traffic management functions and techniques such as:

- traffic parameters
- connection admission control (CAC)
- conformance monitoring and control
- queuing

6.6.2.1　Traffic Parameters

We have seen different types of service classes (CBR, UBR, etc.) available in ATM. Each of these service classes is defined by a set of traffic parameters, which describes the characteristic of the traffic source. These parameters are set to ensure a proper resource allocation to provide a guaranteed bandwidth and adhere to the QoS across the ATM network. The most common traffic parameters are:

- PCR (peak cell rate)
- SCR (sustainable cell rate)
- MBS (maximum burst size)
- MCR (minimum cell rate)
- CDVT (cell delay variation tolerance)

6.6.2.1.1 Peak Cell Rate (PCR)

This is defined as the maximum instantaneous rate that the user will transmit. It can be calculated as the inverse of the minimum time interval between cells. If the time interval between two cells is 1 μs, then the $PCR = 1/(1 \times 10^{-6} \, s) = 10^6 \, cells\, s^{-1}$.

6.6.2.1.2 Sustainable Bit Rate (SCR)

This is defined as the average rate of the cells, when measured over a period of time. It is also known as the mean cell rate.

6.6.2.1.3 Maximum Burst Size (MBS)

This is defined as the maximum number of cells that can be transmitted at the peak cell rate.

6.6.2.1.4 Minimum Cell Rate (MCR)

This is defined on the basis of the minimum cell rate that is required by the user. It is a rate that is negotiated by the end-system and the network such that the cell transmission rate never falls below the minimum specified/negotiated value.

6.6.2.1.5 CDVT (Cell Delay Variation Tolerance)

This is an error margin and defines the acceptable variation in the cell transmission time interval, i.e. defines the upper bound in the variation in cell delay.

Parameters like source type, PCR, SCR, MBS, etc. define the type of traffic and are known as 'traffic descriptors'. Before a connection is made a *traffic contract* is negotiated based on factors such as traffic descriptors, CDVT, QoS class, conformance definition, etc. These parameters define the traffic type. However, the quality of the traffic is also an important aspect of ATM. The ATM forum has defined six parameters for this purpose. Of these six parameters, the first three listed below can be negotiated while the last three cannot.

- CLR (cell loss ratio)
- CTD (cell transfer delay)
- CDV (cell delay variation)
- CER (cell error ratio)
- SECBR (severely errored cell block ratio)
- CMR (cell mis-insertion ratio)

6.6.2.1.6 Cell Loss Ratio (CLR)

This is the ratio of the number of cells lost (during transmission) to the total number of transmitted cells. The cells may be lost due to error, congestion or even significant delays in their arrival.

6.6.2.1.7   Cell Transfer Delay (CTD)

When a cell is getting transferred from the source to its destination, a delay may occur due to propagation, queuing, etc. CTD can be defined as an average time taken by the cell to travel from source to destination inclusive of delays.

6.6.2.1.8   Cell Delay Variance (CDV)

Variation in the delay of cell transfer is measured by the CDV parameter. CDV can be defined as the difference between a measure of cell transfer delay and the mean cell transfer delay on the same connection.

6.6.2.1.9   Cell Error Ratio (CER)

The ratio of the cell(s) delivered with an error to the total number of cells.

6.6.2.1.10   Severely Errored Cell Block Ratio (SECBR)

The ratio of SES blocks (of cells) received to the total number of blocks transmitted.

6.6.2.1.11   Cell Mis-Insertion Ratio (CMR)

Owing to some errors in headers, the cells may end up at the wrong destination. It is the total number of mis-inserted cells observed during a specified time interval divided by the time interval duration (equivalently, the number of mis-inserted cells per connection per second). Mis-inserted cells and time intervals associated with severely errored cell blocks are excluded from the calculation of the cell mis-insertion rate.

ATM traffic parameters are related to the services offered by the ATM network. This is shown in Table 6.2.

6.6.2.2   Connection Admission Control (CAC)

CAC algorithms determine if a new connection is to be accepted or rejected (similar to AC in Chapter 5). It accepts a connection request only if sufficient resources are available and if it will not affect the QoS of existing circuit. Some factors that are considered for a new connection request are:

- Traffic parameters of new connections and QoS requirements.
- Existing traffic contract and connection.
- Bandwidth, both allocated and unallocated.

Table 6.2 ATM service categories.

Parameter traffic type	Traffic parameter	QoS parameter
CBR	PCR	CDV, CTD, CLR
RT-VBR	PCR, SCR, MBS	CDV, CTD, CLR
NRT-VBR	PCR, SCR, MBS	CLR
ABR	MCR	
UBR	PCR	

6.6.2.3 Conformance Monitoring and Control

Conformance definition is responsible for indicating when the traffic contract is broken, e.g. the network is not able to provide a bandwidth that is agreed, or a user exceeds the requested bit rate. This is done through two mechanisms: policing and traffic shaping. Policing is a usage parameter function (UPF) mechanism which ensures that during the connection, the network uses the traffic contract defined for the connection to check that it stays within the contracted service. If there are non-conforming cells, then the network takes appropriate action on them, e.g. setting the CLP bit of the non-conforming cells, thus making the cells eligible for discarding. This discarding is done to prevent any non-conforming cells from affecting the QoS of the conforming cells of the connections. The traffic shaping function modifies the traffic flow and changes the characteristics of the user cell streams to achieve improved network efficiency and to get the lowest cell loss.

6.6.2.4 Queuing

To maintain the network optimum performance, the ATM switch performs a series of cell treatment mechanisms, such as queuing, buffering, cell servicing, and congestion control, thus maintaining the desired QoS. Queuing occurs in the following cases: when two cells arrive at the same time and are going to the same destination then queuing will occur; and when cells of higher bit rate pass through a virtual connection with lower bandwidth, congestion takes place. Buffering of cells occurs when two or more conforming cells are destined to the same output at the same time. Cell servicing, such as dropping of cells, occurs when non-conforming cells with the CLP bit set to one arrives which causes congestion.

6.6.3 Network Element and Interface Configuration Parameters

Configuration of the network elements requires specification of a lot of parameters, which are related to the interfaces, ATM, IP, etc. Since ATM is used as a switching and multiplexing technique in 3G networks, every base station would require a unit that can handle ATM connections. This unit can be integrated with the base station (as shown in Figure 6.2) or can be a standalone unit. In either case, parameter planning for this unit would be required. Parameter planning/setting would be required for the interface units/hardware (PDH/SDH), ATM terminations/cross-connections, and those related to the IP addressing and synchronisation of the ATM cross-connect or AXC unit.

6.6.3.1 Interface Unit Parameters

The interface unit parameters are related to the hardware and the PDH/SDH interface units that are used for transmission of ATM on the PDH or SDH. Hardware unit parameters basically are related to the type of system hardware that is being used, i.e. whether it is ETSI based (E1) or ANSI based (T1), etc. PDH/SDH interface parameters are related to the configuration of the PDH or SDH terminals that are being used. These also include some testing parameters apart from the ones needed to configure these terminals. As the structure of SDH is more complicated than that of PDH, more parameters are required for its configuration, e.g. related to the physical, multiplexer, regenerator and virtual container sections.

6.6.3.2 ATM Termination/Cross-connection Parameters

This constitutes the largest group of parameters to be configured. It contains parameters that are required for the identification and termination of the virtual channels, virtual paths and their cross-connections. Also, parameters describing the type of traffic that is being carried by these virtual channels and paths are configured. The 'rules' given by the ATM forum or ITU recommendation for virtual channel and path identifiers should be followed, e.g. number of VCIs reserved or maximum/minimum number of bits that can be used for VCI/VPI, etc. Every vendor/operator follows its own numbering scheme that should be respected. Cross-connection in the ATM network can be at a physical level, VP level and VC level. (In GSM transmission networks, the cross-connections take place only at a physical level.) If expansion is expected soon, VP level cross-connections should be planned as only a few reconfigurations are required even when new features are added. VC level cross-connections are quite complex especially if only one VP is being used. One example is shown in Figure 6.10.

6.6.3.3 IP Addressing Parameters

IP addressing is an important feature of parameter planning and detailed planning of the 3G transmission networks. Both private and public IP addresses should be clearly defined. Private IP addresses are the ones that are used for internal communications and cannot be changed while public IP addresses are the ones that are *seen* by other network elements and can be changed. Parameters related to the IP addresses and routing tables should be defined.

6.6.3.4 Synchronisation Parameters

Parameters related to the synchronisation of ATM cross-connections, such as definition of the clock source or synchronisation source, should be defined along with the priority of these sources if there is more than one clock.

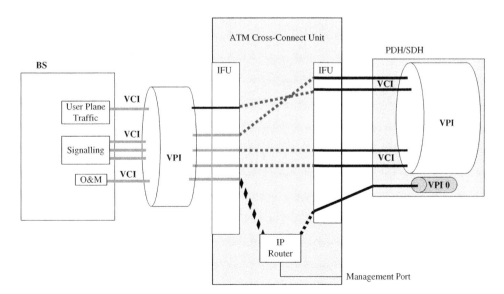

Figure 6.10 ATM termination and cross-connection.

6.6.4   ATM Planning

For ATM planning, a detailed understanding of ATM technology is necessary. ATM planning involves aspects such as:

- Defining the type of traffic.
- Defining the VP and VC connections.
- The number of VPs and VCs, i.e. VPIs and VCIs.
- Defining VP and VC cross-connections.
- Defining physical, VP and VC connection parameters.

Some of these aspects are shown in Figure 6.11: grey lines show the physical paths, thick black lines show the VPCs, and thin black lines show the VCCs.

All these aspects of ATM planning are complicated, especially for a large network. In a network roll-out phase, this process becomes complicated and tedious. For this reason, good ATM planning tools are used.

6.6.5   Synchronisation Plan

The design of the synchronisation plan for a 3G network is quite similar to that for a 2G network, i.e. the clock moving from the MSC to the BSC, from the BSC to the BTS, and from the BTS to the other network elements. The external clock, usually the PRC

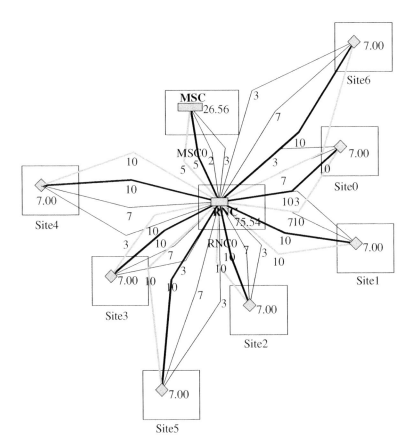

Figure 6.11 ATM planning.

Figure 6.12 Synchronisation hierarchy in 3G transmission networks.

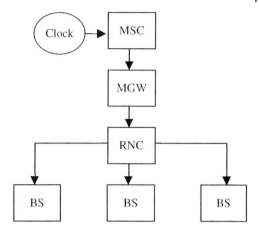

(primary reference clock), is applied to the MSC, which then distributes it further to the MGW, which sends it further to the RNC, and finally it is distributed to the base stations and other network elements. One important thing to remember even in 3G networks is that there might be some network elements that may belong to the 2G networks. In such cases, the ATM parameters need to be observed very carefully, as the 2G network elements are used to the TDMA traffic and work with an assumption that traffic received would be synchronous, which is not true for the ATM (Figure 6.12).

6.6.6   Network Management Plan

The last step in the design of the transmission network is the management plan. The management plan consists of the management communication of the DCN (data communication network) which will use the next hierarchal layer over the ATM, i.e. the IP layer. The plan will also have information about management of the transmission equipment, such as the PDH radio and SDH equipment. While doing DCN planning, the designer has to take into consideration many aspects but mainly the capacity of the I_{ub} interface, which may range from 32 to 128 kbps. Another aspect is the topology. Same topologies that are used in the network design can be used for DCN planning. This may include tree or chain topology. The connections can be based on IP over ATM or purely IP. In either case, the IP addressing has to be such that there is always scope for future expansion, both in terms of network elements and technology. Of course, equipment limitations have to be taken into consideration when planning the IP addressing.

6.7   Ethernet Radio

More than 60% of the world traffic moves on microwaves. Till now we have seen back haul for 2G and 3G networks that was based on TDM. Consumer needs have changed significantly from the late 1990s to the early 2000s. This means change in technology as well. The last few years have seen technology change to Ethernet and being applied to UMTS and LTE networks. Instead of two options (PDH/SDH), radios now have three options (PDH, SDH and Ethernet connections).

Before discussing the Ethernet, let us see where we use it. It is used in our everyday office life when we connect to Local LAN cable. This is based on Ethernet standard IEEE802. The technology was developed to handle higher bandwidth and multi-media. The communication is data blocks/packets and addressing uses a 48-bit MAC address. An evolution of the Ethernet is the Gigabit Ethernet (GbE). As the name suggests, the data transmission rate is in gigabits per second.

Why should the Ethernet be used? In simple terms, it is because it has the advantages of being less costly, flexible. and easy to use. Ethernet is used in microwave technology. The challenges posed by fibre and other cable systems are easily mitigated by the microwave technology using in Ethernet. The capacity benefits that fibre brings are also captured by Ethernet, i.e. up to 1 Gbps and beyond. Challenges do remain in terms of propagation-related performance, spectrum availability, etc. Ethernet used in cellular networks is 'carrier ethernet' that is based on standardisation done by MEF22. Ethernet over WAN developed as 'carrier ethernet' and has capacity to run on any access technology.

Ethernet options available include packet microwave access, optical-technology-based PON/GPON/10GPON access (explained in Chapter 1) and Ethernet over SDH. Packet microwave is an ALL-IP-based network, using packet switching technology. A packet microwave system delivers a high quality service and supports protocols such as MPLS. In Ethernet over SDH, the Ethernet frame is transparently carried over SDH systems.

As discussed above, we are witnessing an Ethernet backhaul. These are IP-based systems. In the GSM, GPRS, and EDGE systems, TDM-based backhaul was used. Both consumer demand and bandwidth requirements were less. However, as the scenario changed in latter half of the first decade in the twenty-first century, the consumer requirements changed and so did bandwidth requirements with 3G and HSPA requiring approximately 30 and 42 Mbps, respectively. Both 4G/LTE and 5G have even higher requirements. The microwave used is IP which gives advantages including increased bandwidth and adaptive modulation leading to higher capacities.

There are three functions of microwave technology in such scenarios: aggregation, network, and transmission. Microwave will aggregate streams from various sources, forwarding services from source to destination, and work on transmission of the received information using adaptive modulation technique. Thus, there are three components of packet microwave:

- IWF (inter-working function)
- standard Ethernet interfaces
- carrier Ethernet switch

Another aspect to know of is hybrid microwave. This operates both the TDM and packets. The CSPs have existing GSM/GPRS networks that work on TDM and LTE networks that require carrier Ethernet. The resultant microwave frame combines both the TDM and packet traffic on its transmission network. For this to happen, a TDM processing unit and Ethernet switch are used that cater for both the TDM and packet traffic.

The Ethernet used for microwave networks is evolving and will go beyond 1 Gbps. Both packet and hybrid microwave technologies meet the requirements to deliver high speed, high capacity multiple access technologies while being cost efficient and innovative.

6.8 3G WCDMA Transmission Network Optimisation

6.8.1 Basics of Transmission Network Optimisation in 3G Networks

We have already seen the transmission network optimisation process for GSM transmission networks. The fundamental process remains the same except that we now have to focus on two more aspects: parameter setting and QoS. The process starts with defining the KPIs and the data collection process followed by analysis of this collected data in terms of capacity, quality and parameter settings. The final optimisation plans are then made to meet the desired QoS (Figure 6.13).

6.8.2 Process Definition

Optimisation in 3G networks is done mainly to optimise the capacity and quality against the backdrop of the QoS desired by the applications used by mobile subscribers. In GSM transmission network optimisation, there are very few KPIs, mainly related to the quality of the microwave network. However, as ATM is used in 3G transmission networks, the number of parameters increases tremendously, thus increasing the number of KPIs. The main KPIs are related to the ATM performance, apart from the microwave KPIs discussed in Chapter 3. As seen in the sections above, ATM QoS parameters will be the ones that will be mainly monitored and measured and these parameters act as KPIs for the transmission network optimisation in the 3G networks.

Data collection mainly involves the existing network data, e.g. site configuration, network topology, link budget calculations, commissioning data, monitored data (i.e. power levels monitored at the time of site commissioning), and the performance of the KPIs. Most of the data collected is similar to the types we have previously seen in Chapter 3.

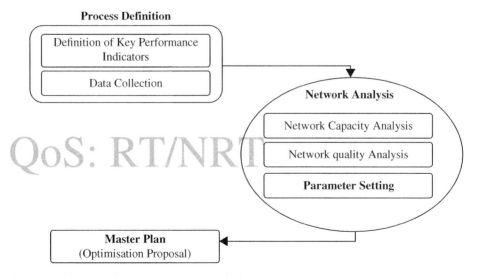

Figure 6.13 Optimisation process in 3G transmission network.

6.8.3 Network Analysis

Network analysis consists of two main steps, capacity analysis and quality analysis, with parameter settings playing a major role in maintaining QoS for any change done in the network.

6.8.3.1 Capacity Analysis

In a GSM transmission network, the capacity calculation and assessment is relatively straightforward. If the site configuration is known in terms of TRXs, then the required capacity for the link could be easily calculated. However, in capacity analysis it is not so direct. We have already had an overview of capacity/traffic calculations. Capacity calculations are dependent upon the type of traffic and cannot be calculated directly from the number of TRXs or the number of application managers used in a base station.

Apart from the regular input for this optimisation such as site configuration, link capacities, etc., the type of traffic, traffic parameters and the quality that is to be achieved would be important inputs. In 3G transmission network optimisation, capacity would be dependent upon inputs such as traffic (flowing in network) and quality (required), a feature that was generally seen only in the radio network (both in GSM and WCDMA radio networks).

An increase in traffic would be the most probable reason for analysis of capacity. Usually when the link capacity is dimensioned in the pre-planning phase, future needs are taken into account. In GSM transmission network optimisation, an increase in traffic would lead to an increase in the number of TRXs or number of sites. However, in 3G networks, capacity analysis would be required more frequently as not only traffic increase but also the type of traffic (RT, NRT, etc.) that increases, would be important factors in this process. An increase in traffic can lead to an increase in the number of SPs/AMs and/or TRXs in a base station. Any increase in these would lead to an increase in the link capacity. We have already seen that a base station site with a small number of users generates substantial traffic. However, even if the increased number of users is small, if the packet data increases, i.e. the small number of users generates PS traffic, then both the number of AMs and link capacity will need to be increased. An increase in the link capacity would mean that RNC capacity would need to be re-examined.

Thus, capacity of the transmission network can be increase by:

- Increasing the number of signal processors/application managers and/or TRXs.
- Increasing the number of base station sites.
- Increasing the number/capacity of RNCs.

6.8.3.2 Quality Optimisation

ATM was chosen for the B-ISDN services so that the QoS could be guaranteed. Thus, ATM will be a focus area in optimisation. Performance of the physical layer would be another area of scrutiny during this process. Thus, there will be two main areas, i.e. ATM and the physical layer, to be analysed in quality optimisation in 3G as compared with only one in 2G, i.e. the physical layer.

6.8.4 Analysis of the ATM Layer

The performance analysis of two ATM layers would be of immediate interest: AAL2 and AAL5 (for CS and PS data, respectively). The performance of these two layers can be monitored with the help of counters and performance indicators on management systems. These counters/measurements are mainly observed to calculate the error ratios and delays. In most cases, delay is the reason behind the degradation. The factors that may cause degradation in quality due to delay are as given in the following.

6.8.4.1 Propagation

Variation in the propagation conditions may force the physical media to cause delays, which is responsible for transporting the bits in ATM cells between ATM nodes and switches. Another probable cause may be the performance of the physical media. We have seen in Chapter 3 that propagation affects the performance of the microwave links. Any degradation of the ESR and SESR will result in degradation in performance of the ATM layer as well.

6.8.4.2 Traffic

ATM connections are designed for a given set of parameters. If the traffic load increases beyond what is expected, it may result in cell delays. Moreover, due to this traffic increase, a higher QoS cannot be guaranteed.

6.8.4.3 Architecture

Architecture of the network elements, such as switches, can have a deep impact on the performance of the network. Factors such as buffer capacity, ATM cross-connection switching capacity and speed, etc. will affect the QoS.

We have seen already above that there are three negotiable traffic (QoS) parameters in ATM: CLR, CTD, and CDV. An optimum performance of these three parameters would lead to the optimum performance of the ATM layer. The performance objectives to be met by the ATM layer are described in recommendation ITU-T I.356. It gives the provisional QoS class definitions and network performance objectives for parameters such as CDV, CLR, CMR, SECBR, and CTD.

One important parameter that would need to be understood at this stage is CDVT. CDVT defines the upper bound in cell delay. This parameter is responsible for definition of the policing function in the network, i.e. making sure that the traffic (or cells) reaches its destination within a required delay tolerance value, thus ensuring a QoS. Basically, any cell that does not respect the service contract is thrown out or discarded. For this purpose, some algorithms are used. In general, CDVT can be calculated as the inverse of PCR. Transmission planning engineers should refer to recommendation ITU-T I.371 for more information on the appropriate values of CDVT. Functions like *frame discard* and *partial packet discard* are used to avoid congestion and improve the throughput performance. Traffic shaping is one of the mechanisms that will affect the cell stream characteristics to ensure a more predictable incoming traffic with lower cell loss. Thus, both the efficiency and quality of the cell stream is maintained.

Performance of the physical layer is another key issue. There are no objectives specified for SES_{ATM} or ATM availability (except the definitions). SES_{ATM} is defined as a second when $CLR > 1/1024 = 9.8 \times 10^{-4}$ or $SECBR > 1/32 = 3.1 \times 10^{-2}$ as from ITU-T Recommendation I.357. Based on simulations, the equivalent worst-case BER for SES_{ATM} is 3.4×10^{-5} and the limiting parameter is SECBR.

Performances and unavailability of ATM connections are related to this threshold value. Some examples are shown Tables 6.3 and 6.4.

Table 6.3 Access network performance objectives.

Radio capacity	Nearest SDH container	ITU-R Rec. F.1189-1 Performance Objectives for a National Access System: from G.828							
		ESR	SESR	SES/month	BBER	BER$_{ESR}$	BER$_{SES}$	BER$_{BBER}$
PDH 2E1	VC-12	0.0034	0.00017	442	1.7E-05	1E-09	1.7E-04	2E-08
PDH 4E1	VC-2	0.0043	0.00017	442	1.7E-05	7E-10	1E-04	5E-09
PDH 8E1/16E1	VC-3	0.0064	0.00017	442	1.7E-05	8E-10	4E-05	3E-09
STM-1	VC-4	0.0136	0.00017	442	1.7E-05	1E-10	2E-05	6E-10
		$C = 8.50\%$		Fixed block allocation				
		$L = 500\,km$		Access length rounded to nearest 500 km				

Source: Reproduced with permission of ITU.

Table 6.4 Access network performance objectives.

From I.356					
Parameter	Allowance (%)	HRX 27 500 km	Access network	Scaling ref.	Equivalent BER
CLR	24	3.00E-07	7.20E-08	10.67	1.70E-11
CER	9.50	4.00E-06	3.80E-07	383.40	9.91E-10
SECBR	9.50	1.00E-04	9.50E-06		
SES$_{ATM}$	—	—	—	—	3.40E-05

Source: Reproduced with permission of ITU.

6.8.5 Parameter Setting

The network analysis of the 3G transmission network is much more tedious compared with the analysis done in the 2G transmission network. The inputs of the whole analysis process can be divided into three major parts. Dimensioning parameters form the first input. The second inputs come from the actual implemented network, such as the capacity of the links and the media used, along with the synchronisation and network management settings. The third major input is in the form of the ATM parameters and the radio network parameters (shown in Figure 6.14).

The dimensioning parameters play a very important part in the whole process. By the time the network is dimensioned based on factors such as existing knowledge and equipment limitations at the time the actual network is launched, lots of factors and parameters change. As the network is designed using the original dimensioned parameters, it becomes necessary to use the inputs, which help in describing the particular behaviour of these parameters.

The general data consists of the actual implemented and commissioned data, such as the topology, link budget calculations, interference analysis, capacities of the links, etc.,

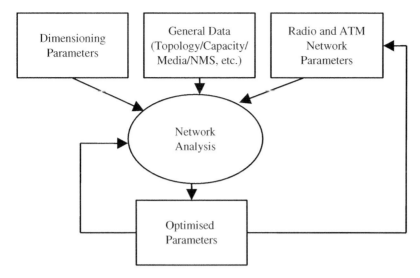

Figure 6.14 Process to study the impact of radio parameters on the transmission network.

along with the actual received power, system curves and the recoded fade margins. Also, for the long hops, the fade margins can be recorded for a certain period of time. Availability performance data can also be recorded for a longer period of time.

The radio network parameters are again subdivided into the dimensioned and the optimised radio parameters. Thus, effects of the optimised radio parameters can be seen on the transmission network parameters. In this way, later on down the process, the relationship between certain radio parameters and the transmission network performance can be found, allowing the setting of the values of these parameters such that the performance of both the radio and transmission network is optimal giving the best E2E (end-to-end) QoS for the network.

The network analysis is based on the inputs of the dimensioning parameters, general data, and the radio/ATM network parameters. The analysis and the subsequent corrections to these parameters would have to be again recycled and the radio/ATM parameters may require fine-tuning. An important aspect would be the balance needed between the radio and the ATM/transmission network parameters. Once this process is done, the optimised parameters would have to be looped back again to the network analysis so as to cross-check whether they are giving the desired network performance.

7

3G Core Network Planning and Optimisation

7.1 Basics of Core Network Planning

Core network planning in a 3G (WCDMA) system consists of both circuit core and packet core planning.

7.1.1 Scope of Core Network Planning

The scope of core network planning includes dimensioning of network elements such as MGW, MSC (and VLR) and HLR (and AC/EIR) from the CS side, while the PS core will include SGSN, GGSN along with interfaces (Figure 7.1).

7.1.2 Network Elements in the Core Network

Network elements in the 3G core network remain almost the same as seen in the previous chapters on core networks, except that a couple of new concepts are added: MSC and MGW (media gateway). These are briefly explained below.

According to the 3GPP Rel4, both the control and user plane traffic gets separated, with MSC and MGW handling the control plane and user plane traffic, respectively.

MSC, also known as MSC Server in 3GPP Rel4, handles the call and mobility control function for CS traffic generated by the registered subscribers, incoming calls from other networks, mobile-originated calls, and mobile-terminated calls. Apart from this, it is also responsible for signalling conversion (user-to-network signalling converted to network-to-network signalling). The MSC (server) is also responsible for controlling the MGW. The MSC (server), as in GSM, contains the VLR as well as other elements such as a group switch (GSW), etc.

The MGW is responsible for user plane traffic handling. The main function of this includes switching voice and data towards the required destinations, converting the ATM traffic into time division mode and vice versa. It is able to handle/switch both the switch and data traffic. $I_{u\text{-}cs}$ signalling termination also takes place on the MGW. Both the versions of IP (v4.0 and v6.0) are supported by the MGW.

Fundamentals of Network Planning and Optimisation 2G/3G/4G: Evolution to 5G,
Second Edition. Ajay R. Mishra.
© 2018 John Wiley & Sons Ltd. Published 2018 by John Wiley & Sons Ltd.

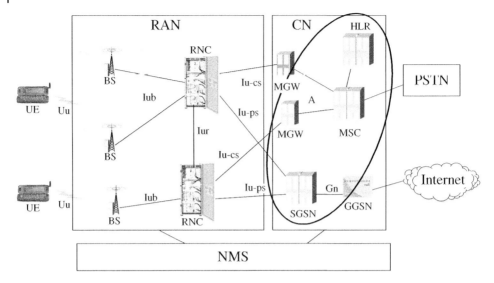

Figure 7.1 Scope of core network planning in a 3G system (WCDMA).

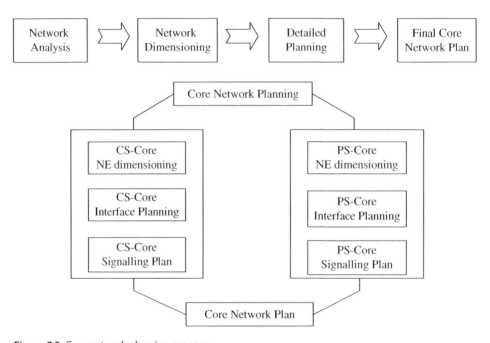

Figure 7.2 Core network planning process.

7.2 Core Network Planning Process

The process of network planning for both the CS core and PS core essentially remains the same as described in previous chapters (Figure 7.2). However, as there are changes in the radio network elements, transmission network elements and interfaces, core network planning also gets affected. Thus, the network elements, interfaces and signalling plans are the main areas of focus in the core network plans. In the following sections, we will go through the process and changes in the CS and PS cores.

7.2.1   Circuit Switch – Core Network Planning

Although CS core network planning involves both switch and signalling network planning (as in GSM), in a WCDMA 3G network it is quite different from in GSM core network planning. This difference is mainly due to the changes mentioned above, i.e. new network elements and interfaces. The process begins with network analysis and dimensioning, followed by detailed planning to produce the final CS core network plan.

7.2.1.1   Network Analysis

Information collection is an indispensable part of the process as in GSM core network planning. The information required would include the existing/designed network topology (both radio and transmission), subscriber database (present and forecasted), topography of the region, traffic measurements, etc. Information such as quality and performance targets would give direction to the analysis and design of the core network. In GSM core network planning, the MSC or switch was one important network element around which the whole design process revolved. However, in 3G, as the functionalities extend to MGW, the whole process will revolve around these two network elements, i.e. MSC and MGW, there inter-connection and the planning of the signalling network.

7.2.1.2   Network Dimensioning

Circuit core planning in 3G networks is similar to that of the 2G networks. Existing subscriber base and the growth rate would be an important factor determining the dimensioning of the network. As seen before, dimensioning should be done in such a way that when implemented, the network should be able to handle the traffic for the next few years. For dimensioning, usually network-planning tools are used (see Appendix A). However, in 3G networks, a twofold approach is taken for dimensioning the CS core network, i.e. the network is dimensioned with respect to user plane and control plane. As we have seen in the sections above, the MGW and MSC (or MSC server) are two elements that are responsible for the user plane and control plane traffic. Thus, dimensioning of the CS core would include dimensioning of the number of MGWs, number of MSCs, and signalling related to them.

Dimensioning the number of MGWs is dependent upon a few factors, and mainly includes the traffic distributions, subscriber database, location of elements like the MSC, RNC and external (other network) elements. The number of MGWs can also be calculated by the amount of area that needs to be covered by a given MGW. The location of the MGW is another focus area. The MGW can be distributed in the network (near the RNCs) or can be co-located near the core site. A distributed structure would result in transmission saving but on the other hand it would also lead to increase in the number of MGWs. Also, an increased number of MGWs would result in an increased number of inter-MGW connections, which increases the complexity of the network making it difficult to maintain or expand. A concentrated structure would decrease the equipment costs but would increase the transmission costs (readers can draw an analogy from Chapter 3 where the concept of BSC location is discussed). Also, in a

From	MGW	External Network
MGW	X	800
External Network	1600	X

Figure 7.3 Traffic generation (in Erlangs).

concentrated structure, the complexity is less. However, in network roll-out plans, both types of structures are used simultaneously depending upon the requirements.

Dimensioning of the number of MSCs (or MSC servers) is directly dependent on factors such as subscriber base, their calling behaviour and the number of RNCs. The reason is that each MSC contains VLR in a limited capacity, i.e. can handle a specified number of subscribers and the number of attempts to log on to the network. Every RNC can be connected to only one MSC.

Once the subscriber base is known for the network (based on which number of MGW and MSC is calculated), the traffic that is expected to flow in the CS-network is calculated. This is similar concept as explained in Chapter 4 before. The traffic matrix would contain the traffic flowing in own network due to the calls generated internally and (expected) calls generated in other networks. One small example is shown in Figure 7.3. Unlike a similar figure shown in Chapter 4 (Figure 4.3), the traffic matrix contains MGW as the network element instead of MSC.

7.2.2 Packet Switch – Core Network Planning

7.2.2.1 Network Analysis

The packet core network is similar to that seen in GPRS network planning (see Chapter 5). The main elements include SGSN, GGSN, border gateway (BG), domain name servers (DNS), LIG, switches and routers, charging gateways (CGs), firewalls (FWs), and the G-interfaces as shown in Figure 7.4. (Note: Figure 7.4 forms part of Figure 5.1.)

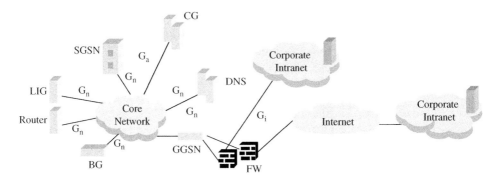

Figure 7.4 Elements and interfaces to focus on in 3G packet core network planning.

7.2.2.2 Network Dimensioning

We have already seen most of these elements in Chapter 4, however, there are some differences in 2.5G and 3G packet core networks. One is obviously the data rate, which is higher in 3G. The evolution of networks also brings new challenges with the increase in the number of PDP contexts. In early releases of GPRS, there is only one PDP context per access point for one subscriber. However, with the new releases and QoS differentiation, the network is able to handle several PDP contexts per access point for one subscriber. This has an impact on the GSN dimensioning as discussed in Chapter 5 (in transmission network dimensioning). As the data increases in 3G in terms of type, rates, etc., the SGSN and GGSN dimensioning gets more complex than in previous chapters. Predictions for number of subscribers, voice and data would be important in this SGSN (and GGSN) dimensioning, as they would give an indication on the PDP context and data traffic that will be needed in the dimensioning process.

Although overheads were taken into account in 2.5G networks, in 3G networks they play a more crucial role. In fact, the impact of overheads is quite substantial in nearly all three parts of planning, i.e. radio, transmission, and core. In the packet core network, overheads need to be calculated as they affect the capacity and subsequently the quality of the network. Overheads in a packet core network can be calculated on the basis of the protocol stack shown in Figure 7.5 (this is part of Figure 5.4). The overheads due to each layer need to added during the dimensioning phase. Layer 1 or L1 is the physical interface while layer 2 or L2 is the ATM (or Ethernet) layer. The percentages of overheads due to each of these layers depends upon the traffic.

Dimensioning of BGs and FWs is also part of core network planning. Any given 3G network will have diverse types of data traffic. Also, this network would interact with different kinds of networks. Thus, FWs become more important in 3G networks. BG dimensioning is dependent upon the number of 'own' subscribers in external networks and external subscribers in the 'own' network. Firewall dimensioning may not be a part of the initial network plan, but would be required in the later stages to protect the network from 'intruders'.

Core network planning engineers should remember that other factors, such as redundancy of equipment, are highly recommended in the core network but cost implications should be considered as well.

Figure 7.5 Protocol stack in a PS network.

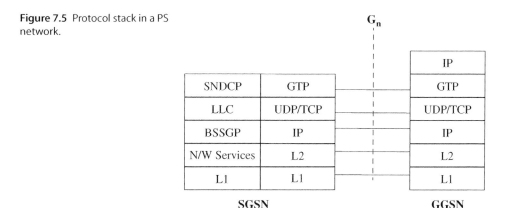

7.3   Detailed Network Planning

7.3.1   Circuit Switch Network

The detailed plans for the CS core will include plans related to routing, signalling and parameter settings. All these three plans are based on the core network plan that is an output of the network dimensioning (though it may change in the detailed planning phase).

The network plan consists of the CS core network elements such as MSC, MGW and the inter-connection of these elements amongst themselves and the other elements in the cellular network. A CS voice network plan, as shown in Figure 7.6, consists of the connection between the MGW and MSC. Also, this plan will indicate the capacity required for the links. Apart from this, the routing plan for both the voice and signalling is required. This plan contains primary routes and secondary routes for carrying the voice and signalling traffic. A load sharing factor is also decided. The location of the MGWs/MSCs, inter-connection between them, connection between the 'own' network and external network, etc., should be decided. In the core network, elements can be inter-connected physically or logically or in both ways. In the present scenario of 3G core networks, physical connections can be made using LANs, IP trunks, etc., depending upon the type of connection switches (or facility) the core network element has. Aspects such as planning of IP addressing schemes usually accompany this.

The naming and numbering conventions remain similar to those of the GSM core network planning, with destinations, sub-destinations, and circuits groups clearly defined. Also signalling points, signalling links, signalling link sets, etc., are defined in the detailed plans. Differentiation between the subscribers of different networks can be done by IMSI analysis.

Signalling plans are based on the concepts discussed in Chapter 4.

Example 1   *Naming of the circuit groups*
Assume that a network is coming up for the operator Cite-Telecom.
   Circuit groups can be named as: CTMSC2
   where CT is the name of the operator (Cite-Telecom), MSC gives the destination code, and the 2 stands for the exchange terminal.
   This is based on the scheme decided by the core network engineers and network operator taking into consideration equipment specifications.

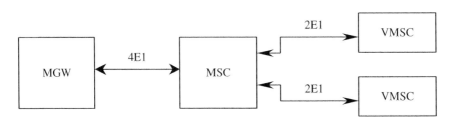

Figure 7.6 Simplification example of a CS voice network plan.

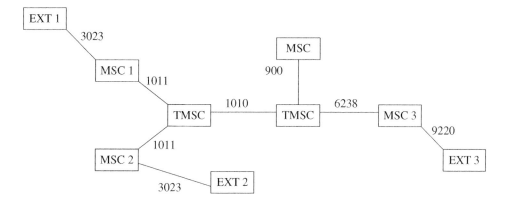

Figure 7.7 Traffic routing.

Planning of the transit layer (a concept similar to that of the transit layer planning in Chapter 4) is another concept to be taken care of during the detailed planning process. We have seen the necessity of the transit layer before, however, in the 3G CS core networks the transit layer is planned for MGWs and MSCs instead of just MSCs as in GSM core networks. Though the planning fundamentals remain the same, the need for a transit layer applies in the ATM backbone network unlike in the IP backbone network where the transit layer concept is not needed. One such example, with the traffic, is shown in Figure 7.7.

The final aspect of detailed planning in a 3G CS core network is parameter planning. Parameters are defined for configuration of the exchange terminals, traffic routing, signalling, circuit groups, destination, and sub-destinations, etc. Though the number of parameters defined is substantial (depending upon the equipment capability, etc.), some commonly used parameters are mentioned below.

7.3.1.1 Configuration of Exchange Terminals

Parameters related to the configuration of exchange terminals include the definition of ports for the termination of the PCMs and those related to defining the synchronisation for the core network element. The limitation of the equipment and naming convention should be respected when giving the numbers for it. A synchronisation hierarchy should be clearly defined, i.e. from where the MSC/MGW, etc. would receive the master clock.

7.3.1.2 Routing Parameters

Routing parameters give the direction of the traffic from the switch. Other parameters are related to outgoing calls for their signalling registration, control parameters, and the starting point for a call.

7.3.1.3 Destination and Sub-destination Parameters

Parameters related to the call destination, and the route it takes to reach the destination, are required. The call may be routed through one or more than one sub-destinations which need to be defined in the naming conventions. Also, the parameters related to load sharing have to be clearly defined. Based on these parameters, the call will take the shortest route to reach the desired destination. The parameters generally include the name destinations (sub-destinations), route identifiers, types of calls, etc.

7.3.1.4 Circuit Group Parameters

Circuit group names, types and identifiers should be defined in the parameter lists. Also, parameters associated with the functioning of the circuit groups such as the signalling registers, SPCs, control, destination, etc., are defined.

7.3.1.5 Call Parameters

Parameters related to calls need to be defined. These parameters include:

- Network-related parameters such as country code (e.g. 34 for Spain), codes required for making national and international calls.
- Definitions of IMSI (such as indicators, range) and associated PLMN.
- Call identity parameters such as calling and called party status (e.g. called subscriber is roaming or in conversation mode, etc.).
- Network- and switch-related parameters like mobile country code and the mobile national code for the operator (e.g. 12 for a particular operator), etc.
- Parameters required for charging include parameters such as the call type, e.g. if the subscriber is using the network for a roaming call or for sending a SMS, if it is a free call or chargeable call, if the incoming call is from an own-network subscriber or from an external-network subscriber, etc.

7.3.1.6 Signalling in the CS Core Network

RANAP signalling exists between the RNC and the MGW, i.e. on I_{u-cs}, and is carried to the MSC. The signalling protocol on the I_u interface is shown in Figure 7.8. RANAP messages are embedded in MTP3SL messages. SS7 signalling network services over ATM network require signalling AAL (SAAL) to deliver MTP3 signalling messages between network elements. SAAL is responsible for the correct transfer of signalling messages on an ATM-based SS7 signalling link. Each of these sublayers is considered when doing the traffic calculations and parameter settings.

The signalling load calculation forms an important part of CS core network planning. The factors affecting the RANAP signalling load are the amount and type of traffic generated by the subscribers and the number of subscribers. Based on these, the amount of bandwidth required for signalling is calculated. When calculating the signalling load

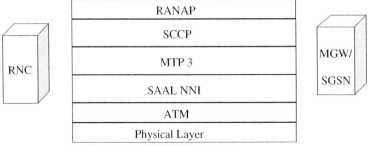

Figure 7.8 SS7 signalling on an I_u interface.

towards the MSC (from MGW, also known as 'SIGTRAN'), overheads must be calculated. The signalling parameters group consists of signalling points, signalling point codes, signalling links, signalling link sets, etc. Also included are the signalling routing definitions.

7.3.2 Packet Switch Network

As shown in Figure 7.4, the core network in 3G is quite similar to that discussed in Chapter 5. Here we will discuss two aspects of detailed PS core network planning, namely:

- Connecting the (PS) core network elements to each other, i.e. planning for the interfaces.
- IP addressing and traffic routing.

As in any of the domains for cellular network radio/transmission/core, the equipment specification would play an important role not only in dimensioning but also in detailed planning. In a packet core network, SGSN and GGSN capabilities are the most important defining factors in the detailed planning. For detailed planning, it is important to understand the functioning of these two network elements.

As shown in Figure 7.9, SGSN contains four main blocks: interface unit, router, GPLC/GTPU, and management. The interface unit is required for handling the data connections in the direction of the RNC on one side and the GSN (SGSN/GGSN) on the other side. A GPLC card is also present on either side with the interfaces and is used for receiving, sending and managing the packet data. The router in the SGSN performs the functions of both a router and a processor. In fact, this is the unit that 'controls' the SGSN. All the units in the SGSN are usually Ethernet or ATM interfaces and are connected to the router. The traffic movement is guided by the IP addresses. The GGSN is much simpler than the SGSN and it can be termed a simple router. The SGSN is also

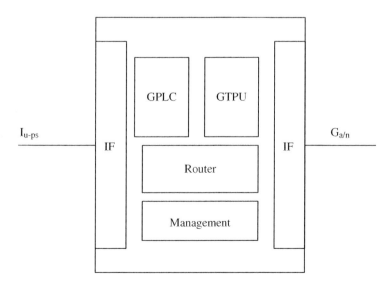

Figure 7.9 Simplified block diagram of SGSN.

responsible for the mobility management related functions. The functionality of the GGSN router is only routing without much processing to be done, unlike in SGSN where the router performs processing functions as well. The main function of the GGSN is to generate the PDP contexts and to set up the GTP tunnel. Once this is done, all the GGSN has to do is to maintain them. Again, the IP address plays an important part in the process, as it is required for the generation of the PDP contexts.

In previous chapters, we have already seen the planning aspects of the I_{u-ps} interface and the SGSN. However, there are some more interfaces, i.e. G_n and G_a, that need to be planned apart from IP addressing and routing. As other interfaces, e.g. G_i (interface between the GSN and external networks), in the core network involve external element characteristics, they are outside the scope of this book.

7.3.2.1   GSN Inter-connection

GSN inter-connection or the G_n interface planning solutions are part of the packet core network detailed plans. Traffic calculations are required so that traffic can be distributed equally between GGSNs. One of the ways to send the traffic is in a round-robin fashion wherein the traffic is sent to the next available GGSN. Another way could be a DNS deciding to which GGSN the traffic is routed through to the SGSN. Also, the amount of traffic is another factor to consider when deciding the routing for the traffic between the GGNs as it can be on an equal- or unequal-sharing basis.

G_a interface planning involves one more network element, i.e. CG, apart from SGSN and GGSN. Information exchange or messages transfer between the GSN and CGs is done using GTP-based protocols. Inter-connections between these network elements are planned based on the capabilities of the network elements and network configuration. One of the possible ways to inter-connect these elements is to use virtual LANs along with the planning of IP addressing, redundancy of paths and associated signalling, etc.

7.4   Core Network Optimisation

The 3G core network optimisation process is similar to that seen in GSM /GPRS/EDGE core network optimisation in previous chapters. As has been the case with the 3G network optimisation processes in radio/transmission, the QoS becomes an important factor against the backdrop of which the optimisation needs to be done. The KPI definition, analysis and the preparation of the final optimisation plan remain almost the same as that already discussed in Chapters 4 (for the CS core) and 5 (for the PS core). The final optimisation plan consists of suggestions for changes in the existing network elements (e.g. increased capacity of the network elements), new network elements and their locations, definition of allocation of transit elements and layers, new simplified number plans, etc. Also covered is optimising the interfaces within the network such as I_{u-cs}, I_{u-ps}, G_n, etc., and interfaces to external networks, e.g. G_i (Figure 7.10).

However, one aspect changes, i.e. timing of the optimisation. In the GSM networks, core network optimisation did not usually begin instantly after the network launch. It took some time (from a few months to few years) before it was realised that core network optimisation was needed unlike in the radio network where the optimisation cycle started almost parallel to the network planning. In some cases of 2.5G networks and in

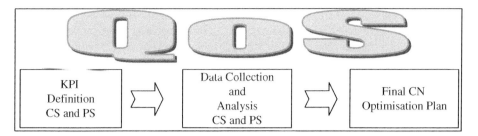

Figure 7.10 3G core network optimisation process.

almost all cases of 3G networks, core network optimisation would be an inherent part of network optimisation. The reason for this is that the QoS in 3G is considered from end-to-end rather than from an individual network subsystem (e.g. radio/transmission/core) perspective, as shown in Figure 7.11. As the quality of the packet data is so important, the following section focuses on the QoS of packet data and ways improve it.

7.5 End-to-End Quality of Service

The mobile subscribers' perspective of the quality of the network is what matters in any mobile network, especially in 3G where the QoS is the backbone of any network design and implementation. All three network subsystems contribute to enhancing the QoSs delivered. How is the required QoS achieved? For a given application, a group of parameters are identified. These are also called QoS parameters. The objectives for this particular group parameter (for this particular application) should be met in order to achieve the desired quality standards. One important aspect of optimisation is to keep this parameter group as precise and small as possible. An increased number of parameters leads to increased complexity in the network and the optimisation process. Each parameter group has control mechanisms so that the quality can be controlled as and when desired. The reason is that the quality of any given application should be negotiable. As we have seen previously, four traffic classes are defined in UMTS networks: conversational, streaming, interactive, and background. Various applications fall under each category, but technically speaking these four classes can be differentiated by two factors:

- maximum permissible delay
- maximum allowed errors (BER)

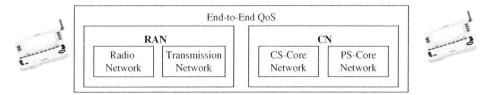

Figure 7.11 E2E QoS structure.

These traffic classes and the QoS associated with them are responsible for handling the packet flow (PDP contexts) in the network. The important aspects related to the PDP contexts include the type of PDP context, associated network elements (e.g. GSN in the case of a PS core network) and the expected QoS of a PDP context. Applications belonging to different traffic classes, such as WWW, Telnet, video conferencing, FTP, etc., would need different processes to achieve the desired QoS. Whenever the mobile subscriber activates a PDP context, the parameter group can be negotiated both in terms of quality (of service desired) and quantity (number of PDP contexts desired by the user). The network should be optimised in such a way so as to meet the most stringent demands made by the mobile subscriber in terms of quality and quantity. This means that the correct parameters have to be assigned by the network to a given application and to the availability of the resources for handling the requested application. This loops back to the initial discussion of optimising the quantity of network elements and the interfaces between them.

In the initial phase of optimisation, when the traffic may not be known, measurements can be done to observe the quality of the network. For knowing the quality of the web-related services, parameters such as the call (session) access rate and call (session) establishment time, dropped call (session) rate, etc., can be observed which in turn are dependent upon the PDP context establishment. Also, the data capacity both in the uplink and downlink directions can give an estimate of the end-to-end (E2E) QoS of the network from the WWW perspective. Similarly, the quality of all applications important to the network should be measured, two of which are FTP and WAP file transfers.

E2E QoS is an area that is still being explored in 3G networks. It will remain an area of focus for some years because each application will need a different QoS. Applications and their requirements will keep on changing with the further development of 3G networks.

Part III: 4G

LTE Network Planning and Optimisation

8

4G Radio Network Planning and Optimisation

The 4G-LTE (long term evolution) network evolved from a 3GPP perspective from Release 99 (March 2000) with UTRA in FDD and TDD (2.84 Mcps) modes. This was followed by Release 4 (March 2001) containing TD-SCDMA and Release 5 (March 2002) that produced HSDPA with IMS. In 2005, Release 6 came up with HSUPA with MGMS. DL MIMO (multiple-input multiple- output), optimised real-time services, e.g. VoIP, Push-2-talk came in Release 7. It was as early as 2008 that Release 8 introduced LTE for the first time. Some of the important aspects covered were higher data rates, low latency, introduction of MIMO, etc. Releases 9, 10, 11, etc. (further to Release 8) continued the standardisation of the radio networks.

8.1 Basics of Radio Network Planning

The process remains similar to that discussed in previous chapters. However, as the network architecture has changed substantially (more simplified), apart from the fact that it is now a data-oriented network, the radio network planning and its focus changes with respect to GSM/WCDMA networks.

8.1.1 Scope of Radio Network Planning

As shown in Figure 8.1, the scope of radio network planning will cover, from an equipment perspective, eNodeB planning and, from an interface perspective, the X2 interface, in addition to taking into consideration the U_u interface. First, let us consider an overview of LTE fundamentals.

8.1.2 LTE System Requirements

Some of the most prominent system requirements in the LTE network are as follows (based on Release 8):

- DL/UL: 300/75 Mbps
- latency: 10 ms
- air-interface (DL/UL): OFDMA/SC-FDMA
- antennas: MIMO

Fundamentals of Network Planning and Optimisation 2G/3G/4G: Evolution to 5G,
Second Edition. Ajay R. Mishra.
© 2018 John Wiley & Sons Ltd. Published 2018 by John Wiley & Sons Ltd.

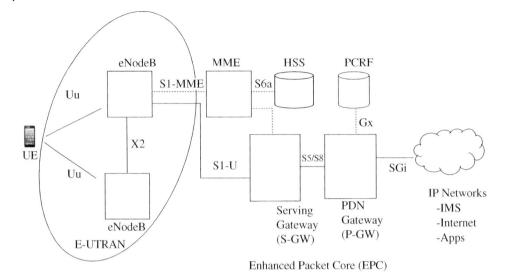

Figure 8.1 Scope of 4G/LTE network planning.

- mobility: $450 \, \mathrm{km \, h^{-1}}$
- spectrum: 1.4, 3, 5, 10, 15, and 20 MHz
- SAE (system architecture evolution): All IP network
- backward compatibility with GSM/EDGE/UMTS systems

8.1.3 LTE Radio Fundamentals

The basic propagation characteristics remain similar to those seen earlier in this book in chapters related to GSM/EGPRS and WCDMA radio networks. Some new and interesting concepts have arisen in LTE and we will consider them here.

8.1.3.1 LTE FDD and TDD

LTE is being deployed across the world on both the TDD and FDD domains. We have already discussed both modulations – time division and frequency division – in earlier chapters. Let us consider duplexing more deeply. The term fundamentally means using one channel for two-way communications. It is of two types: half duplex and full duplex. An example of the former case is the wireless system used by police, where one person speaks and another listens, and vice versa. In the latter case, both parties can talk and listen at the same time, so it is a two-way communication system. There are two types of full duplex communication systems: frequency division and time division.

In the FDD mode, there are two frequency channels for communication with a guard band to avoid interference while in the TDD mode, a single frequency is used where transmitted and received data is 'sent' using different timeslots as shown in the Figure 8.2. This means that FDD uses a paired spectrum while TDD uses an unpaired spectrum.

Let us compare the two technologies: FDD is used where UL and DL data rates are symmetrical, while TDD is used where asymmetrical data rates exist. It is possible to change the capacity dynamically in the TDD mode but this is not possible in the FDD

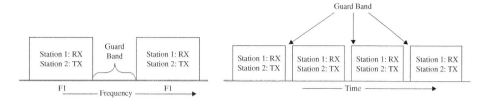

Figure 8.2 FDD versus TDD mode.

mode. In the FDD mode, a guard band is needed, while in the TDD mode a guard time/ period is needed. The guard time/period will limit the capacity of the system, unlike the guard band in FDD where the capacity is not impacted. As FDD needs a paired spectrum and TDD an unpaired spectrum, the amount of spectrum needed in FDD is more than in TDD. This directly impacts the costs greatly especially since prices of an auctioned spectrum can turn the tide in favour/against an operator. Another cost implication is the use of diplexers. In the FDD mode, diplexers are used to separate the frequencies which is not a requirement in the TDD mode. As there are separate frequencies for UL and DL, continuous transmission takes place; this is not the case in TDD where there is discontinuous transmission (due to utilisation of a single frequency for UL and DL). Discontinuous transmission can degrade the performance of the power amplifier in the transmitter.

8.1.3.2   FDD Versus TDD Mode
Both FDD and TDD networks have distinct advantages. The network deployment is not just based on technology, but also on some other important factors such as equipment availability, cost, frequency spectrum, and device availability. According to GSA reports (January 2016), there have been 480 LTE networks commercially launched in 157 countries. Of these, 409 operators deployed FDD mode, 50 operators deployed TDD mode, and 21 operators deployed both FDD and TDD modes.

8.2   LTE Air Interface

8.2.1   OFDMA

The underlying technology used by LTE is orthogonal frequency division multiplexing (OFDM). This technique has existed since 1966 but has only recently found relevance in the cellular world. This modulation system is also used in other technologies such as WLAN, DVB, etc. There are two distinct elements in OFDM: frequency division multiplexing (FDM) and orthogonality. In FDM, one communication channel will carry a combination of many signals wherein each signal (subcarrier) has a different frequency within the main channel. At the receiver end, a demultiplexer is used along with a band pass filter to receive the correct frequency. In orthogonality, signals are orthogonal to each other. For this reason, even in tight spectral conditions, subcarriers can be transmitted without any interference from each other. Due to orthogonality, there are no filter requirements at the transmitter and receiver. Hence, it not only reduces cross talk but also negates the presence of cross talk.

GSM and UMTS networks are based on TDM and CDMA technology, respectively. UMTS networks have a problem related to the impact of multipath fading and to counter that OFDM was chosen in LTE networks in the downlink. In UMTS, the signal was spread over the complete 5 MHz bandwidth (called the spreading phenomenon) but here the data is transmitted over many narrow band carriers of 180 kHz each. In LTE, data is transmitted over many narrow subcarriers. This technology which is based on FDM (frequency division multiplexing), also meets other requirements such as spectrum flexibility and cost-efficient solutions for wide carriers with high peak data rates.

In summary, the benefits of OFDM in LTE are:

- A single frequency can be used in the network leading to SFNs (single frequency networks).
- It offers robust mechanisms against interference.
- High spectral efficiency and low latency.
- Complex equalisation filters not needed to cope with severe channel conditions.

However, one major disadvantage of OFDM is the time sequences that are produced after transformation of the complex symbols into a small set of subcarriers have a high peak to average power ratio (PAPR). The effect of high PAPR is the high cost of equipment design and higher power consumption. Hence, OFDM cannot be used in UL. single carrier frequency division multiple access (SC-FDMA) is used in UL in order to reduce the PAPR.

8.2.2 SC-FDMA

This is a modified form of OFDM and is seen as discrete Fourier transform (DFT)-coded OFDM. To generate a frequency domain signal, fast Fourier transform (FFT) takes a random signal and multiplies it with complex exponentials over a range of frequencies. This is followed by summations of products and plots resulting in coefficients of frequencies. These coefficients are called the spectrum and signify the amount of frequency in that signal. The analogue version of FFT is applied in the digital domain by DFT. So, in DFT-coded OFDM, the time domain data signals are converted using DFT to the frequency domain and then processed using OFDM modulation. In simpler terms, in OFDM, each data symbol is assigned a separate subcarrier, while in SC-FDMA, each data symbol is carried by multiple subcarriers. SC-FDMA can be considered aso frequency spread (DFT spread) OFDM as well. Thus, SC-FDMA has all the strengths of OFDM along with a low PAPR making it ideal for the UL direction.

8.3 LTE Frame Structure

These are of two types based on topology, either FDD or TDD. There are two frame structures: type 1 for LTE FDD (Figure 8.3) and type 2 (Figure 8.4) for LTE TDD. In type 1, one LTE radio frame is of 10 ms, and each frame is divided into 1 ms while each subframe is 0.5 ms. In type 2, there are two half frames of 5 ms each, and each of these is subdivided into five frames each 1 ms long. These frames are used to send system information while techniques such as frequency hopping happen at the subframe level.

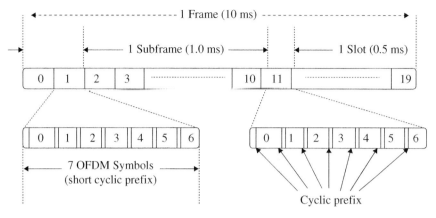

Figure 8.3 Type 1 frame structure.

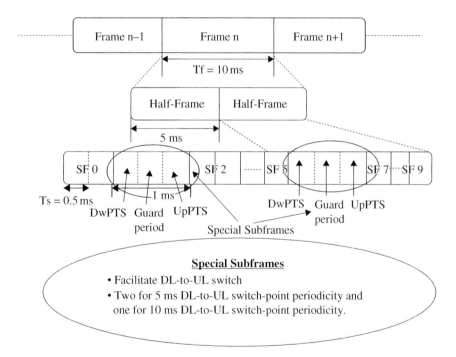

Figure 8.4 Type 2 frame structure.

Type 1 frame structure:

1 radio frame = 10 ms subframe.
1 subframe = 1 ms each.
No. of slots per subframe = 2.
1 slot = 0.5 ms.
No. of orthogonal symbols per slot = 7.
No. of samples in one frame (10 ms) = 307 200.
No. of timeslots in one radio frame = 307 200.

At the start of a symbol, a cyclic prefix is used. There are two types of cyclic prefixes – normal and extended. On using an extended cyclic prefix, the number of symbols will get reduced to six. The sampling rate for both TDD and FDD is the same.

Type 2 frame structure:

1 radio frame = 10 ms.
Two half frames = 5 ms (each).
No. of time slots in half frame (5 ms) = 8.
No. of fields in half frame = 3.
Length of each field (DwTP, GP, and UpPTS) = 1 ms.

Here GP is the guard period, the DwPTS field carries synchronisation, user data and the downlink control channel for transmitting scheduling and controlling information, and the UpPTS field transmits the PRACH and SRS (sounding reference signal).

8.4 LTE Protocol Stack

The OSI model is a conceptual model that has seven layers. Of these seven layers, three layers (1–3) are media layers while the top four layers (4–7) are host layers (as explained earlier in Chapter 1).

8.4.1 Bearers in LTE

There are three kinds of bearers in LTE: radio bearer, S1 bearer, and S5/S8 bearer. The radio bearer transports packets between UE and eNodeB. The S1 bearer transports packets between eNodeB and S-GW and the S5/S8 bearer transports packets between eNodeB and P-GW.

The air-interface protocol stack in LTE is based on the first three layers: physical, data link, and network. There are three kind of protocols (signalling, user plane, and transport plane) and two planes (data and control). The fundamental function remains the same as we have seen in previous chapters. The air-interface is divided into two levels: access stratum (AS) and non-access stratum (NAS) (Figure 8.5).

8.4.1.1 Layer 1
8.4.1.1.1 Physical Layer
Layer 1 or the physical layer carries information over the air-interface. The information is received from the MAC transport channels, i.e. offering data transport services to higher layers. The major functions carried out by the physical layer include modulation, synchronisation (frequency and time), HARQ, rate matching, power control, MIMO processing, error detection, and RRC measurements.

8.4.1.2 Layer 2
8.4.1.2.1 MAC
MAC lies in layer 2 and exists both in UE and eNodeB. One of the main functions of the MAC layer is mapping between logical and transport channels. Other functions include error corrections (HARQ), transport format selection, priority handling between the UE

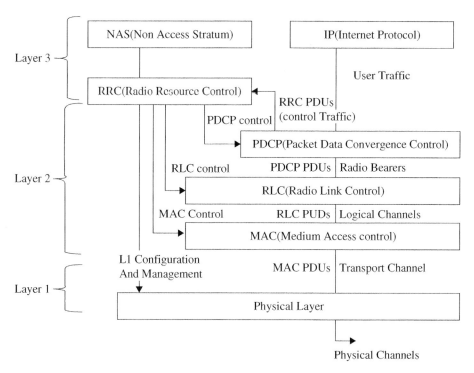

Figure 8.5 LTE protocol stack.

(dynamic scheduling), scheduling reporting and multiplexing/demultiplexing of MAC SDUs of one of the different logical channels into/from transport blocks delivered to/from the physical layer on transport channels.

8.4.1.2.2   *Radio Link Control (RLC)*
A layer 2 protocol, RLC is located above the MAC and below PDCP layer. It transfers the upper layer PDUs. RLC operates in three modes: transparent mode (TM), unacknowledged mode (UM), and acknowledged mode (AM). Its functions include error detection, error correction (ARQ), concatenation, segmentation/resegmentation, and reassembly.

8.4.1.2.3   *Packet Data Convergence Protocol (PDCP)*
This is a layer 2 protocol, lying between RRC and RLC. It is used for transfer of data. Other functions include header compression/decompression and ciphering/deciphering.

8.4.1.3   **Layer 3**
8.4.1.3.1   *Radio Resource Control (RRC)*
RRC is a layer 3 protocol that exists in UE and eNodeB lying between NAS and lower level layers. Some of the functions of RRC include paging, key management, mobility functions, establishment management, maintenance and release of RRC connection between UE and E-UTRAN, and broadcasting system information to the NAS and AS.

8.4.1.3.2   NAS

This is the highest stratum of the control plane between the MME and UE. Apart from supporting the mobility of UE and supporting session management procedures, some of the other key functions of this layer include authentication, security control, mobility handling, and evolved packet system (EPS; radio bearer, S1 bearer and S5/S8 bearer combined are known as EPS bearer) bearer management.

8.4.1.4   Signalling Protocols

The control functions are carried by the means of signalling. MME controls the eNodeB using the S1 application protocol while the base stations interact using the X2 application protocol.

8.5   LTE Channel Structure

As discussed in previous chapters, there are three types of channels (physical, transport and logical) apart from the traffic channels. We have seen generic characteristics of the channels before but let us look at some of the important LTE channels (Figure 8.6) and their functions.

8.5.1   Physical Channels

The physical channels for both the uplink and downlink have different requirements and have different operations.

8.5.1.1   Downlink

PDCCH (Physical Downlink Control Channel): The main purpose of this physical channel is to carry mainly scheduling information.

PBCH (Physical Broadcast Channel): This carries information known as MIB (master information block) that helps the UE to access the networks.

PHICH (Physical Hybrid ARQ Indicator Channel): This gives information related to HARQ, thus indicating if the information received is correct or not. HARQ has been discussed in earlier chapters.

PCFICH (Physical Control Format Indicator Channel): This indicates to the UE the format in which information is received.

8.5.1.2   Uplink

PRACH (Physical Random Access Channel): This is the only non-synchronised transmission made by UE in LTE and is used for the random access procedure known as RACH.

PRACH includes RA (random access) preambles generated from Zadoff–Chu sequences. Of the five formats defined, four are for FDD modes. The preamble is selected based on the maximum estimated call range. PRACH carries RA preamble that is used to initiate the RA. The RA procedure is used for achieving UL synchronisation between UE and eNode B. It is also used to obtain resources, e.g. for RRC connection requests.

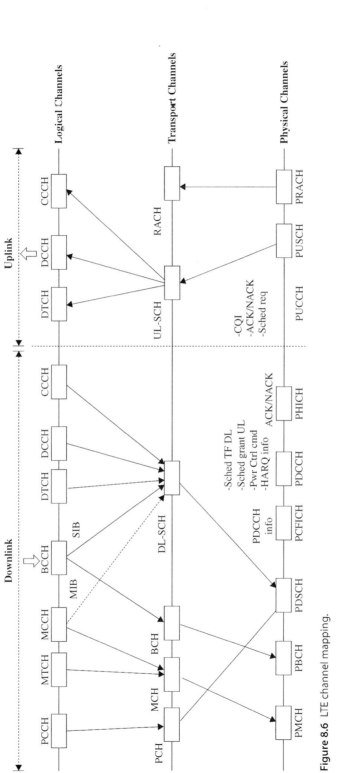

Figure 8.6 LTE channel mapping.

PUSCH (Physical Uplink Shared Channel): As for the PDSCH channel in uplink, this is used for uplink data transmission by the UE.

PUCCH (Physical Uplink Control Channel): This provides various control signalling requirements including scheduling requests, CQI information, and ACK/NACK.

8.5.2   Transport Channels

There are four downlink transport channels and two uplink transport channels.

8.5.2.1   Downlink

BCH (Broadcast Channel): This maps to BCCH and, as the name suggests, has requirements such as MIB to be broadcast to the entire coverage areas.

PCH (Paging Channel): This carries paging information. This is mapped to PDSCH, and can be dynamically used for traffic/control channels.

MCH (Multicast Channel): This is used for MBMS services, transmits MCCH information and is mapped to PMCH.

DL-SCH (Downlink Shared Channel): This is used by many logical channels, and is the main channel for downlink data transfer.

8.5.2.2   Uplink

RACH (Random Access Channel): This is used for the RACH procedure and requirements.

UL-SCH (Uplink Shared Channel): This is similar to DL-SCH but is used for uplink data transfer.

8.5.3   Logical Channels

There are seven logical channels in downlink and three logical channels in uplink.

8.5.3.1   Downlink
8.5.3.1.1   *Control Channels*
BCCH (Broadcast Control Channel): This provides broadcasting system control information, e.g. MIBs/SIBs to all mobiles connected to the eNodeB.

PCCH (Paging Control Channel): This is used for paging information for the location of the UE in the network.

CCCH (Common Control Channel): As the name suggests, this is common to multiple UE and is used to transfer control information between base stations and UE.

DCCH (Dedicated Control Channel): This is used for transmitting control information and is dedicated to a UE.

MCCH (Multicast Control Channel): This is used for transmitting information for multicast reception.

8.5.3.1.2   *Traffic Channels*
DTCH (Dedicated Traffic channel): This channel is dedicated to the UE.

MTCH (Multicast Traffic Channel): This is used for transmitting information for multicast data.

8.5.3.2 Uplink

8.5.3.2.1 Control Channels

DCCH (Dedicated Control Channel): This is used for transmitted control information and is dedicated to a UE.

CCCH (Common Control Channel): This is common to multiple UE and is used to transfer control information between base stations and the UE.

8.5.3.2.2 Traffic Channels

DTCH (Dedicated Traffic Channel): This channel is dedicated to the UE.

It will be seen that many of the LTE channels bear similarities to those used in previous generations of mobile telecommunications.

8.5.4 Channel Estimation

Channel estimation is done by UE to decide the original signal. Reference signals are the ones which are used by UE to perform a downlink channel estimation. Based on that, only UE does the channel measurements and takes the decision on whether to stay on the carrier or do some mobility.

8.5.4.1 Reference Signal Sequence

LTE has both DL and UL reference signal sequences. In the DL, the signals are two types: reference and synchronisation. Based on the quality of the signal, prediction of certain aspects can be done; the signals are known as reference signals. In the DL, there are a few types of reference signals: UE specific, cell specific, CSI, position, etc. In the UL, two types of reference signals are supported: demodulation and sounding. The former is associated with transmission of PUSCH/PUCCH and the latter is not.

8.6 Multiple Input Multiple Output Antenna Technique

We have seen earlier how more than one antenna is used for reception. In this technique, two antennas are used at the receiver end and signals received by both are combined using a combiner leading to higher signal strength. This is called diversity processing and is used to counter the fading phenomenon. If more than one antenna is used for transmission and more than one antenna is used for reception, it is known as the MIMO antenna technique. Here the multiple input is used at the transmitter and multiple output for the receiver. MIMO is used to increase the data rates. This is done by increasing the number of antennas for which the multiple data streams can be set up. So, if there is 2×2 system, i.e. two transmitting and two receiving systems, then the total number of signals received at the receiver is four (Figure 8.7). This scheme was added in 3GPP Release 10. Within MIMO, there is an open and closed loop spatial multiplexing technique. In the open loop spatial multiplexing, there was no mechanism of feedback from UE while in the closed loop method, the pre-coding matrix indicator is sent from the IE to the eNodeB in order to optimise transmission. Another important technique is beam forming where the linear array allows antennas to transmit in a focused direction. The main advantage of this is to reduce interference and increase the capacity for a particular UE in whose direction the antennas are focused. This technique has become important in 5G technology.

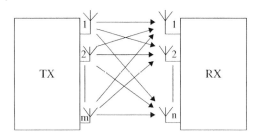

Figure 8.7 MIMO.

8.6.1 Beam Forming

This is specific to the LTE TDD mode. However, beam forming is a technique associated with antenna arrays – phased array antennas and adaptive array antennas. This fundamentally the ability of an antenna array to place a radiation pattern in a particular way/ direction. This is achieved by each antenna of the array being applied relative to the amplitude and phase shift, and resultant signals are then added and transmitted in a particular direction to achieve beamforming.

8.7 Network Elements in a LTE Radio Network

8.7.1 User Equipment (UE)

The UE in LTE is similar to the UE/MS in GSM or 3G networks. Of course, smart phones are now dominating the market. The fundamental functions of the UE include mobility management, call control, session management and identity management. One has to understand a fundamental shift: the term mobile station is also known as UE, however, the term UE in LTE is used not only for a mobile station but also for other user devices communicating with the base stations.

There are a few main components of the UE: USIM, mobile termination (MT), terminal adapter (TA), and terminal equipment (TE). We have discussed USIM in previous chapters. All the communication-related functions are handled by MT. MT is responsible for sending/receiving/terminating the RRC. Application-specific service protocols are terminated at the TA. All services terminate at the TE.

3GPP 36.306 has defined 14 categories of UE. These categories are needed so that the base stations can communicate correctly with the UE after determining its performance.

The top eight categories of UE are given in Table 8.1. (Note: Category 8 exceeds the requirements for IMT-Advanced.) The various categories have varied characteristics, e.g. MIMO 4×4 is supported by category five.

8.7.2 Base Station (eNodeB)

Let us briefly consider the evolution of the base station architecture so that we can understand the LTE base station better. We have seen in previous chapters the GSM and UMTS base stations. However, let us understand the evolution of the base station according to 3GPP releases from the perspective of some of the major changes.

Table 8.1 Data rates (Mbps) for LTE UE categories.

	LTE UE category							
	1	**2**	**3**	**4**	**5**	**6**	**7**	**8**
Downlink	10	50	100	150	300	300	300	1200
Uplink	5	25	50	50	75	50	150	600

GSM base stations were connected to the BSC through E1s (A_{bis} interface) which is channelised through time division multiplexing. In Release 99, the bearer is IP/ATM rather than TDM while new air compression techniques are used which provide greater spectral efficiency. Both these technologies are compatible. Release 5 incorporated changes on the transport layer, i.e. IP became the prime bearer. Release 5 was compatible with Release 99 and GSM as well. Further releases (Releases 6 and 7) delivered increased user connection rates, supporting services related to increased data rates. More details on 3GPP releases can be found in Appendix D.

The main aim of LTE technology is to be 'everywhere'. To decrease the cost of controlling radio networks centrally through BSC or RNC in GSM and UMTS technologies, 3GPP introduced SAE (system level architecture) which removes the need for BSC or RNC. Thus, the radio network in LTE consists only of UE and base stations.

Some of the main units of eNodeB include the base band (BB) processing unit, the multiplexing unit (MUX), and the control unit (CU). There are two other important units: the radio processing unit and the demultiplexing unit. The signal from the UE goes to the radio processing unit, which then interacts with the multiplexing unit. The signal then goes through the BB unit towards the core network through the CU. The radio processing unit performs functions such as amplification and analogue to digital conversion. This signal is then further processed by the BB unit through the physical layer, the MAC layer, the RLC layer and DSPs (digital signal processors). The CU performs functions that involve the transmission interface, call connection, and control/measurements. Thus, eNodeB performs RRM functions towards the UE. In LTE, as there is no RNC, the radio processing and control functions are terminated at the eNode B; the handovers are hard handovers.

8.8   Key Phenomena in LTE

8.8.1   Interference in LTE

Interference impacts almost all elements related to coverage, capacity and network quality. Interference impacts TDD networks the most. As seen in previous chapters, interference can take place between the same and different carriers, and also between the same and different cells.

8.8.1.1   Inter Cell Interference
While the RNCs and BSCs control interference in the UMTS and GSM networks, respectively, in the LTE network this is done by the eNodeBs themselves. By using the X2 layer, eNode Bs can 'talk' to each other and schedule the resources in order to control

interference. This technique is known as ICIC (inter-cell interference coordination) and is defined in Release 8. By using the ICIC technique, the base stations are able 'talk' to each other using the X2 interface about the interference status of neighbours. Based on this information, radio resource are allocated to the respective UE. Thus, the inter-cell interference (ICI) is reduced/finished. The X2 message includes information such as HII (high interference indicator), RNTP (relative narrowband transmit power), and OI (overload indicator). Based on this, there are four frequency planning schemes: conventional frequency planning, fractional frequency reuse, soft frequency reuse, and soft fractional frequency reuse. These are explained elsewhere in the chapter.

When filters are not used, the interference between two eNodeBs is the most serious as it can cause the most impact on the network performance. This is caused when one of the base stations is in transmission mode while another is in reception mode. This can be countered by providing a guard band of 5 MHz. In FDD as there is sufficient frequency separation, there is no need for a guard band, while for collocated TDD-FDD systems, a 5 MHz guard band and separate antenna systems should be used.

8.8.2 Scheduling

We have discussed the scheduling mechanism in earlier chapters. However, let us consider it from an LTE perspective. Here we have the scheduling request (SR) mechanism. The trigger for the SR is the buffer status report (BSR). This report contains more information than the SR. The SR sends single bit information on the data that needs to be transmitted through the UE while the BSR contains details. The UL and DL scheduling are separated and independent and able to take advantage of the different channel conditions using techniques such as round-robin, C/I (max) and proportional fairness. The factors that affect scheduling include the radio conditions, traffic volume, and QoS requirements of the UE. Some of the key parameters include resource allocation (bearer quality, interference, channel occupancy), QoS parameters (e.g. SPI), and channel quality (HRAQ CQI reports). From a protocol perspective, the scheduler works in the RLC and MAC layers.

8.8.3 Quality of Service

QoS is the basis of almost all network designs in the telecom industry – both wired and wireless (3GPP, IEEE, IETF, etc.). QoS in a LTE network is discussed in terms of EPS bearer – the virtual connection between the UE and P-GW. One UE can have multiple EPS bearers, e.g. VoIP, internet. When the UE latches to the network, it creates what is known as a default EPS bearer, and remains active until the UE is detached from the network. When the UE uses the VoIP service, it is known as a dedicated EPS bearer. The QoS of the default EPS bearer is non-GBR (non-guaranteed bit rate) while for the dedicated EPS bearer it is non-GBR and GBR (Figure 8.8).

The key parameters associated with QoS are: QCI (QoS class identifier), GBR (guaranteed bit rate), ARP (allocation and retention priority), and MBR (maximum *guaranteed* bit rate). While QCI, GBR and ARP are per EPS bearer, MBR is per group of EPS bearers. QCI is an integer from 1 to 9 where the QoS performance of each IP packet is indicated through each of the integers, i.e. there are nine different QoS performance characteristics. The QCI contains information such as resource type (GBR/non-GBR), and packet information such as packet delay/error loss.

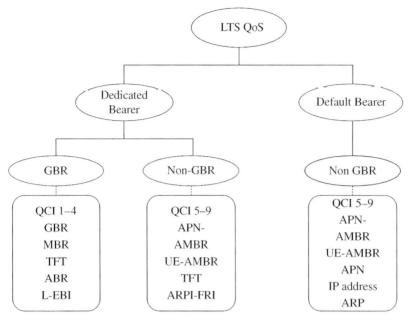

Figure 8.8 QoS in LTE.

8.9   Radio Network Planning Process

The fundamental process of radio network planning remains similar to that seen in GSM or UMTS radio network planning. The network planning starts with a pre-planning phase.

8.9.1   Pre-Planning Phase

The most important element of the pre-planning phase is dimensioning. The most important inputs come from the customer, frequency spectrum availability, existing network inputs (if it is not a greenfield operator), geography information, etc. Based on this, link budget calculations are done resulting in primary site count and locations. The basic elements include:

- Link budget calculations and coverage analysis.
- Traffic and capacity analysis.

8.9.1.1   LTE Link Budget

As seen in previous chapters, link budget calculation is done to calculate the maximum path loss. Based on this path loss calculation, the number of cell sites can be calculated. Once the site locations are primarily finalised (on the tool as site searches is a parallel on-going process especially in a greenfield operator). Path loss calculations are then used along with propagation models (e.g. Hata, COST-231 Hata or Erceg-Greenstein) to find the actual areas that would be covered after the LTE network launch. Some of the

Uplink link budget	800 MHz_20 MHz_20W - FDD	
UE output power	23,0	dBm
Resource blocks (RBs)	5	
Power per RB	15,9	dBm
Thermal noise	−174,0	dBm
RBS noise figure	1,8	dB
User bitrate	0,27	Mbps
SNR	−3,0	dB
RBS sensitivity	−122,6	dBm
Antenna gain (RBS + UE)	15,0	dBi
Installation loss (RBS + UE)	0,2	dB
Penetration loss	16	dB
Fading margin	6,7	dB
Max. pathloss unloaded	130,6	dB
Utilization	30%	
Interference margin	0,2	dB
Max. pathloss	130,5	dB
Range	1,31	km
Site-to-site	2,0	km

Downlink link budget		
RBS output power	20,0	W
Resource blocks (RBs)	100	
Power per RB	23,0	dBm
Thermal noise	−174,0	dBm
UE noise figure	7,0	dB
User bitrate	5,4	Mbps
SNR	−4,1	dB
UE sensitivity	−118,5	dBm
Antenna gain (RBS + UE)	15,0	dBi
Installation loss (RBS + UE)	2,2	dB
Penetration loss	16	dB
Fading margin	6,7	dB
Max. pathloss unloaded	131,6	dB
Utilization	30%	
Interference margin	1,1	dB
Max. pathloss	130,5	dB
Range	1,31	km
Site-to-site	2,0	km

Figure 8.9 LTE link budget example: 800 MHz.

key inputs (as seen previously) for the link budget include transmitter and receiver power, antenna gains (both transmitter and receiver), cable and connector losses, impact of cell loading, interference margins, and propagation models. An example of a LTE link budget is shown in Figure 8.9.

We have discussed certain link budget parameters that are similar to LTE link budget calculations, e.g. propagation or equipment parameters. Let us now look at certain LTE specific parameters such as beam forming, cell edge rate, and MIMO antenna gain.

Beamforming is applicable for the TDD mode. Two modes of beam forming have been defined by 3GPP: single data mode and dual (mux) data mode. The former can improve only coverage while the latter has an impact on coverage and throughput.

The cell edge rate is the data speed at the edge of the cell. The higher the data speed, the lower is the coverage and vice versa. Some of the aspects that impact cell edge rate include the MIMO schemes chosen, the modulation mode, the eNodeB power amplifier, and the number of eNodeBs at the cell edge.

MIMO antenna gain is the gain brought in by using MIMO antenna technology. This is composed of various gains such as the array gain, the spatial mux gain, and the power combining gain. This helps to improve system capacity and coverage.

Next, we consider traffic analysis. In GSM it was quite simple with voice as the key factor. In EGPRS, there was a certain amount of data and with UMTS, the data quantity increased substantially. However, LTE is completely data only where voice also travels as data. Also, data utilises the maximum capacity of the network. The ability of the network to carry the traffic load is also calculated. After the traffic calculations are done, dimensioning of interfaces is done.

Some of the key inputs of dimensioning include the number of subscribers, subscriber behaviour of service utilisation, throughput of the cell site, spectrum and channel bandwidth, data rates (utilisation behaviour of subscribers), and cell radius.

Outputs of dimensioning include the cell radius, coverage plans, capacity planning, and number of cell sites. This will help in determining the deployment cost of the network.

There are some differences between the UMTS and LTE link budget calculations. Usually LTE is deployed in networks where there is an existing UMTS network (although this is not mandatory) and hence LTE itself does not provide any positive change in coverage area. In the UL, the LTE link budget will see smaller interference margin and no soft handover gains.

From the LTE link budget, we obtain the cell size. Dividing the total coverage area with the area covered by one cell site, we get the number of sites needed. However, in LTE dimensioning, the number of cell sites needed is also calculated from the capacity. It is based on the number of subscribers, traffic modelling, and site capacity. The types of traffic data are: voice, real time data (e.g. video call), and non-real time data (NRT). Cell throughput depends on factors such as the cell range (from the path loss), channel bandwidth (1.4–20 MHz), cell load, MIMO (open/closed loop spatial multiplexing and diversity), and scheduling. Unlike in earlier networks where capacity was limited by the base station capacity and the number of base stations in a geographical area, the capacity in in LTE is impacted by factors such as interference, modulation, coding, and scheduling.

During the dimensioning phase, the best possible method to calculate the cell throughput is by mapping the SINR (resulting from simulation) into bit rates (MCS). Other factors to consider include traffic analysis – number of subscribers in the area (subscriber density), traffic mix and profile, peak data rates, and average data rates during the day and during busy hours. Just like in GSM and UMTS, capacity planning is done for urban, suburban and rural areas.

Another important term used in capacity dimensioning is the overbooking factor or OBF. Complete dimensioning cannot be done at peak rates as the cost of the network will increase and resources will be wasted. OBF can be defined as the average number of subscribers that can share a given channel unit. Mathematically, OBF is the product of the ratio between the peak and average data rates and the utilisation factor. The utilisation factor is usually less than 85% to guarantee QoS.

Thus, the overall data rate is the product of the total number of subscribers, peak data rate, and OBF. The total number of cell sites needed to fulfil the capacity requirements is the ratio between the overall data rate and site capacity.

Thus, the total number of sites required will be the higher of two numbers – one from the output of coverage and the other from capacity calculations.

8.9.2 Detailed Network Planning

The main elements of detailed planning are: configuration planning, coverage and capacity planning, parameter planning, and frequency planning.

8.9.2.1 Configuration Planning

To achieve the desired coverage and capacity (inputs from dimensioning), the base station configuration needs to be planned. Other inputs include site surveys results for geographical inputs for propagation modelling. Based on the configurations that are different for coverage-oriented planning and capacity-oriented planning, the coverage and capacity calculations are done again.

8.9.2.2 Coverage and Capacity Planning

We have already seen the process for coverage and capacity planning in previous chapters. The important aspects of the detailed planning are the 'practical inputs' such as exact site locations, specifications of the equipment (radio base stations, UE, antennas, etc.), parameter planning, and simulations. One of the most utilised simulations is the Monte Carlo simulation. Network technology and mobile phone technology have both changed. By the advent of LTE networks, the markets were full of smart phones and users had access to all kinds of applications such as Twitter and Facebook. This meant more data, and continuous connectivity requirements. The signalling traffic substantially increased. Thus, the QoS became more tightly linked to capacity and coverage than ever before. Some of the aspects that need to be considered in capacity planning are:

- capability of the UE
- user behaviour
- spectrum and bandwidth available
- RSRP (coverage)
- interference-related parameters
- scheduling
- latency
- timeslot management
- equipment capability
- handover

While many of these areas are quite familiar from previous chapters, there are few that play a significant role in LTE; one of them is interference. Overall throughput decreases in the cell with an increase in the number of users. Thus, interference mitigation needs to be done to have good cell throughput. Another is the scheduling algorithm which is explained elsewhere in the chapter. Timeslot management, including guard band requirements between uplink and downlink, needs to be managed. Nine configurations defined by 3GPP are used.

How is capacity dimensioned? The average throughput is calculated based on bandwidth and spectrum efficiency. This is done through simulations. Based on the total number of subscribers, calculation of the total throughput requirements that depend on average throughput (BH) is done. Now the total number of sites can be calculated. Based on this, the S1 and X2 interface capacity is determined. Factors such as handover and frequency of handovers also play an important role in X2 interface capacity planning.

For coverage, some key aspects include interference, required SINR, and spectral efficiency.

We have discussed interference in detail in this chapter. Another important aspect is the required SINR. This is one of the main performance indicators for LTE, fundamentally for the cell edge. It is dependent on two factors: propagation channel models and coding schemes. Another element that has an impact on cell edge is spectral efficiency. When deriving spectral efficiency, link level simulations are not taken into account and both MAC and RLC protocol overheads are taken to be negligible.

We have seen the process of site surveys, identification, etc. in earlier chapters in this book. Hence, we focus on two aspects: the fundamentals of parameter planning and simulation.

8.9.2.3 Parameter Planning

8.9.2.3.1 SINR (Signal to Interference and Noise Ratio)

SINR is probably one of the most important parameters. It is defined as the ratio of average signal power received to the summation of the average interference power and noise power. The average interference power is the power of measured signals (channel interference) from other cells in the same system.

8.9.2.3.2 PRB (Physical Resource Block)

The unit of allocation of subcarriers to a user for an air-interface (OFDM) for a set amount of time in LTE is known as the PRB. One PRB contains 12 subcarriers. Each of these subcarriers has 7 OFDM symbols, thereby making it 84 modulation symbols.

The UE is connected to the most suitable cell. This suitability is determined after UE monitors the system information and paging of the cell it is connected to. The selection/reselection process is based on RS (reference signal) measurements: RSRP, RSRQ, RSSI.

8.9.2.3.3 RSRP (Reference Signal Received Power)

According to 3GPP, RSRP is defined as the linear average over the power contributions (in watts) of the resource elements that carry cell-specific reference signals, across a specific bandwidth. Calculated against the useful part of the OFDMA symbol, UE has to measure this for aspects such as cells selection/reselection and handover functions. As LTE UEs have a minimum of two receive antennas, the UE can measure RSRP and choose the stronger signal of the two receive antennas.

8.9.2.3.4 RSRQ (Reference Signal Received Quality)

This is the ratio of RSRP and total received signal and noise power (normalised to 1PRB bandwidth). In the UMTS system, it is equivalent to UMTS CPICH E_c/I. it is calculated as a product of the number of PRBs and the ratio of RSRP and RSSI.

8.9.2.3.5 RSSI (Received Signal Strength Indicator)

As the name suggests, RSSI is the measurement of the received radio signal power. Basically, it tells if there is sufficient signal strength to have a good connection. So, what is the difference between RSSI and dBm? While dBm is an absolute value (mW), RSSI is a relative value. According to 802.11, RSSI can be scaled between 0 and 255 and each vendor can define its own maximum value of RSSI. RSSI depends on both signal strength and bandwidth.

8.9.2.3.6 PCI (Physical Cell Identity)

PCI is used for two main purposes: neighbour cell handover measurement and as a resource allocation parameter. Neighbour cell handover measurement reports are done by the UE. For this purpose, PCI is used. PCI is encoded in the physical layer signal transformation and it should uniquely identify the neighbour cell within the network. Care should be taken to make sure that the UE is not able to measure and report two cells with the same PCI. This is possible by making sure that the PCI reuse distance is long enough. There are two ways to assign PCI: one is based on resource element to resource element interference; and other is resource element to PDSCH/PDCCH interference. The PCI is similar to the scrambling code in UMTS. However, scambling codes range from 0 to 511, and PCI ranges from 0 to 503. 3GPP required the value of PCI/3 in each cell to be 0, 1, and 2 in each base station.

8.9.2.3.7 TA (Tracking Area)

Tracking area parameter planning is similar to location area and routing area planning in GSM/UMTS networks. When the UE is in the active state, its location is known by the network at cell level. However, when in the idle state, its location is known at TA level. A group of cells or base station can be defined under a TA. The update of the UE in the idle state is done at the MME by sending a TAU (tracking area update) message.

8.9.2.4 Simulations

One of the most used simulations is Monte Carlo simulation. Monte Carlo simulation was developed during the Second World War, and is used in various fields including finance and accounting. This helps the user to make decisions in areas that have extreme variations in results. In LTE, these simulations are used to identify the aggregate interference by randomly placing the user in a predefined deployment scenario. We have already studied the concepts of scheduling, admission control, and power control in previous chapters. Using the algorithms related to these features, the UE transmit power and eNodeB is simulated to calculate the interference. This is then used to predict the capacity of the network. However, this does not use small scale fading and mobility.

8.9.2.5 LTE Frequency Planning

8.9.2.5.1 Spectrum Management

We have already discussed TDD and FDD modes earlier in this chapter. ITU has decided in consultation with regulatory bodies the frequency bands for LTE technology (3GPP Release 8/9) shown in Tables 8.2 and 8.3. In FDD mode, the frequency bands are paired and have separation between the frequencies. In TDD mode, frequencies allocated are unpaired as both the uplink and downlink share same the frequency band.

Bands 33–43 are used for TDD mode while other frequency bands, especially 1–21, are used for FDD mode. The channel bandwidth assigned to the network is operator-dependent. The lower the frequency, the higher the coverage.

8.9.2.5.2 Spectrum Refarming

This will be required for most operators globally as there are many that have evolved from GSM and UMTS to LTE. Thus, the spectrum that has been used for GSM/UMTS may be refarmed/reused for LTE. The LTE spectrum has been sold at staggering prices across the world. Frequency refarming is a great way to enhance the spectrum bands for the operator to roll out LTE. However, deploying GSM frequencies for LTE coverage may lead to interference. Here, frequency planning plays an important role.

8.9.2.5.3 Channel Spacing

The channel spacing between two adjacent carriers is defined as:

$$\text{Channel spacing} = \left(BW_{ch1} + BW_{ch2}\right)/2$$

where BW is bandwidth. It is dependent on factors such as frequency blocks and channel bandwidth. This can be played on to improve system performance. Carrier centre frequency is always an integer multiple of 100 kHz.

Table 8.2 LTE frequency bands.

LTE band number	Uplink (MHz)	Downlink (MHz)	Width of band (MHz)	Duplex spacing (MHz)	Band gap (MHz)
1	1920–1980	2110–2170	60	190	130
2	1850–1910	1930–1990	60	80	20
3	1710–1785	1805–1880	75	95	20
4	1710–1755	2110–2155	45	400	355
5	824–849	869–894	25	45	20
6	830–840	875–885	10	35	25
7	2500–2570	2620–2690	70	120	50
8	880–915	925–960	35	45	10
9	1749.9–1784.9	1844.9–1879.9	35	95	60
10	1710–1770	2110–2170	60	400	340
11	1427.9–1452.9	1475.9–1500.9	20	48	28
12	698–716	728–746	18	30	12
13	777–787	746–756	10	−31	41
14	788–798	758–768	10	−30	40
15	1900–1920	2600–2620	20	700	680
16	2010–2025	2585–2600	15	575	560
17	704–716	734–746	12	30	18
18	815–830	860–875	15	45	30
19	830–845	875–890	15	45	30
20	832–862	791–821	30	−41	71
21	1447.9–1462.9	1495.5–1510.9	15	48	33
22	3410–3500	3510–3600	90	100	10
23	2000–2020	2180–2200	20	180	160
24	1625.5–1660.5	1525–1559	34	−101.5	135.5
25	1850–1915	1930–1995	65	80	15
26	814–849	859–894	30 / 40		10
27	807–824	852–869	17	45	28
28	703–748	758–803	45	55	10
29	n/a	717–728	11		
30	2305–2315	2350–2360	10	45	35
31	452.5–457.5	462.5–467.5	5	10	5

n/a, Not applicable.

8.9.2.5.4   Conventional Frequency Planning

LTE employs OFDMA systems, wherein the spectrum is split into orthogonal frequency bands (subchannels). Each of the sets of subchannels is allocated to different UE. The allocation is based on bandwidth requirements. This has low symbol rate and hence low ISI (inter-symbol interference).

Table 8.3 LTE TDD frequency bands.

LTE band number	Allocation (MHz)	Width of band (MHz)
33	1900–1920	20
34	2010–2025	15
35	1850–1910	60
36	1930–1990	60
37	1910–1930	20
38	2570–2620	50
39	1880–1920	40
40	2300–2400	100
41	2496–2690	194
42	3400–3600	200
43	3600–3800	200
44	703–803	100

Earlier in the chapter, we have seen interference impacting the performance of the network. Owing to this, frequency planning assumes importance in the network. As we have seen before, frequency planning can be: conventional, fractional frequency reuse, partial frequency reuse and software frequency reuse.

In conventional frequency planning, the frequency reuse factor is 1. This offers the worst interference especially at cell edges as the frequency spectrum is used in all cells without any restriction. A way to reduce this interference is to reuse the frequency three times. Here, the frequency band is divided into three parts, all orthogonal to each other. Thus, the frequency is adjacent sector and is different thereby reducing the interference. However, this comes at the cost of capacity loss. In GSM systems, adjacent cells had different frequencies reducing the interference and decreasing the spectrum efficiency. UMTS networks are single frequency networks, where the reuse factor is 100%. The spectral efficiency is high and so is interference.

8.9.2.5.5 Fractional Frequency Reuse (FFR)
In LTE, the FFR technique is used. Here the same set of frequency is used but all base station allocation is done using FFR. The idea is to partition the cell into two parts: centre area and edge area. Then use the hybrid of reuse 1 and 3, respectively. Another way to do this is let the centre area use all frequencies and the whole of the edge area uses only one-third of the available frequency subchannels. This scheme improves the interference levels at the cell edge area. An example of FFR is shown in Figure 8.10.

FFR can be of two types: partial frequency reuse (P-FFR) and soft frequency reuse (S-FFR). There are two differences in the two types of FFR. In P-FFR all sectors (frequency reuse factor = 1) use a common frequency band at equal power while the power in sub-bands is coordinated among neighbouring cells. In S-FFR, though some frequency sub-bands transmit full and some transmit reduced power, each sector transmits the whole frequency band. SFR is the recommended in frequency planning and can be

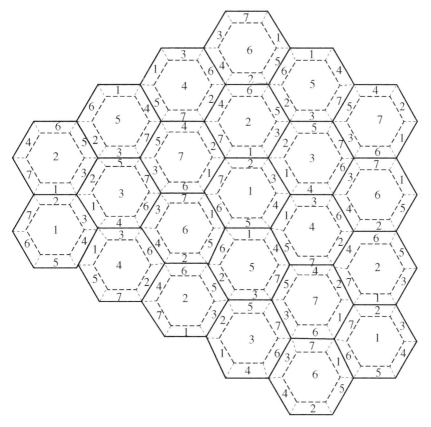

Figure 8.10 LTE frequency planning.

used in both FDD and TDD methods. The whole carrier bandwidth can be divided into three sections – one for each sector. This would reduce interference as each of the sectors uses one-third of the carrier bandwidth. This also results in higher throughput for the users.

When working on TDD, there are a few things to be kept in mind during frequency planning. These include considering the carrier bandwidth, co-frequency, the guard band, DwPTS, GP and UpUTS. In TDD mode, synchronisation also plays an important role for interference mitigation purposes.

8.10   LTE Radio Network Optimisation

LTE pre-launch optimisation can be divided into three different tuning stages.

8.10.1   Initial Tuning

The main activities in this stage are the single cell functionality test (SCFT) based on a static test as well as drive tests to check LTE parameters for every single cell of a LTE site, i.e. RSRP, RSRQ, SINR, PCI, maximum throughput or data rate, intra eNodeB handovers (clockwise and anticlockwise), handover interruption time, ping latency, etc. Besides recording these measurements through drive test software, SCFT tests also focus on finding any eNodeB hardware and configuration-related issues. So, all the issues related to parameters or KPI thresholds need to be resolved in the initial tuning-based optimisation to achieve an optimum level of operation of a single eNodeB before proceeding to the next level of cluster optimisation.

8.10.2   Cluster Tuning

Once the SCFT for each and every cell in the network has been performed up to a desired level of QoS, sites/eNodeBs are grouped in clusters and cluster-based tuning is initiated on planned drive routes among all sites in a particular cluster and among clusters to check the status of the LTE parameters, e.g. average throughput, inter-site handovers, and to resolve any boundary issues. The aim of the cluster-level optimisation is to check all the important roads – arterial, VVIP, throughways and highways – in order to achieve optimal level of LTE performance within these clusters.

8.10.3   Market Level/Network Tuning

When the optimisation of all the clusters within a network is completed, network or market area level tuning is undertaken through agreed drive routes to collect network level parameters and check all the different types of LTE handovers, namely inter-cell or intra-eNodeB, inter- eNodeB (via X2 or S1), and intra or inter MME (while the drive routes will intersect the MME boundary). The border issues are also analysed and resolved in the network level tuning (e.g. to mitigate any sort of interference from external networks being observed by the border sites). As the level of KPI thresholds (e.g. accessibility, retainability, integrity and handover) are somewhat relaxed on moving from SCFT to cluster and then on to network level tuning, the same needs to be checked and calculated based on the network level LTE parameters collected during such network drive tests to achieve the required QoS of the LTE network.

   A couple of examples from network optimisation are shown in Figure 8.11 and Figure 8.12.

8.10.4   Self-organising Networks

The concept of self-organising networks was developed by NGNM and standardized by 3GPP. The key element behind the development of self-organising networks was to impact (lower) CAPEX and OPEX of the network operations while at the same time improve user experience and network performance. As the name suggests, the parameters in LTE networks are configured in real time based on the performance measurements information/reports received. This automation helps to use the resource optimally and forms the basis of optimisation in LTE. There are three key functions of self-organising networks: self-configuration, self-optimisation, and self-healing.

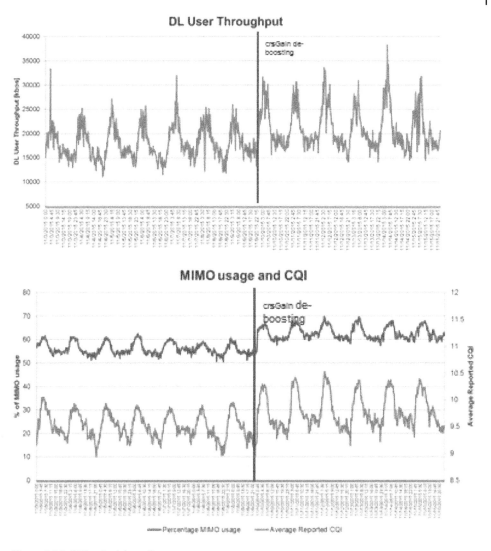

Figure 8.11 CRS gain deboosting.

8.10.4.1 Self-configuration

This is fundamentally plug-and-play and comes into action after installation of the base station taking care of all software-related deployment functions. The main purpose behind self-configuration is to minimise human intervention resulting in faster and error-free deployment.

The eNodeB establishes connection with DHPC/DNS servers after self-test, followed by establishment of secure tunnels. After this, eNodeB will interact with the confirguration server. Mainly based on Release 8, self-configuration is all about dynamic plug-and-play of eNBs. The key elements include: PCI (physical cell ID assignment), cell download, inventory, and neighbour relation.

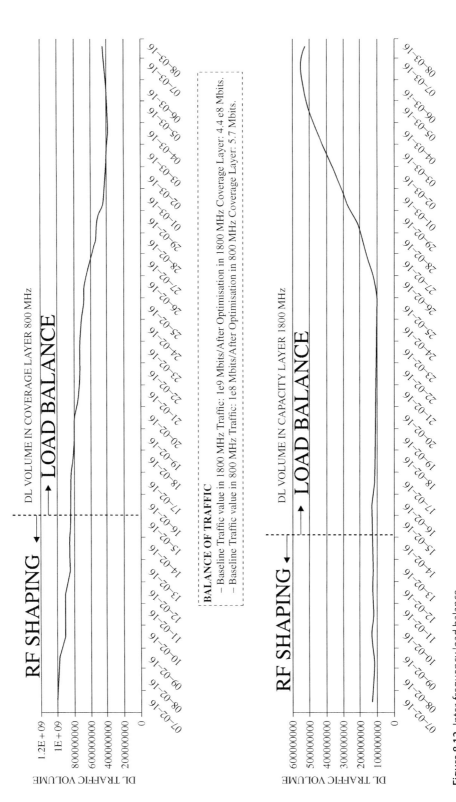

Figure 8.12 Inter-frequency load balance.

8.10.4.2 Self-optimisation

This is all about performance and quality. Based on the performance data, self-organising network function enables optimisation on the nodes. The features described in self-optimisation come from Release 9. Some of the key elements include: PCI, handover, RACH, ICI, load balancing, and ANR (automatic neighbour relation).

8.10.4.3 Self-healing

As the name suggests, this is about the network's own procedure correctness. A generic example could be a software failure and the system's ability to reinstall the software. On account of this set of procedures, many problems could be rectified in the network. This self-healing action takes place after alarms are generated. Both self-diagnosis and self-healing are part of these procedures. This is available in Release 10 and includes: ICI coordination, optimisation related to coverage and capacity, outages, self-recovery, test drive minimisation, and energy savings.

8.10.4.4 Self-organising Network Architecture

There are three types of self-organising network architecture: distributed, centralised, and hybrid.

8.10.4.4.1 Distributed

This architecture is dynamic due to its ability to scale and adapt. This is possible because algorithms run at network nodes and communication between nodes is direct. However, as the optimisation is also at the node level, the total network optimisation might not be too stable.

8.10.4.4.2 Centralised

All the algorithms run at the network management level so there is more control and coordination over the nodes. Network optimisation takes place centrally and for the whole network. The multi-vendor elements can be added in the centralised domain; this is not possible in distributed self-organising networks. Owing to the centralised architecture, the response times and power requirements are also higher.

8.10.4.4.3 Hybrid

This has the advantages of both the distributed and centralised self-organising networks. Features such as quick response time and centralised coordination from distributed and centralised self-organising networks, respectively, are there; so are the drawbacks such as multi-vendor integration and lower scalability/adaptability.

8.10.5 Key Performance Indicators

LTE performance parameters are divided into five classes:

- Accessibility – user access to the requested services.
- Retainability – capability of network to retain user for providing desired services for time requested.
- Mobility – related to handover, network ability to provide services to moving user.
- Availability – percentage of time cell is available to give services to user.
- Integrity – service quality for the user.

The first three are network-related KPI classes while the last two are service-related KPI classes.

Whenever a new site is on air, single site verification is performed to identify the challenges related to coverage, throughput, signal quality, and handover success rate.

The entire network is divided into many clusters to get a bigger picture of the network performance. As soon as the sites within the clusters are meeting requirements that the clusters are ready, then a cluster drive is performed.

The drive test will include assessment of accessibility (call set-up success rate), mobility (intra frequency/inter frequency/inter radio access technology handovers), retainability (dropped calls), integrity (throughput, latency, block error rate, jitter, etc.).

As drive test activity on a daily basis is very time consuming and costly, performance indicators are used to optimise the network which plays a vital role. With the help of performance indicators, the optimisation engineer can have a clear idea of how the network is performing with respect to design.

Although practical implementation and design may not match 100%, based on performance indicators, the network can be optimised to get the best possible outcome with existing resources.

As an air-interface is unpredictable and not constant, radio network optimisation should be a continuous activity throughout the life cycle of any network to meet the thresholds of KPIs. Introducing new user equipment with new software can also create compatibility issues in the network which can create interference to existing subscribers and lead to poor user experience. Capacity crunch can lead to blocking of new subscribers to services offered by the network operator.

Based on design strategy, new features are implemented in a radio network to improve performance. Capacity crunch, which increases capacity expense and operational expense of the service provider, can be reduced to an extent by activating the best suitable features available with that equipment vendor.

8.11 LTE Advanced

LTE Advanced or LTE-A is an evolved version of LTE. Speeds that were expected to be delivered in LTE will be delivered in LTE-A. It is also known as 4G+ in some countries. LTE-A or 4G+ is an amalgamation of various technologies including carrier aggregation (CA), support for relay nodes (RNs) and multi-antenna techniques (MIMO). So, what does LTE-A or 4G+ offer? 3GPP Release 8 allows a data rate of 300 Mbps in UL and 75 Mbps in DL in a bandwidth of 20 MHz, while Release 10 or LTE-A allows a peak data rate of 1 Gbps at a bandwidth of 100 MHz. Releases 11 and 12 provide technology improvements in areas of CA, MIMO and spectrum efficiency.

Some key elements of LTE-A or 4G+ are discussed in the following.

8.11.1 Carrier Aggregation

The term 'carrier aggregation' is nothing to do with carriers or telecom operators but, simplistically speaking, it is all about the ability of a mobile device to transmit and receive data in/from two different frequency bands at same time. This result is a peak data rate of 1 Gbps and 500 Mbps in UL and DL, respectively. While it supports both the TDD and FDD version, it is also backward compatible to LTE devices (Releases 8 and 9).

So, how is this data rates possible? The bandwidth can be achieved by aggregating five carriers of 20 MHz each. We have seen before that the carrier bandwidth is 1.4, 3, 5, 10, 15 and 20 MHz in LTE. A maximum of five carriers can be aggregated. Carrier aggregation can be of two types: contiguous inter-band, and contiguous and non-contiguous intra-band. In former, the frequency arrangement is such that communication between carriers is done by a contiguous band greater than 20 MHz; the communication is achieved using two different carrier frequency bands. This results in higher throughput. In non-contiguous intra-band, communication is done using multiple carriers in the same frequency band (used for multiple operator scenario). A point to remember here is that in Release 10 there are two component carriers in the DL and only one in the UL (hence no carrier aggregation in the UL); in Release 11 there are two component carriers in the DL and one or two component carriers in the UL when carrier aggregation is used. So, what happens at the physical layer in carrier aggregation? LTE and LTE-A have OFDMA-based radio networks. After extension of the bandwidth, the synchronisation signal carries out a cell search which is transmitted on the centre frequency of each cell where each cell has an arranged centre frequency of 100 kHz raster. PBCH is also multiplexed in the same way. Both HARQ and AMC are performed separately one each carrier. They are then combined using PDSCH to form a single carrier.

8.11.2 MIMO

We have covered MIMO elsewhere, but from a LTE-A perspective, eight antennas in DL and four antennas in UL are supported.

8.11.3 Coordinated Multi-point Transmission and Reception (CoMP)

The CoPM feature is from Release 11 and is used to improve the network performance at the cell edge. This enhancement is done by avoiding base station interference and call disconnections at the cell edges. In CoMP, as the name suggests, a number of TX (transmit) points provide coordinated transmission in the DL, and a number of RX (receive) points provide coordinated reception in the UL. A set of TX/RX points used in CoPM are co-located TX/RX antennas that provide coverage in the same sector. Two or more TX points under joint transmission can transmit on the same frequency and in the same subframe. The same can happen for reception. The received data is combined to improve the received power and service quality.

8.11.4 Relay Nodes

HetNet is a key technology in LTE-A and 5G networks. Network planning is efficiently done using RNs. The RNs are low power base stations that establishes connection with the RAN using a donor cell. This helps in enhanced cell edge capacity and coverage. This is possible using demodulating and decoding to signal, and recoding and remodulating before transmitting.

9

4G Core Network Planning and Optimisation

9.1   Introduction

We have seen in Chapter 8 the basics of 4G radio network planning and optimisation. Unlike in Universal Mobile Telecommunications System or UMTS networks, base stations are directly connected to the core network. The core network in LTE is not only quite enhanced (enhanced packet core) but also sets the foundation for technologies of the future, including that of 5G. Here in LTE, we have only the packet core that is known as the Evolved Packet Core (EPC).

9.2   Basics of EPC Network Planning

Core network planning has always been an important part of network planning. In LTE, this has become even more important. The scope of LTE EPC network planning is shown in Figure 9.1.

9.2.1   Scope of EPC Network Planning

Beyond the eNodeBs, everything falls under core network planning. The important elements that are involved include MME, S-GW, HSS, and P-GW. Before going into detail, let us consider some of the key concepts that define the LTE core network.

9.2.2   EPC Network Elements

As we have seen in a previous chapter, 3GPP has evolved to a three-layered architecture from a four-layered architecture. 3GPP focused on three main aspects for EPC development: the core network should support high data rates and low latency; the core network should have the ability to support 3GPP, 3GPP2, etc. systems; and, finally, the core network should have the ability to handle IP traffic.

As discussed previously, EPC consists of packet and user domains. The packet domain consists of MME, SGW, PDN-GW while the user domain consists of HSS.

We have been given an overview of the functions of network elements of EPC in Chapter 1. Let us now consider some of these elements in more detail.

Fundamentals of Network Planning and Optimisation 2G/3G/4G: Evolution to 5G, Second Edition. Ajay R. Mishra.
© 2018 John Wiley & Sons Ltd. Published 2018 by John Wiley & Sons Ltd.

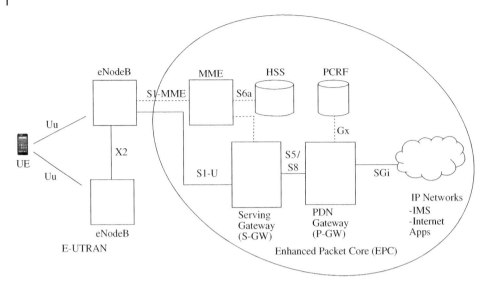

Figure 9.1 Scope of LTE EPC network planning.

9.2.2.1 Mobility Management Entity (MME)

Some of the key functions of MME involve authentication, bearer and traffic management and interworking support. Non-Access-Stratum (NAS) Signalling is a pure signalling entity inside the EPC. SAE uses tracking areas to track the position of idle UEs. The basic principle is identical to location or routing areas from 2G/3G. MME handles attaches and detaches to the SAE system, as well as tracking area updates. It generates and allocates temporary IDs for UEs. MME is also responsible for the signalling coordination to set up transport bearers (SAE bearers) through the EPC for a UE.

9.2.2.2 Serving SAE Gateway

This connects the eNodeB and PDN-GW. The serving gateway is a network element that manages the user data path (SAE bearers) within EP. It therefore connects via the S1-U interface towards eNB and receives uplink packet data from here and transmits downlink packet data on it. The serving gateway is like a distribution and packet data anchoring function within the EPC. It relays the packet data within the EPC via the S5/S8 interface to or from the PDN gateway. A serving gateway is controlled by one or more MMEs via the S11 interface. At a given time, the UE is connected to the EPC via a single Serving-GW.

9.2.2.3 Packet Data Network (PDN) SAE Gateway

The PDN gateway provides the connection between the EPC and a number of external data networks, It is comparable with GGSN in 2G/3G networks. A major functionality provided by a PDN gateway is the QoS coordination between the external PDN and EPC. The PDN gateway can be connected via S7 to a PCRF (Policy and Charging Rule Function). If a UE is connected simultaneously to several PDNs this may involve connections to more than one PDN-GW.

9.2.2.4 Home Subscriber Server (HSS)

The HSS evolution has been from HLR (see Chapter 2). However, the concept of HSS was introduced in the UMTS Release 5. There is always the possibility of one or more than one HSS in a home network. Some of the functions of HSS include mobility management (CS, PS, core network), subscriber authentication, security, etc. Other functions include communication with SIP servers, etc. With LTE/SAE the HSS will get additional data per subscriber for SAE mobility and service handling. Some changes in the database as well as in the HSS protocol (DIAMETER) will be necessary to enable HSS for LTE/SAE. The HSS can be accessed by the MME via S6a interface.

9.2.2.5 Policy and Charging Rules Function (PCRF)

This is not considered as an EPC element but as a control plane element. This gives a dynamic control over charging, bandwidth and network usage. In simpler terms, PCRF is used to apply the policy and charging control rules of the communication services provider. Thus, it is the element which decides whether or not a new connection is to be allowed and if a new bearer is to be created. This is connected to the PDN-GW through a S7/Gx interface and to IMS through a Rx interface.

9.2.3 EPC Network Interface

9.2.3.1 S1 MME

This is the interface between E-UTRAN and MME. It is used for control plane protocol. S1-Flex allows several MMEs to be connected to a single eNodeB. Owing to S1 flex, pooling of MMEs is also possible.

9.2.3.2 S1-U

This is the interface between the E-UTRAN and the serving gateway and is responsible for inter-eNodeB handovers.

9.2.3.3 S3

This is the interface between SGSN and MME and helps the mobility between inter-3GPP networks.

9.2.3.4 S4

This is the interface between SGSN and the serving gateway. This gives user plane support between GPRS and S-GW.

9.2.3.5 S5

This is the interface between S-GW and P-GW. The user plane tunnelling and tunnel management function between the two is given by S5. Also, it provides multiple PDN gateways and is used for S-GW relocation (related to UE mobility).

9.2.3.6 S6a

This is the interface between MME and HSS and is used for the transfer of subscription, authentication and authorisation.

9.2.3.7 Gx

This is the interface between PCRF and P-GW. It provides the transfer of QoS policy and charging rules from PCRF to P-GW.

9.2.3.8 S11

This is the control plane interface between MME and the serving gateway, used for EPS management.

9.2.3.9 SGi

This interface is between P-GW and he Internet and is equivalent to the G_i interface in GPRS explained in Chapter 2.

9.3 EPC Network Planning: Key Concepts

The inputs for a network plan that is given to the customer come from various sources. Radio planners give inputs such as base station IDs, radio network parameters and performance counters; IP and transmission engineers give inputs such as IP and transport access plans, parameters, and counters; and core planners give MME IDs, MME-SCTP plans, etc.

The fundamental network planning process remains like the ones we have seen in earlier chapters. However, for efficient designing of EPC, a few concepts need to be understood.

9.3.1 Latency and Delay

9.3.1.1 Latency

Latency can be defined as the delay between the cause and the effect of some physical change in the system being observed. In a technical way we can say that latency is the time taken by a bit to travel from the transmitter to the receiver or the destination. Throughput is inversely proportional to delay.

9.3.1.2 Processing Delay

This is the time taken by the router to process the packet header. While processing the packet, the router needs to check the bit level errors which can come in a packet during the transmission.

9.3.1.3 Serialisation Delay

This is the fixed delay required to clock a voice or data frame onto the network interface. It is directly related to the clock rate on the trunk. At low clock speeds and small frame sizes, the extra flag needed to separate frames is significant.

9.3.1.4 Queuing Delay

Processing delays in high-speed routers are typically on the order of microseconds or less. After nodal processing, the router directs the packet to the queue where further delay can happen (queuing delay). The queuing delay is a variable delay and is dependent on the trunk speed and the state of the queue.

9.3.1.5 Propagation Delay

This can be defined as the time taken by the signal to travel from sender to receiver. Also, it can be defined using the formula: time = distance/speed, i.e. the ratio between the link length and the propagation speed over the medium.

9.3.2 EPC Deployment Model

In this section we will discuss the design considerations/deployment model with respect to the EPC. CSP (Communication Service Provider) can use various combinations of centralisation and decentralisation of nodes based on the network configuration. The majority of 3G core deployments use a centralised architecture in which a centralised GGSN serves multiple SGSNs at distributed locations but for LTE it is a bit flexible. There are four types of deployment models.

9.3.2.1 Completely Centralized

This mode is fully centralised as every node, i.e. MME, SGSN, GGSN, SGW and PGW, is kept in the same place. This is mainly an architecture in which all of the EPC functions are centralised.

9.3.2.2 Completely Distributed

This is an architecture in which all of the EPC functions are distributed and generally deployed together.

9.3.2.3 Centralised Bearer/Distributed Control

In this case, the SGW, PGW and GGSN is centralised whereas the MME and SGSN are distributed purely as signalling entities.

9.3.2.4 Centralised Control/Distributed Bearer

In this case, the signalling or the control blocks are centralised whereas the other blocks such as SGW, PGW and GGSN are kept and deployed in a distributed manner. This deployment model is considered for better user experience.

9.3.3 Transport Network

As shown in Figure 9.2, there are three different layers plus the site in the transport network.

9.3.3.1 Core/Super Backbone

This is the main tunnel which comes out from the core (without any partitions). The function of the super backbone is to establish proper connectivity with all the other regions/entities. The routing mechanics being used by the super backbone is MPLS VPN which is a family of methods for using multiprotocol label switching (MPLS) to create virtual private networks (VPNs). MPLS VPN is a flexible method to transport and route several types of network traffic using an MPLS backbone.

9.3.3.2 Aggregation Layer

The main function of the aggregation layer is for agglomeration of the access and pre-aggregation layer, i.e. super backbone.

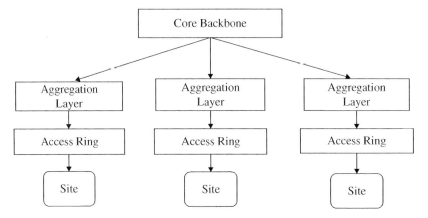

Figure 9.2 Transport network.

9.3.3.3   Access Ring
This is the branched part of the aggregation layer which eventually connects to the site.

9.3.4   Security Framework

The security framework is based on 3GPP TS33.401. It consists of five function domains:

- network access security
- network domain security
- user domain security
- application domain security
- visibility and configurability of security

There are three areas of security: LTE authentication, non-access security, and access security. LTE authentication is mutual authentication between the UE and LTE network. This is done through a procedure known as EPS AKA (Authentication and Key Agreement). Non-access security is responsible for the security of messages between UE and MME, and uses integrity key, ciphering key, etc. Access security is responsible for the security of messages between UE and eNB.

9.4   EPC Network Dimensioning

The number of mobile devices, connections and data consumption are expected to continue to increase. Operators are expected to ensure that their networks are able to meet this increasing demand. This, in turn, will result in more revenue and higher market share.

Keeping this in mind, core network dimensioning is necessary to handle the huge amount of data traffic and high number of subscribers. Also, a proper set of inputs is vital for dimensioning to yield nearly accurate results.

Core network dimensioning is an integral part of the general wireless planning process. Dimensioning is necessary to identify and calculate the necessary capacity, user data and signalling traffic, and to modify strategy for coverage, capacity and quality.

Dimensioning of an LTE EPC network involves the following elements:

- traffic (bh) and signalling dimensioning
- equipment dimensioning

9.4.1   Traffic (BH) and Signalling Dimensioning

A few parameters are needed to dimension the expected traffic that will flow in the EPC networks. They include the expected smart phone users that will give inputs of the expected data traffic that will be generated. The analysis should include the services supported based on the QCI. It will also be wise to look into other sources of data generators, e.g. data cards. The data generated can be streaming, interactive, internet or simple download. Based on the classification of session size and number of sessions in the busy hour, traffic at busy hour can be calculated. For dimensioning signalling procedures, the number of signalling messages between network elements and throughput per interface in the control plane are calculated. These are based on signalling size, transmission time and throughput, and throughput and capacity per interface.

9.4.2   Equipment Dimensioning

This involves dimensioning five key elements of the LTE EPC network: MME, S-GW, P-GW, HSS, and PCRF. The number of MME can be calculated based on the number of transitions – attached users, idle/active users, number of procedures, and capacity of MME. For S-GW and P-GW, a similar concept applies, expect that bearers are used instead of users. For HSS, the inputs are the number of active users and the HSS capacity. For PCRF, the inputs are the number of procedures and the PCRF capacity.

9.5   Detailed EPC Network Planning: Key Concepts

Detailed network planning for LTE EPC will involve the traditional processes discussed earlier for core networks. However, for LTE, 5G and beyond, some elements and concepts such as IMS, VoLTE, virtualisation, etc., will play an important role to make networks 'behave' qualitatively or rather as 3GPP expected.

9.5.1   Dynamic Tracking Area List

With the functionality of the dynamic TA list implemented, the TA list is dynamically produced for the UE. The list selection is based on the TAU history, the UE mobility pattern within thee TAU signalling and, of course, less manual configurations.

9.5.2   MME-HSS Selection

The attach procedure is performed by MME by doing IMSI analysis. This is based on subscriber IMSI. This results in PLMS of IMSI and the destination which further results in the S6a interface to subscriber HSS. Furthermore, the destination realm is added to AIR (Authentication Information Request)/ULR (Update Location Request). Then the message is sent to HSS using the S6a interface. The same interface is used to enable signalling between elements. For selecting the MME for a UE, eNB uses the GUTI (Global Unique Temporary Identity) protocol. This contains GUMMEI (Global MME Identity) and M-TMSI (MME-Temporary Mobile Subscriber Identity).

9.5.3 DNS (Domain Name System)

9.5.3.1 DNS Records

DNS is a system that is used for correspondence between the host name and an IP address on an IP network. There are two kinds of records: 'A' record and 'AAAA' record. Both of them map host names to the IPs. However, the former is for looking up IPv4 records while the latter is for looking up IPv6 records. NAPTR (Name Authority Printer) is a powerful tool that allows DNS to be used to look up services. The SRV resource record allows a single domain with static load balancing to be used by DNS administrators. This helps in moving services from one host to another while allowing some hosts to be designated as primary servers.

9.5.3.2 DNS in LTE

According to ETSI TS129303 V14.3.0, once a UE is registered in a LTE network, it is assigned a serving and PDN gateway with the help of the DNS system. S-NAPTR methodology is used for GW network address resolution. This procedure is used when any core network element has the Fully Qualified Domain Name (FQDN) of an entity and needs to find one or more services of that entity. The S-NAPTR procedure returns either one SRV record set or one A/AAAA record set. UE can roam in and out of a network while using the same PDN connection. The UE may or may not have 3GPP capability. MME uses a DNS query to find the correct S-GW and P-GW. The S-GW that is close to eND is found using TAI as the input parameter, while APN is used as the parameter to find P-GW. In general, DNS is used for dynamic selection of MME, PDN, SGW, HSS and SGSN in the network.

9.5.4 Bearer and QoS (Quality of Service)

There are two types of bearer traffic in a LTE network: SDF and EPS. While the former refers to a group of IP flows associated with a service of the user, the latter points to IP flows of aggregated SDFs that have the same QoS class. The EPS bearer is of two types: default and dedicated.

The bearer in the EPC can be classified as:

- SAE/access bearer: between the eNB and SAE GW.
- S5/S8 bearer: between P-GW and S-GW.
- External bearer: between P-GW and the application layer.

Within the same bearer, all data should have the same QoS. As seen in previous chapters, for default bearers N-GBR is allocated while for dedicated bearers GBR is allocated.

9.5.5 Bearer Parameters

Some of key SDF and EPS bearer parameters are: ARP, QCI, GBR, and MBR.

9.5.5.1 ARP (Allocation/Retention Protocol)

This is the resource allocation priority parameter assigned during the SAE bearer set-up. There are three components: priority level (1–15), pre-emption capability, and pre-emption vulnerability. This is used during call admission. After establishment, it has no impact on packet movement.

9.5.5.2 QCI (Quality Class Identifier)

As the name suggests, this indicates different QoS performance characteristics. An integer number is assigned to each bearer to show the QoS category (1–9) assigned to it. It is pre-configured in the service-provider network.

9.5.5.3 GBR (Guaranteed Bit Rate)

This indicates the minimum bit rate guaranteed for a bearer.

9.5.5.4 MBR (Maximum Bit Rate)

This indicates the maximum bit rate for a bearer while at the same time discarding any excess traffic by using a rate policing function.

9.5.6 Handover

The handover in LTE can be:

- Inter LTE: The UE MME and/or S-GW is changed.
- Intra-LTE: None of the above are changed after handover.
- Inter-RAT: This is between various radio technology networks.

The three phases of handover are: preparation, execution and completion. In the preparation phase, both the base and target eNBs prepare for handover. During the actual handover phase, the UE gets connected to the target eNB after getting itself detached from the base eNB. This happens after resources are allocated to the UE from the target eNB. Once the handover happens, the resources towards the base eNB are released. Both the UL and DL traffic is delivered through the target eNB.

9.5.6.1 Roaming

A basic functionality of any mobile network, in LTE SAE, is that MME provides the S6a interface to HPLMN HSS and supports PDN gateway selection for assisting roaming. The S8a interface between S-GW and P-GW is used for the same purpose.

9.5.6.2 Policy and Charging Control

This is controlled by PCRF and executed by PCEF. Each network has only one logical PCRF database. It is basically a repository of policy and charging control rules. Each rule is associated with a SDF template applied to matching data flow.

9.5.6.3 Authentication, Authorisation and Accounting (AAA)

As the name suggests, this authenticates the user (for IP address allocation, based on Access Network Identity), while accounting is used for session tracking. These procedures support providing consistent quality of experience across the networks and technologies (roaming, resource sharing, etc.).

9.6 IMS (IP Multimedia Subsystem)

IMS was first introduced and designed by 3GPP Release 5 and has been further enhanced in 3GPP Releases 6 and 7. It is now being adopted by other standard bodies such as ETSI. This is one of the next generation network architectures which is used for providing multimedia services, both fixed and mobile. This technology started off as a technology under 3GPP but is now a part of the wired technology as well and built on SIP (Session Initiation Protocol) which enables clients to invite others into a session. This is access dependent and supports multi-access types such as GSM, WCDMA, CDMA 2000, WLAN, wire line broadband and other packet data applications. This is used for controlling the communication in IP-based NGNs (Next Generation Networks) and also provides real-time services on top of the UMTS packet-switched domain. This technology consists of session control, connection control and an application services framework along with subscriber and service data. IMS is intended to be an open systems architecture, a full proof architecture which simplifies and speeds up service creation. The location of IMS in a LTE EPC network is shown in Figure 9.3.

9.6.1 Applications of IMS

IMS is used in different fields such as presence services, full duplex video telephony, instant messaging, unified messaging, multimedia advertising, multiparty gaming, video streaming, web/audio/video conferencing, and push-to services, such as push-to-talk, push-to-view, and push-to-video. It delivers person-to-person real-time IP-based multimedia communications and enables applications in mobile devices to establish person-to-person connections.

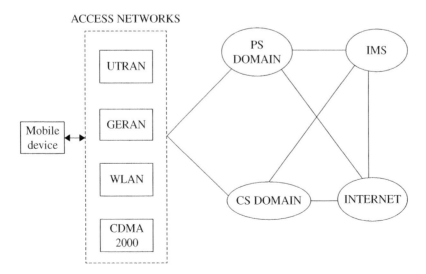

Figure 9.3 IMS location in LTE EPC.

9.6.2   IMS Architecture

As shown in Figure 9.4, IMS architecture is split into three main layers:

Service or application layer: This layer comprises application and content servers to execute value added services for the user. The service layer is where all of the actual services 'live' and also perform different functions such as configuration storage, identity management, billing, and control of voice and video calls.

Control layer: This layer comprises network control servers for managing call or session set up, modification and release, and also contains a full suite of support functions such as charging and O&M.

Connectivity or transport layer: The connectivity layer comprises routers and switches, both for the backbone and the access network. The transport layer is responsible for the abstraction of the actual access network from the IMS architecture, and acts as the intersection point between the access layers and the IP network above it.

9.6.3   IMS Core Site

The IMS solution consists of Connection Processing Solutions (CPS) and IP Multimedia Register (IMR) products and may be co-located with other core systems such as the media gateway and mobile switching centre (Figure 9.5). There are many protocols which are used in implementation of the IMS site providing features such as security, efficiency and resilience. IMS network elements support a dual stack, i.e. both IPv4 and IPv6, where IPv4 is used for network management and charging.

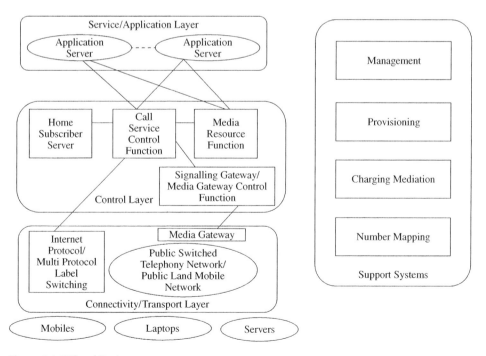

Figure 9.4 IMS architecture.

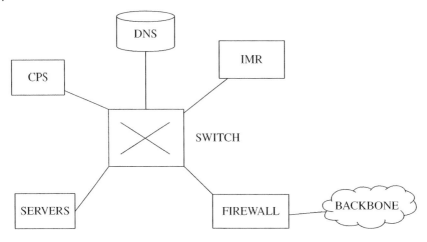

Figure 9.5 IMS core site.

9.6.3.1 Domain Name Server (DNS)
This is the hierarchical, distributed database which stores information for mapping Internet host names to the IP address and vice versa and other data used by internet applications. This server is a standalone element. There are two servers on the DNS: one is the master server which is the ultimate source of information in an IMS domain; and the other is the slave server which loads the zone contents from the master server using a replication process known as 'zone transfer'. It also answers to cluster node requests and receives zone transfers from the master DNS server. A DNS cache is used in every node for reduction of ethernet traffic and minimises DNS query time.

9.6.3.2 Firewall
This is the most effective means of controlling the flow of IP traffic between two networks or servers. This is used at the boundaries of the network and to implement network 'islands' and security domains within a wider range. This also applies different security rules for incoming and outgoing traffic on an interface.

9.6.3.3 Access Gateway
This provides an interface between the radio network and the IP-based network.

9.6.3.4 Access Network
This is the radio portion of the network.

9.6.3.5 Breakout Gateway Control Function
This controls the resource allocation to IP sessions.

9.6.3.6 Call Session Control Function
This provides the control and routing function for IP sessions.

9.6.3.7 Foreign Agent
This provides registration information to the home agent and also forwards packets from the mobile to the home agent.

9.6.3.8   Home Agent

This tracks the current foreign agent serving the mobile and forwards packets to it.

9.6.3.9   Home Subscriber Server

This can take the place of a HLR in an all-IP network and contains AAA functions and other databases.

9.6.3.10   Media Gateway

This provides an interface for bearer traffic between IP and PSTN.

9.6.3.11   Media Gateway Control Function

This provides signalling interoperability between IP and PSTN domains.

9.6.3.12   Policy Decision Function

This assigns network bandwidth and resources in real time and also assigns resources according to demand and QoS requirements.

9.6.3.13   Position Determining Entity

This provides assistance by way of location determination algorithms while some mobiles can determine the position independently.

9.6.3.14   SIP Application Server

This represents a platform for SIP application development and operation.

9.6.4   Functions and Interface in IMS

IMS functions and reference points are shown in Figure 9.6.

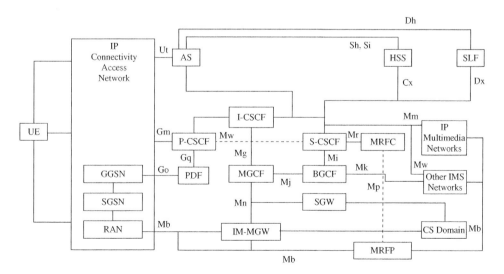

Figure 9.6 IMS functions and reference points.

9.6.4.1 Call Session Control Function (CSCF)

This provides the session control for both applications and terminals and also includes securing SIP message routing and monitoring.

9.6.4.2 Serving Call Session Control Function (S-CSCF)

This is the heart of an IMS system which provides the SIP signalling, routing, translation and interaction with other services and charging and also supports authentication of users after registration.

9.6.4.3 Interrogating Call Session Control Function (I-CSCF)

This helps in routing the SIP request to an assigned S-CSCF.

9.6.4.4 Proxy Call Session Control Function (P-CSCF)

This is a SIP proxy that is the first point of contact for the IMS terminal, which is located either in the visited network or in the home network. It also authenticates the user and establishes the IP security association with the IMS terminal.

9.6.4.5 Media Gateway Control Function (MGCF)

This controls the connection for media channels in an IMS-MGW mode and communicates with the CSCF.

9.6.4.6 IP Multimedia Subsystem-Media Gateway Function (IMS-MGW)

This provides support to bearer control, media conversion, and payload processing.

9.6.4.7 Media Resource Function (MRF)

This implements functionality to manager and process media streams.

9.6.4.8 Multimedia Resource Function Controller (MRFC)

This is responsible for controlling the media stream resources.

9.6.4.9 Multimedia Resource Function Processor (MRFP)

This provides functions and provisions for multimedia resources and these resources are controlled by the MRFC.

9.6.4.10 Subscription Locator Function (SLF)

This is located in the subscriber database which helps in response to queries from the I-CSCF or AS.

9.6.4.11 Breakout Gateway Control Function (BGCF)

This is responsible for routing of the telephony sessions which are initiated in the IMS moving towards the circuit-switched network.

9.6.4.12 Application Server (AS)

This is placed in the user's home network and provides value-added IMS services.

9.6.4.13 Home Subscriber Server (HSS)

This is the master database which includes information handling, authentication, and mobility management.

9.6.4.14 Signalling Gateway Function (SGF)

This is responsible for signalling conversion between SS7 and IP networks.

9.6.4.15 Policy Decision Function (PDF)

This allocates the IP bearer resources entering the packet switched network (Table 9.1).

Table 9.1 Reference points.

Interface name	IMS entities	Description	Protocol
Cr	MRFC, AS	Used by MFRC to fetch documents like scripts and other resources from an AS	HTTP over declined
			TCP/SCTP channels
Cx	I-CSCF, S-CSCF, HSS	Used to communicate between I-CSCF/S-CSCF and HSS	Diameter
Dh	SIP AS, OSA, SCF, IM-SSF, HSS	Used by AS to find a correct HSS in a multi-HSS environment	Diameter
Dx	I-CSCF, S-CSCF, SLF	Used by I-CSCF/S-CSCF to find a correct HSS in a multi-HSS environment	Diameter
Gm	UE, P-CSCF	Used to exchange messages between UE and CSCFs	SIP
Go	PDF, GGSN	Allows operator to control QoS in a user plane	COPS, Diameter
Gq	P-CSCF, PD	Used to exchange policy decision related information between P-CSCF and PDF	Diameter
ISC	S-CSCF, I-CSCF, AS	Used to exchange messages between CSCF and AS	SIP
Ma	I-CSCF, AS	Used to directly forward SIP requests which are destined for public service identity hosted by the AS	SIP
Mg	MGCF, I-CSCF	MGCF converts ISUP to SIP signalling and forward SIP signalling to I-CSCF	SIP
Mi	S-CSCF, BGCF	Used to exchange messages between S-CSCF and BGCF	SIP
Mj	BGCF, MGCF	Used to exchange messages between BGCF and MGCF in the same IMS network	SIP
Mk	BGCF	Used to exchange messages between BGCFs in different IMS networks	SIP
Mm	I-CSCF, S-CSCF, external IP network	Used for exchanging messages between IMS and external IP networks	SIP
Mn	MGCF, IM-MGW	Allows control of user-plane resources	H.248
Mp	MRFC, MRFP	Used to exchange messages between MRFC and MRFP	H.248

(Continued)

Table 9.1 (Continued)

Interface name	IMS entities	Description	Protocol
Mr	S-CSCF, MRFC	Used to exchange messages between MRFC and S-CSCF	SIP
Mw	P-CSCF, I-CSCF, S-CSCF	Used to exchange messages between CSCFs	SIP
Rf	P-CSCF, I-CSCF, S-CSCF, BGCF, MRFC, MGFC, AS	Used to exchange offline charging information with CCF	Diameter
Ro	AS, MRFC	Used to exchange online charging information with ECF	Diameter
Sh	SIP AS, OSA-SCS	Used to exchange information between SIP AS/OSA-SCS and HSS	Diameter
Si	IM-SSF, HSS	Used to exchange information between IM-SSF and HSS	Mobile application part
Sr	MRFC, AS	Used by MRFC to fetch documents like scripts and other resources from an AS	HTTP
Ut	UE, AS (SIP-AS, OSA-SCS, IM-SSF)	Facilitates the management of subscriber information related to services and settings	HTTP, XCAP

9.6.5 Protocol Structure in IMS

The main protocol, namely SIP (Signalling Initiation Protocol), is used in the IMS on the basis of requirements such as flexibility and security and provides establishment, modification and termination of the multimedia sessions between two devices. There are other functions which are handled by SIP such as service control, billing, QoS authorisation, and resource management. There are five types of logical entities in an SIP network: user agents, proxy servers, redirect servers, location server, and registrars. SIP works in conjunction with the SDP to initiate multimedia sessions which describe session initiation and session announcement. SIP is used to establish the sessions by the following general steps:

- *Session initiation:* The user's device signals are needed and the user's network location is identified and a unique session identifier (SIP URI) is assigned.
- *Session description:* The SDP protocol is used. This delivers a description of the session to the user device.
- *Session management:* RTP and RTSP are used and once the session is accepted by the user's device, media and other content are then directly exchanged between the end points.
- *Session termination:* Either party in the session can request session termination once the data or media exchange is complete.

Some key aspects of IMS are as follows:

- *Resilience:* This is the ability of the network to withstand faults and failures and provide a high degree of availability for services. This is required at network element level, site level, transport level, and disaster recovery level. The target is to avoid a single point of failure at the site. One way to achieve high availability in site solution is to use the Virtual Router Redundancy Protocol (VRRP) which provides network redundancy for IP networks.
- *Virtual LANs (VLANs):* Nodes are grouped together in VLANs, so that the nodes in different VLANs do not see each other. For the IP routing process, one VLAN can have multiple IP subnets. Each VLAN is one IP sub-net which helps maintenance and configuration. There are two ways to use VLANs: port and IP-based. It is recommended to use at least 04 VLANs with IMS for O&M traffic, charging traffic, lawful interception traffic, and all the other signalling traffic.
- *Multiprotocol label switching (MPLS):* This technology is used for speeding up network traffic flow and makes it easier to manage a network for QoS and involves setting up a specific path for a given sequence of packets. This is called a 'multiprotocol' because it works with the IP, ATM and frame relay network protocols. MPLS-based virtual private networks (VPNs) can be used for carrying the different types of traffic separated across the IP/MPLS backbone.
- *IPSec:* This provides security for transmission of sensitive information over unsecured networks. This operates in two modes: one is the transport mode which places the IPSec header after the original outer IP header, before the upper layer protocol; and the other is the tunnel mode which encapsulates the entire IP header and datagram, presents an IPSec header and then creates an outer IP header to tunnel the packet. There are two main protocols provided by IPSec: Authentication Header (AH) which protects a packet against modification during transit; and Encapsulation Security Payload (ESP) which is the encryption part of IPSec.

9.6.6 IMS Security System

IMS is essentially an overlay to the PDS and has a low dependency on the PDS. PDS is deployed without the multimedia session capability due to which separate security association is required between the multimedia client and IMS before access is granted to multimedia services. IMS authentication keys and functions on the user side may be stored in a secure memory location on the UE. These authentication keys and functions are logically independent and are used for PDS authentication. There are five different security associations and different needs for security protection for IMS (numbered 1–5) (Figure 9.7). These functions are as follows:

1) This association provides mutual authentication between the UE and the S-CSCF. The HSS collective delegates the performance of subscriber authentication to the S-CSCF. HSS is responsible for generating keys and challenges.
2) This association provides a secure link and a security association between the UE and a P-CSCF for protection of the Gm reference point.
3) This association provides security within the network domain internally for the Cx interface.
4) This association provides security between different networks for SIP capable nodes and is only applicable when the P-CSCF resides in a virtual network.
5) This association provides security within the network internally within the IMS subsystem between SIP capable nodes and is applied when P-CSCF resides in a home network.

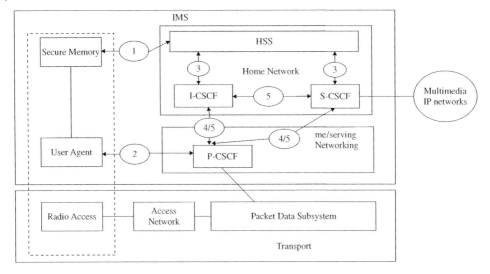

Figure 9.7 IMS security.

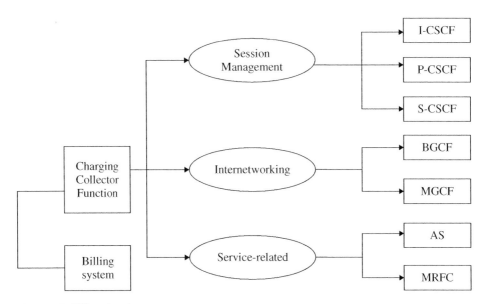

Figure 9.8 Offline charging.

9.6.7 IMS Charging

Offline charging: This is applied to users who pay for their service periodically. All SIP entities use the diameter Rf interface to send accounting information to a CCF located in the same domain. CCF collects all the information and builds a CDR which is sent to the billing system of the domain (Figure 9.8). While each domain has its own charging system, billing information between various domains is exchanged so that roaming charges may apply.

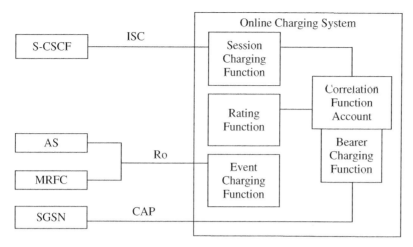

Figure 9.9 Online charging.

Online charging: This is shown in Figure 9.9. Online charging is known as credit-based charging which is used in prepaid services or real-time credit control of postpaid services.

9.6.8   Service Provisioning in IMS

IMS enables services by defining the services, creating related service data, and passing an incoming request to application servers, etc.

9.6.8.1   Registration in IMS

The architecture allows for the S-CSCF to have different capabilities or access to different capabilities. The network operator shall not be required to reveal the internal network structure to another network and to expose the explicit IP addresses of the nodes within the networks. S-CSCF is able to retrieve a service profile to the user who has an IMS subscription and check the registration request against the filter information. HSS supports the possibility to restrict a user from getting access to an IP multimedia core network subsystem from unauthorised visited networks. During IMS registration, as shown in Figure 9.10, the UE may indicate its capabilities and characteristics in terms of SIP user agent and also update its capabilities by initiating a re-registration when the capabilities are changed on the UE. P-CSCF may subscribe to notifications of the status of the IMS signalling connectivity after successful initial user IMS registration and cancel any active subscription.

The SIP protocol is used to establish sessions and further exchange data. First, the client has to register some server and is authenticated during registration and based on triggers the register is sent to various application servers in the IMS to establish that the client is online. After the PDP context is established, the client sends a register request to P-CSCF which selects/resolves the I-CSCF node address in the home client home IMS. I-CSCE is the entry point in a home S-CSCE which communicates with HSS to retrieve the authentication parameter. This results in sending the second register request with authentication parameters.

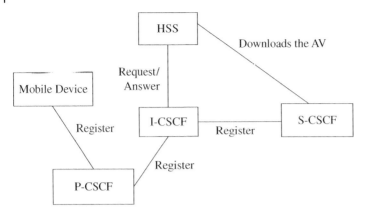

Figure 9.10 IMS registration.

9.6.8.2 De-registration in IMS

This may be mobile initiated or network initiated which may be initiated by register timeout.

Mobile-initiated de-registration: De-registration is accomplished by a registration with an expiration time of zero seconds. The procedure is as follows: The UE decides to initiate de-registration. The UE sends a new register request with an expiration value of zero seconds and sends the register information flow to the proxy. The proxy sends the register information flow to the I-CSCF. The P-CSCF network identifier is a 'string' which identifies at the home network. The I-CSCF sends the Cx-Query information to the HSS which determines that the PUI is currently registered. The Cx-Query is sent from the HSS to the I-CSCF which determines the address of S-CSCF through a name address resolution mechanism and sends the de-register information flow to S-CSCF. S-CSCF sends the de-registration information to the service control platform and performs whatever service control procedures are appropriate.

Network-initiated de-registration: This mechanism should be at the SIP protocol level in order to guarantee access independence for the IM CN subsystem which can initiate the de-registration procedure for the following reasons:

- Network maintenance: Forced re-registration from users, cancelling the current contexts of the user spreads among the IP multimedia core network subsystem nodes at registration and imposing a new IM registration solves this condition.
- Network/traffic determined: This supports a mechanism to avoid duplicate registration or inconsistent information storage. This case may occur at the change of the roaming agreement parameters between two operators and imposing new service conditions to roamers.
- Application layer determined: The service capability offered by the IP multimedia core network subsystem to the application layer has parameters which specify whether the subsystem registration is to be removed or those from one or a group of terminals from the user.
- Subscription management: The operator must be able to restrict user access to the IP multimedia core network subsystem upon detection of contract expiration, removal of IM subscription, and fraud detection.

In the case of changes in service profile of the user, it may possible to actively change the S-CSCF by using the network initiated de-registration by a HSS procedure.

9.7 Voice-Flow in LTE

Voice flow in LTE is shown in Figure 9.11. Voice services can be catered for in LTE networks by the following means: IMS or existing GSM/UMTS networks or VoLTE. We have discussed IMS in previous sections. The LTE network can process only IP (or PS) calls. Some technology inducement needs to be done in the LTE network for it to provide non-IP voice calls. There are three methods to do this:

- VoLGA (Voice over LTE via Generic Access)
- CSFB (Circuit Switch Fall Back)
- VoLTE (Voice over LTE)

9.7.1 VoLGA

This basically means using the generic 2G/3G networks for voice call flow. For this to happen, a new network element, the VoLGA network controller, is introduced. So what role does it play? As in the GSM network, the BSC communicates with the MSC. However, as this was not cost efficient, it failed to make its mark in the network.

9.7.2 CSFB

Most of the networks globally have evolved from 2G/3G networks to LTE networks. Especially in the developing world, 2G/3G networks are still in use. Using the CSFB, the existing 2G/3G network capacity is used by redirecting the UE onto a CS cell/network.

Figure 9.11 Voice flow in LTE.

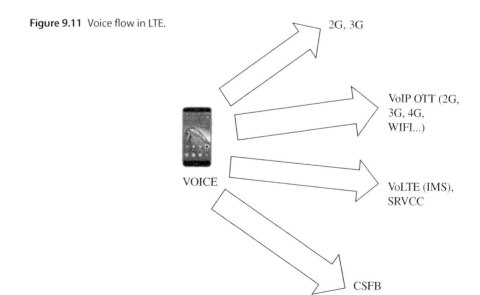

CSFB was specified by 3GPP in Release 8 and can be implemented by a software upgrade. Fundamentally, the voice and SMS are delivered to LTE devices using circuit switch technology.

9.7.3 VoLTE

Once VoLTE is implemented in networks, other solutions will cease to exist. It is an IMS-based solution and is a standard for high speed wireless communication for mobile/data terminals. There are many benefits of using VoLTE including provision of reliable services, battery efficiency, rapid call establishment, and spectrum efficiency. This means the capacity of VoLTE is higher than GSM and UMTS. The speech Codec mentioned in 3GPP is AMR-NB. However, the recommended codec is AMR-WB which is also known as HD voice. According to the GSMA IR.92 document, supplementary services expected to be supported by VoLTE include ad-hoc multi party conference, call forwarding, barring, etc. For VoLTE, IMS needs to be implemented in the LTE network.

Let us consider the following situation: a subscriber is engaged in a voice call in an LTE area. The subscriber moves out of LTE coverage but for the call to continue, the user needs to fall back to the GSM/UMTS network. This is done by using Single Radio Voice Call Continuity (SR-VCC), introduced in 3GPP Release 8. For SR-VCC to function, the IMS framework needs to be in place and the devices should have the capability to talk to both LTE and GSM/UMTS networks at the same time. Also, software upgrades are needed in the MSS, IMS, and LTE/EPC systems.

9.8 Software Defined Network (SDN)

The network has always been very traditional and required very specific skills. Network devices such as firewalls, routers, and switches are proprietary hardware and are sold by network vendors. All the tasks in a traditional network are separated by different planes: control plane, data plane, and management plane. The control plane is responsible for routing information such as the ARP table. The data plane is responsible for forwarding traffic using the information from the control plane. The management plane is responsible for managing and accessing the network devices. From the SDN point of view, the data plane and control plane are of relevance.

Configuring the traditional network is a slow, manual process (sometimes using scripts) using CLI. Usually it takes hours for the network team to configure a network device for new servers or applications and it is done during the maintenance windows.

Hence, businesses are looking at virtualisation that speeds up and automates the network configuration.

9.8.1 SDN Architecture

The main goal of SDN is that the network should be open and programmable, so that it can evolve with the speed of software. If a business wants a specific type of network behaviour, it should be able to develop or install an application to do what it needs. SDN addresses the fact that the static architecture of a traditional network is not flexible to

Figure 9.12 SDN overview.

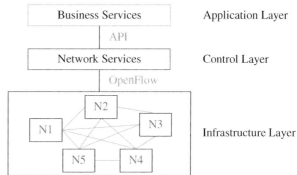

meet today's business needs and agility requirements. As shown in Figure 9.12, SDN architecture suggests a centralised approach disassociating the forwarding devices of network packets (data plane) from the routing process (control plane), thus separating the network control plane from the forwarding plane and the control plane controls various network devices. This centralised control plane is the 'brain' of the network and the forwarding plane just follows instructions. This enables a centralised control which is software driven, and therefore programmable, easily configurable, easily manageable and not dependent on the network hardware devices; hence it is agile.

With SDN, applications can be aware of the network. The control plane is logically centralised and summarises the network state of the applications.

The high level logical architecture of SDN, as shown in Figure 9.13, comprises a network operating system, also called the SDN controller, which will have core services interfacing with network nodes and providing programmable interface to the network applications. The key core services are: topology, inventory, statistics, and host tracking. The network forwarding devices receive packets, take action on the packets and update

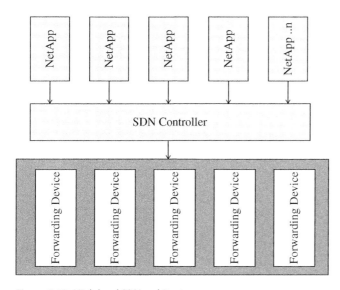

Figure 9.13 High level SDN architecture.

counters. The actions include dropping a packet, modifying a packet header, and sending out packets to a single port or multiple ports. However, the instruction of how to handle a packet comes from the SDN controller. The SDN applications are software programs that communicate the network requirements and desired network behaviour to the SDN controller. Thus, the SDN controller is between the SDN applications and the network infrastructure layer. NBI interfaces communicate between the SDN controller and SDN application layer. CDPI interfaces communicate between the SDN controller and infrastructure layer.

When a packet arrives at a SDN control forwarding device, the forwarding device will parse the packet header and will already know what to do with the packet or query the SDN controller. The network applications will determine what action will need to be done with the packet and pushes the information to the forwarding device using the SDN controller as the translating device. The forwarding device will then take the assigned action on the packet. The forwarding packet also caches the instructions so that for future similar packets, it can take the action without checking with the SDN controller. This process goes on device-by-device until the packet reaches its destination.

The controller has a global view of all network forwarding devices below it. It creates an abstract view of the network to the network applications. The network applications use this information to make key decisions on how to implement network policies. For example, a network security application does not care about all the possible paths between all points in the network and the SDN controller provides a network abstract and shows the forwarding devices as a single large switch and translates the requirements from the SDN application layer to the SDN datapath layer.

The forwarding devices/datapath could be traditional hardware devices but must support the programmable interface line OpenFlow or can be software switches like open vSwitch. Hardware switches provide better performance while software switches provide greater flexibility. The most common protocols for south bound communications are OpenFlow, OVSDB, NetConf, and SNMP. The SDN controller provides Rest API to be consumed by the network applications.

SDN controllers are being sold by many big networking vendors/companies, e.g. Cisco Open SDN controller, Juniper Contrail, Brocade SDN controller, and PFC SDN controller from NEC. Many Open source SDN controllers such as Opendaylight, Floodlight, Beacon, Ryu, etc. are also available.

The key characteristics of SDN Architecture are given in the following.

9.8.1.1 Programmable
Since the network services are separated from the forwarding layer and are deployed as software applications, they are programmable and it is easy to add and update features.

9.8.1.2 Agile
The administrator has more flexibility to manage network traffic flow to meet customer demands.

9.8.1.3 Central Management
The network intelligence is centralised and software-based. This allows it to maintain a global view of the network, appearing to applications and policy engines as a single, logical switch.

9.8.1.4 Configurable

Network management programs are not dependent on the proprietary hardware, so network managers can write programs to manage, configure and operate network resources.

Open standard based and vendor neutral: SDN simplifies network design and operation because instructions are provided by SDN controllers instead of multiple, vendor-specific devices and protocols.

9.9 Network Function Virtualisation (NFV)

A telecoms network consists of an increasing variety of proprietary hardware. The launching of new services demands network reconfiguration and on-site installation of new equipment which requires additional floor space, power, and trained maintenance staff.

In this ever-changing world of demand and technology evolution, it requires greater flexibility and dynamism than hardware-based appliances allow. Hard-wired networks with single function boxes are tedious to maintain, slow to evolve, and prevent service providers from offering dynamic services.

Software applications are dynamically configurable and automated and can be migrated into different environments and new features added and removed. ETSI was founded in November 2012 by seven of the world's leading telecoms network operators. ISG NFV became the home of the Industry Specification Group for NFV. They came up with virtualised network functions which allow networks to be agile and capable of responding automatically to the needs of the traffic and services running over it.

9.10 Virtualising Network Functions

Virtualisation can be considered as the first step in separating the infrastructure services from the physical layer on which that service operates. In virtualisation, the services consumed (compute, storage, or network) are not associated with any physical hardware; instead, the services are a software abstraction layer on any physical resource running the virtualisation software. The service exposes API interfaces which can be consumed for managing and configuring the network, thereby bringing out the full potential of automated provisioning, as shown in Figure 9.14.

With network virtualisation, the virtual network is provisioned in software, with API interfaces that can be consumed by services to bring in agility and speed similar to server virtualisation. The same software tools already provisioning the application's virtual machines can provision both compute and network together and subsequently validate the complete application architecture.

Management, automation and orchestration (MANO) includes the virtualisation and application management layer providing the framework for provisioning virtual networks and managing NFV infrastructures. It also helps components within the network function virtualisation infrastructure (NFVI) to communicate with the existing operational and billing support systems.

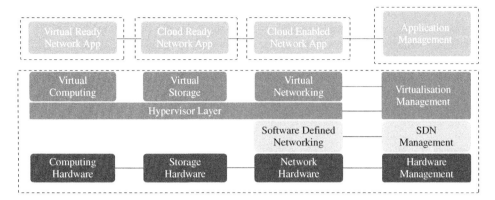

Figure 9.14 Network function virtualisation.

Network function virtualisation will simplify and speed up adding new network functions or applications in the environment and hence speed up the realisation of speed needs. The benefit also includes reduced power consumption and increased physical space in the data centres, thus helping to reduce Capex and Opex costs.

9.10.1 Benefits of Virtualisation

An operator network is composed of a variety of network elements which are a mixture of legacy on premise and virtual environment. All control and data functions of the network elements are bundled together in the device and managed by the supplier, so the deployment of new services or upgrades or modifications to existing services are very rigid and dependent on the supplier and needs a lot of coordination between different suppliers. This leads to operational challenges, increase in operational cost, and inflexibility in business growth with changing market demands.

The key features of SDN NFV are that the operators are looking at separation of the control and data plane, automated orchestration of services and resources, a programmatically control network, scaling of network by using virtualisation, monetising the network infrastructure using multi-tenancy, automation of business processes, and standardisation.

Operators are looking for operational efficiency and business transformation to mitigate the risks and challenges. The operators will achieve operational efficiency utilising an elastic, scalable, and automated network infrastructure. They can utilise the dynamic traffic steering and service chaining functionalities. They will benefit by decreasing the time to market for new and existing features. The environment will provide agile service creation and rapid provisioning opportunities and thus enhance customer satisfaction.

SDN NFV will change the traditional network model and thus bring a paradigm shift in network information technology. There will be a significant change in how operators design, build, manage and deliver their services to customers. This, in turn, will restructure their cost structure and operational processes and will enable them to remain competitive in the industry.

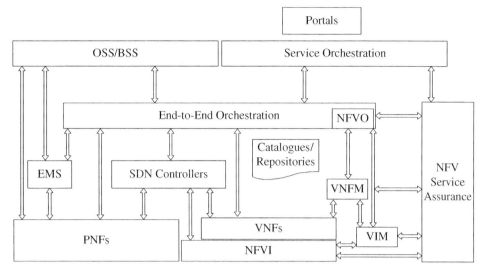

Figure 9.15 Example of Telco using NFV.

Figure 9.15 shows a high-level Telco network architecture using SDN-NFV concept and the following functional blocks:

- NFVI includes all hardware and software components on which virtualised network functions (VNFs) are deployed.
- A VNF is a software implementation of a network function, deployed in the hardware which is capable of running on the NFVI.
- A physical network function (PNF) is the implementation of a network function that relies on dedicated hardware and software for part of its functionality.
- The virtualised infrastructure manager (VIM) is responsible for controlling and managing NFVI compute, storage and network resources. It also includes the physical infrastructure manager (PIM).
- The virtualised network function manager (VNFM) is responsible for VNF lifecycle management (e.g. instantiation, upgrade, scaling, healing, termination).
- End-to-end orchestration (EEO) is the function responsible for lifecycle management of network services. The orchestration function has a number of subcomponents related to different aspects of that orchestration functionality. VNF orchestration is done by the NFV orchestrator (NFVO).
- Catalogues/repositories is the collection of descriptor files, workflow templates, provisioning scripts, etc. that are used by EEO, VNFM, and SA to manage VNFs and NFV/SDN/end-to-end network services.
- Service assurance (SA) collects alarm and monitoring data. Applications within SA or interfacing with SA can then use this data for fault correlation, root cause analysis, service impact analysis, SLA management, security monitoring and analytics, etc.
- The Data Centre SDN Controller (DC SDN Controller) is responsible for managing network connectivity within a data centre.
- The wide area network SDN Controller (WAN SDN Controller) is responsible for controlling connectivity services in the WAN.

- The access SDN Controller is responsible for controlling wireline or wireless access domains.
- The domain/vendor-specific controller is an optional controller that may be required to handle specific vendors or technology domains in the absence of standard interfaces or for scalability purposes.
- Service orchestration is the customer-facing function responsible for providing sa ervices catalogue to the portal.
- Portals include customer portals where customers can order, modify and monitor their services and the Ops portal.
- The Element Management System (EMS) is a legacy system responsible for the management of specific network elements.
- Operations Support Systems and Business Support Systems (OSS/BSS) are responsible for a variety of functions such as order entry, service fulfilment and assurance, billing, trouble ticketing, helpdesk support, etc.

Part IV: 5G

Introduction to 5G Network Planning and Optimisation

10

5G Network Planning

10.1 Introduction to 5G

5G technology was a buzz word in 2016 but now it is at the stage where it is being deployed. So, how is 5G technology being viewed at this stage? In very simple terms, it is seen as a technology having much superior data rates and much lower latency than 4G. Table 10.1 shows that peak data rates in 5G will be beyond 1Gbps with latency lower than 10 ms.

From the technologist's viewpoint, 5G is seen in terms of the following parameters:

- *Data rates:*Data rates are expected to be in the range of 1–10 Gbps.
- *Capacity:*Capacity requirement per user is expected to be 500 GB.
- *Frequency:*While current LTE networks work at frequencies of 4 GHz, higher frequency ranges are expected to be used in 5G networks (6–100 GHz).
- *Latency reduction:*Latency requirements will be under 10 ms and will be as low as 1 ms in certain uses.

This will be in an ecosystem that will have millions of devices, thereby totally transforming the mobile broadband experience. The 5G era will see unprecedented growth in the area of critical communications and in the IoT.

In 5G technology development there are three studies that need to be mentioned: IMT for 2020 and Beyond by ITU-R, the 5G Definition Process by the EU's METIS project, and the 5G White Paper by NGMA.

10.1.1 The World in 2020

The world is expected to witness some startling mobile and device growth by 2020. It will be almost like an explosion. The industry predicts that more than six billion smart phones will be in circulation by 2020. If we add in devices, the figure is expected to be a staggering 50 billion. In 2016, data traffic was around 7 EB (1 EB is one billion gigabytes) per month and this is expected to grow to 45 EB per month by 2020. Another interesting prediction is that data traffic is expected to grow to around 10 GB per month. Looking at the situation from a telecom network perspective, the world will see more than six billion 4G connections and more than 20 million 5G connections across the world (more than one billion 5G subscriptions by 2023). These figures give us an idea of the kind of technology challenges that will arise in the next couple of years.

Fundamentals of Network Planning and Optimisation 2G/3G/4G: Evolution to 5G,
Second Edition. Ajay R. Mishra.
© 2018 John Wiley & Sons Ltd. Published 2018 by John Wiley & Sons Ltd.

Table 10.1 Data rates in 5G.

Technology	Peak rates	Latency (ms)
2G	40 kbps	700
3G	384 kbps	250
4G	42 Mbps	100
5G	>1 Gbps	10–1

10.2   The 5G Challenge

5G technology will be deployed in 2019/2020 with a rapid increase in subscription rate. There will be some deployments in advanced mobile broadband networks.

10.2.1   Devices and Data

As already stated, it is predicted that 2020 will see more than 50 billion connected devices producing more than 45 EB of data per month. This data will be generated by smart phones and the higher number of IoT devices. IoT devices will produce a very small amount of data but will be ably supported by the huge number of devices. Quite a few inputs in current 3G/4G network dimensioning are based on the number of users and this is going to impact scheduling and access control mechanisms apart from the signalling needed for IoT devices.

10.2.2   Network Capacity (Densification, Spectrum, etc.)

Owing to the above reasons, the capacity of the network will be affected. The spectrum used will also go beyond 6 GHz (especially for IoT). The amount of data will lead to a need for capacity planning from an altogether different perspective without drastically impacting the cost. Some of the aspects to be considered could be working on distributed data and signalling methodologies. 5G will also see 'ultra dense networks'. In highly dense areas, ultra dense networks will be deployed to ensure constant connectivity and high data speed. Such networks will be an amalgamation of pico, femto, micro and macro cells in much larger numbers.

10.2.3   E2E Network Architecture

5G as a technology is not coming alone. Virtualisation, SDN, IoT, Cloud, digital transformation, etc. will be an inherent part of the technology or network evolution. Most CSPs/DSPs are being developed with a view to Network Evolution 2020 towards 5G where all these elements are an inherent part of the development. This makes things both interesting and complicated, requiring massive efforts to take the network to the desired function and performance level.

10.2.4   Machine-to-Machine (M2M) Communications

M2M communications enable the networked devices to communicate with each other. 3GPP also refers to this as MTC or machine type communications. The information from a device is sent to a computer (or other device) using an embedded sensor or a communication module. Some M2M applications include security, health, metering, and manufacturing. M2M communication is usually in short data packets ~10–100 s bytes, with higher order latency being permissible.

10.2.5 Device-to-Device (D2D) Communications

D2D communications have existed in the past but with upcoming networks, the devices will be connected to networks, thus making D2D communications an interesting feature in the next generation of LTE-A/5G networks. In simpler terms, it could be considered an evolution of push-to-talk. Examples from the past include walkie-talkies and the use of tetra communication systems. The advantages of using D2D are lower power consumption, better coverage, lower latency, and improved user experience. However, as a network is being used, there is the possibility of increased interference, which needs to be mitigated. Other challenges will include needing a longer battery life and the issue of the usage/sharing of the spectrum between D2D and mobile communications.

10.3 5G Network Architecture

Before we look at the architecture of 5G networks, let us briefly look at the basis for it. Just like 3GPP, we now have 5GPPP.

10.3.1 5GPPP

5GPPP or the Fifth Generation Public Private Partnership will result in 5G standards. It is considered to be the biggest research programme in the world – driven by the European Commission and the ICT industry. It is part of the EU's Horizon 2020 programme. Apart from the EU, there are 34 members involved including vendors, carriers, research bodies, and SMEs. A number of subprogrammes run under 5GPPP including 5G architecture, spectrum, SDN/NFV, and security.

10.3.2 5G Requirements

Let us consider what a 5G network is expected to do. In addition to the features of a 4G network and with a focus on cost and energy efficiency, 5G networks should be able to provide bespoke solutions across various industry sectors. With the IoT, D2D, etc. playing an important role, higher storage and computation speed will be expected from 5G networks. For handling such a varied and diverse ecosystem, the network should be capable of orchestrating the computing and network resources. There will be many times when a single instance of an application (software) will cater to multiple customers (multi-tenancy), with the tenant having the ability to customise a certain part of the application. Services providers will be able to provide services through multiple DSPs. Moreover, as with 3GPP, the 5G networks will need to be future-proof.

10.3.3 5G Use Cases

The three use cases for 5G defined by ITU, as given in IMT-2020, are given in the following.

10.3.3.1 Massive Machine-Type Communication (m-MTC)

There are two types of machine type communications: non-critical and critical. M2M in future years will become much wider, i.e. while a few hundred devices per access node are now connected, in years to come the number will increase to more than a few thousand devices per access node. Of these, the majority will be working on non-critical operations. The key requirements for non-critical applications include low device costs and energy consumption while for critical applications the key requirements are very high reliability and availability with very low latency. Smart wearables, mobile video surveillance, etc. will be available for users.

10.3.3.2 Enhanced Mobile Broadband (eMBB)

In very simple terms, eMBB can be said to be 50+ Mbps and an ultra low-cost network – everywhere including ultra-dense areas. Broadband will be available across the network area including the cell edge. This is all at low ARPU or ARPA. Examples of usage include broadband in high speed trains or mobile hotspots. Services such as Augmented Reality, multi-user interactions, etc. can be delivered with ease. For such use cases, the 5G NR (New Radio) standard in non-standalone operation for control plane anchor will use LTE. 5G NR is expected to give an almost optical fibre like experience to users. 5G NR serves any device – mobile or sensor from mobile, static to vehicle – giving seamless and always connected services. Thus, services such as a 3D cyber office with holograms/HD video, and Cloud services will be at the disposal of the consumer anytime and anywhere.

10.3.3.3 Ultra-Reliable and Low Latency Communication (u-RLLC)

5G will show a significant growth in application where ultra-high reliability and low latency is required, e.g. smart grid. Other use cases under this category would include robotics, vehicle-to-vehicle (V2V) communication, health (life critical applications), and security (drones and public safety).

10.3.4 5G Network KPIs

ITU established the following eight key performance indicators (KPI) to specify, quantify and measure the characteristics of the IMT-2020 (5G) systems.

10.3.4.1 Peak Data Rate (Gbps)

This is the maximum data rate achieved in ideal conditions to a single mobile station after all assignable radio resources for the corresponding link direction are used. Mathematically, the peak data rate is the product of channel bandwidth and peak spectral efficiency. The minimum requirements for peak data rate in downlink and uplink is 20 and 10 Gbps, respectively.

10.3.4.2 User Experienced Data Rate (Mbps)

This is the 5% point of the cumulative distribution function (CDF) of the user throughput. User throughput (during active time) is defined as the number of correctly received bits, i.e. the number of bits contained in the service data units (SDUs) delivered to layer 3 over a certain period (of time). The user experienced data rate for downlink and uplink is 100 and 50 Mbps, respectively.

10.3.4.3 Peak Spectrum Efficiency (bps Hz^{-1})

This is the maximum data received for a single mobile station, that is normalised by channel bandwidth (in bps Hz^{-1}) in error-free conditions after all assignable radio resources for the corresponding link direction are used. Peak spectral efficiency for downlink and uplink is 30 and 15 bps Hz^{-1}, respectively.

10.3.4.4 Device Mobility (km h^{-1})

This is the maximum mobile station speed (in km h^{-1}) at which the desired QoS is achieved. Mobility classes are defined as follows:

- stationary: 0 km h^{-1}
- pedestrian: 0–10 km h^{-1}
- vehicular: 10–120 km h^{-1}
- high speed vehicular: 120–500 km h^{-1}

The high speed vehicular class up to 500 km h^{-1} is mainly envisioned for high speed trains. Table 10.2 gives the mobility classes that shall be supported in the respective test environments.

A mobility class is supported if the traffic channel link data rate on the uplink, normalised by bandwidth, is as shown in Table 10.3. This assumes the user is moving at the maximum speed in that mobility class in each of the test environments.

10.3.4.5 Latency (ms)

With each generation development of cellular technology, latency is becoming more and more important. Various use cases of 5G are very specific about latency, which is defined by some as E2E delay or as TTI (transmit time interval).

As an example, consider that data is to be transferred between two locations A and B that are 1000 km apart and connected by fibre cable.

Table 10.2 Mobility classes.

	Indoor hotspot (eMBB)	Dense urban (eMBB)	Rural (eMBB)
Mobility classes supported	Stationary, pedestrian	Stationary, pedestrian, vehicular (up to 30 km h^{-1})	Pedestrian, vehicular, high speed vehicular

Table 10.3 Traffic channel link data rates normalised by bandwidth.

Test environment	Normalised traffic channel link data rate (bps Hz^{-1})	Mobility (km h^{-1})
Indoor hotspot (eMBB)	1.5	10
Dense urban (eMBB)	1.12	30
Rural (eMBB)	0.8	120
	0.45	500

Distance $(D) = 1000$ km.
Speed of light $(S) = 3*10^8$ m s^{-1}.
Speed of light through fibre $= 31\%$ slower (assume) $= 69\%$ of $S = 207*10^6$ m s^{-1}.
Time taken for data to transfer from A to B $= 1000/207*10^6 = {\sim}5$ ms.
For a round trip $= {\sim}10$ ms.

Latency in 5G networks will range between ${\sim}1$ ms and 10 ms. For some mission critical applications, it may be less than 1 ms. Both control plane and user plane latency has been defined in IMT-2020. Control plane latency refers to the transition time from the most 'battery efficient' state (e.g. Idle state) to the start of continuous data transfer (e.g. Active state) while user plane latency is the contribution of the radio network to the time from when the source sends a packet to when the destination receives it (in ms).

Control plane latency is a minimum of 20 ms while user plane latency is defined as 4 ms for eMBB and 1 ms for u-RLLC.

10.3.4.6 Connection Density (Number of Connected/Accessible Objects per km^2)

This is the total number of devices fulfilling a specific QoS per unit area (per km^2). This is used for the m-MTC scenario. Connection density should be achieved for a limited bandwidth and number of transmission reception points. The target QoS is to support delivery of a message of a certain size within a certain time and with a certain success probability. The minimum requirement for connection density is 1 000 000 devices per km^2.

10.3.4.7 Network's Energy Efficiency

This is the capability of radio interface technology (RIT) to minimise the RAN energy consumption in relation to the traffic capacity provided. A device's energy efficiency is the capability of the RIT to minimise the power consumed by the device modem in relation to the traffic characteristics.

The energy efficiency of the network and the device can be related to support of the following two aspects: efficient data transmission in a loaded case and low energy consumption when there is no data.

Another important term is the sleep ratio. Low energy consumption when there is no data can be estimated by the sleep ratio. The sleep ratio is the fraction of unoccupied time resources (for the network) or sleeping time (for the device) in a period of time corresponding to the cycle of the control signalling (for the network) or the cycle of discontinuous reception (for the device) when no user data transfer takes place. Furthermore, the sleep duration, i.e. the continuous period of time with no transmission (for the network and device) and reception (for the device), should be sufficiently long. The RIT shall have the capability to support a high sleep ratio and long sleep duration.

10.3.4.8 Area traffic capacity (Mbps m^{-2})

This is the total traffic throughput served per geographical area (Mbps m^{-2}). It signifies the correct bits received over a certain period (of time). It is calculated for one use case of one frequency band and one transmission reception point. It is based on average spectral efficiency, network deployment density and bandwidth. If the bandwidth is aggregated across multiple bands, the area traffic density is added over the bands. The target value for area traffic capacity in downlink is 10 Mbps m^{-2} in the indoor hotspot.

10.3.5 Overview of 5G Network

5G will be a truly convergent technology. Cellular technology experts envisaged fully convergent networks for 4G/LTE, but this will be achieved by 5G networks. Almost any and every element would be seen in the 5G network – IoT, MTC, D2D, V2V/V2x, ultra-dense cells, Cloud, smart grid, LTE-A, NR, etc. 5G will be able to seamlessly integrate newly developed millimetre wave technologies with existing Radio Access Technologies.

With respect to the LTE network discussed in earlier chapters, the 5G network will see a new radio (known as NR). The RAN will have both LTE-eNodeB and 5G-NR. Of course, with LTE here we mean both LTE and LTE-A. As in earlier networks, a 5G network would consider both the RAN and the cCore. While RAN would witness 4G and 5G radio, it would also consist of the legacy 3GPP radio access, Wi-Fi and fixed networks. The core domain would see much closer bonding with applications, IoT, and the Cloud.

A broad overview of the 5G networks is shown in Figure 10.1.

10.4 5G Enabling Technologies

The technologies that will impact 5G are discussed in the following.

10.4.1 Densification and Multi-RAT

The deployment of 5G networks will signify a drastic shift from homogenous networks to heterogeneous networks. These RATs include the likes of evolved LTE, Wi-Fi, and LTE in an unlicensed spectrum. The cell density will be higher (due to the higher number of small cells), while providing better spectral density and coverage area. Owing to implementation of M2M/D2D communication, spectrum and energy efficiency will also increase.

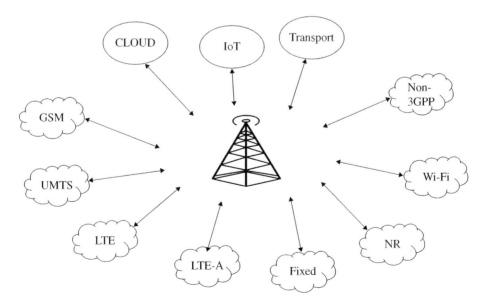

Figure 10.1 5G networks.

10.4.2 Massive Multiple-Input Multiple-Output (Massive MIMO)

For countering the effects of multipath fading and increasing the performance – quality and capacity of the wireless systems – smart antennas are used. There are four kinds of smart antennas: beam forming, spatial diversity and space time spreading, SDMA (Space Division Multiple Access) and MIMO. Beam-forming smart antennas that are half wavelength are used to create spatially selective transmitter or receiver beams helping to reduce the impact of co-channel interference and increase the number of users in the system. When multiple antennas are placed far apart from each other (typically 10λ), the individual received signals are the ones that have experienced independent fading and, at the same time, achieved maximum diversity gain. This smart antenna system increases system robustness. The SDMA system distinguishes between the individual subscribers utilising a subscriber-specific spatial signature, thus using the same frequency band to support many subscribers. And finally, the multiple antenna system MIMO increases the throughput of the system in terms of number of bits that can be transmitted by a subscriber for a given bandwidth. The improvement of the system performance is multi fold and hence is found to be utilised in high rate WLAN, 3G, 4G and 5G systems.

In the massive MIMO system, the number of antennas used is in the range of 32–64. Owing to the use of bigger antenna array, the antenna gain is high. Also, the reliability, robustness, energy efficiency, power saving, and latency reduction is high. However, there are also some disadvantages including signal processing complexity and sensitive beam alignment due to the narrow beam (see Appendix B).

10.4.3 Full Duplex Communication

This has always existed in microwave transmission but equipment such as UE and base stations lack the capability. Full duplex communication in 5G provides the possibility to transmit and receive over the same frequency band. This is possible due to new mobile communication technologies and self-interference (SI) cancellation abilities. There are a few self-cancellation techniques: propagation-domain SI suppression, analogue-domain SI cancellation, and digital-domain SI cancellation. Full duplex communication improves spectral efficiency by almost double and improves high data rates in areas such as D2D, small cell networks, and M2M.

10.4.4 Energy Efficiency

In terms of the efficiency of systems the need for a long lasting battery is understood, especially for sensor devices, it is also important to understand that the mobile industry needs to work on reducing its carbon footprint. This would mean a lower OPEX for mobile operators in longer run. This, in turn, would lead to lower total cost of ownership for the end subscriber as well – truly a win–win situation for all.

10.4.5 Self-Organising Networks

In future networks, the number of small cells will be much higher than that of macro cells; these cells are usually located in the home/indoor environment. These cells/networks would need to be self-configurable, plug-and-play type. They should be intelligent to be able to adapt in the network to be able to reduce the interference.

Self-configuration and self-optimisation features are available in Release 8 and Release 9, respectively. Features for automatic detection and removal of failures and automatic adjustment of parameters are mainly specified in Release 10.

10.4.6 Content Cache

Mobile data traffic growth is huge and will need a large network capacity. An increase in spectrum or spectrum efficiency and an increase in number of sites would not be effective to achieve this. Moreover, 80% of the traffic comes from social media applications such as Facebook and Twitter. As bandwidth and power is limited, the content caching technique is used by DSPs. This technique increases the capability of the system to deliver information over a longer period. This solution lowers content access latency and backhaul traffic loading. This improves customer experience at lower costs.

10.4.7 New Physical Layer: Air-Interface

The 5G network will have multiple RATs. We consider the two main ones here: LTE-A and NR. For best network performance, there has to be tight inter-working between the two. The range of spectrum used is indeed vast (1–100 GHz). The service requirements, KPIs, etc. also need to be addressed. Air-interface design/selection is based on criteria such as throughput, delay, capacity, and spectral efficiency reliability. Thus, due to availability of multiple air-interfaces, there would be a need for harmonisation at various layers including PHY, MAC and/or PDCP. There are two key elements used in harmonisation: orthogonality and synchronisation. The former is about the absence of cross talk while the latter is about clock commonality. LTE-A uses orthogonal frequency division multiplexing (OFDM) which is suited for some but not all applications. The way to handle this is by using existing OFDM with modifications, enhancements, different numerologies and/or coexisting different waveforms to cater for entire ranges of requirements.

Partial orthogonality is being used in 5G air-interface development. Some of the candidate technologies for this are:

- DFTs-OFDM (Discrete Fourier transform-spread-OFDM)
- CP-OFDM (Cyclic Prefix-based OFDM)
- FBMC (Filter Bank Multi-Carrier)
- UF-OFDM (Universal Filtered Multi-Carrier)

However, it seems that the initial phase of standardisation will focus on OFDM and DFTS-OFDM.

It is worth noting that despite all efforts to harmonise the lower layers of the air-interface of NR and LTE-A, there might be limitations due to the non-backward compatibility of 5G NR. Thus, PDCP may be the layer where harmonisation takes place.

10.5 5G Radio and Core Network

Some of the key RAN design requirements include ability to scale, multi-connectivity/ D2D, centralised processing, flexibility, wider spectrum operability, high energy efficient, network slicing, and future-proofing. It should be able to provide diverse services with more stringent QoS at lower costs. Some of the elements to be considered are given in the following.

10.5.1 5G NR

NR Access Technology is not backwards compatible with LTE and is operable from sub-1 to 100 GHz. 5G NR is a global wireless standard for the new 5G air-interface that has evolved in 3GPP Release 15 and beyond having the ability to connect everything (IoT/IoE) and giving a fibre-like experience. 5G NR is based on OFDM technology and its aim is to provide devices (both sensors and mobiles) in a 100% connected environment. We have already seen the benefits of OFDM in previous chapters. Release15 is for 5G NR deployments planned in 2020 while some important elements will be ready for roll-out in 2018/2019. Release 16 is will cover all elements/use cases for IMT 2020 submissions. It will cover all three use cases (eMBB, m-MTC, and u-RLLC).

OFDM-based waveform and scalable numerology brings a diverse spectrum of bands under 5G NR – sub-6 GHz frequencies to spectrum bands above 24 GHz including licensed, shared licensed and unlicensed spectrum.

10.5.2 5G NGC (Next Generation Core)

With every passing generation from 2G to the upcoming 5G networks, the core network has been increasing its role to make the technology succeed. The role has evolved from a bare minimum CS in 2G, to PS in the form of SGSG and GGSN in 2.5G and 3G, to further use in LTE all packet traffic flow, and, finally, in 5G to fulfil requirements and get the desired performance.

As seen before, the biggest difference between LTE and 5G is the convergence of various technologies – multiple devices (from technologies), applications, requirements, and performances. In simpler terms, there are more devices using less data, e.g. in IoT and low latency for critical applications. All this will impact the core network leading to the development of a new core network called 5G NGC or 5G Next Generation Core.

Virtualisation is already being implemented in LTE networks (making them 5G ready), but with NFV, SDN, and Cloud, it will make 5G networks as desired – scalable, flexible, and dynamic. These three features are in the process of being implemented in the 4G/LTE networks worldwide. One very interesting feature that will exist in 5G networks is network slicing.

10.5.3 Network Slicing

Based on principles similar to NFV and SDN, it is a virtual network. By using a common shared physical infrastructure, virtual networks can be created on top of it. And each of these virtual networks can be used for customers/industry/DSP, etc. to fulfil specific requirements. A whole network is seen to be 'cut' into 'pieces' or 'slices' to cater for individual needs; the process is known as network slicing. Slicing is fundamentally done on the core network and is delivered by the RAN network. An example of network slicing is shown in Figure 10.2. A lot of work on network slicing has been done and is reported in 3GPP TR22.891.

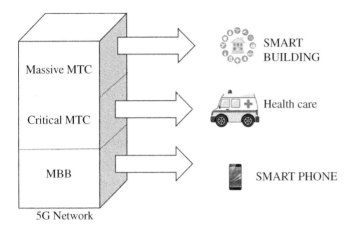

Figure 10.2 Example of network slicing.

Let us now consider a simple example of network slicing: A person is driving a vehicle along a road. V2V communication is working (e.g. for driver/vehicle security) that needs low latency and low throughput while at the same time the driver is enjoying video streaming that needs a high throughput with normal latency. Both applications are running on the same network/DSP/operator. In this case, the same E2E is logically working in two slices for the two applications.

3GPP TR22.891 specifies the following:

- The operator shall be able to create and manage network slices that fulfil required criteria for different market scenarios.
- The operator shall be able to operate different network slices in parallel with isolation that e.g. prevents data communication in one slice to negatively impact services in other slices.
- The 3GPP system shall have the capability to conform to service-specific security assurance requirements in a single network slice, rather than the whole network.
- The 3GPP system shall have the capability to provide a level of isolation between network slices which confines a potential cyber attack to a single network slice.
- The operator shall be able to authorise third parties to creat and manage a network slice configuration (e.g. scale slices) via suitable APIs, within the limits set by the network operator.
- The 3GPP system shall support elasticity of the network slice in term of capacity with no impact on the services of this slice or other slices.
- The 3GPP system shall be able to change the slices with minimal impact on the ongoing subscriber's services served by other slices, i.e. new network slice addition, removal of existing network slice, or update of network slice functions or configuration.
- The 3GPP system shall be able to support E2E (e.g. RAN, CN) resource management for a network slice.

Networks across the world are moving towards virtualisation in core networks – NFVs and SDNs. Separating the control and user planes makes implementation of network slicing in core networks easy. This separation allows independent scaling of resources,

Cloud deployment/migration, functionality choice for various applications, location choice for the user plane, etc. On the RAN side, a configuration set is applied to the user and control plane functions. The coordination between various slices is done by common control functions.

10.6   5G Spectrum

The spectrum has been at the core of all technology evolution from 1G to 4G. Lower frequencies cover more areas, while at the same time equipment, filters, etc. developed in certain frequencies also become important when deciding which frequency bands need to be worked on. Characteristics such as compliance, security, effectiveness, etc. are taken into consideration.

The spectrum for 5G can be broadly divided into three blocks:

- Sub-1 GHz: Used for 5G Broadband coverage, IoT.
- 1–6 GHz: Used for both coverage and capacity applications.
- Beyond 6 GHz: Ultra-high speed mobile broadband.

However, the demand for the total spectrum needed in each country/region would be different. The requirements are calculated based on the procedure defined in ITU-R M.1768. Four basic steps are involved:

- definition of services
- market expectations
- technical and operational framework
- spectrum calculation algorithm.

An insight into the national spectrum requirements of some countries is given in Table 10.4.

10.6.1   WRC-15

WRC-15's action on the low- and high-range spectrum for 5G are shown in Table 10.5 and Table 10.6, respectively. The factors to be considered by WRC-19 include:

- IMT provides telecommunications services on a global scale, and demand for IMT is growing.
- IMT systems help support global economic and social development.
- Ultra-low latency and very high bit rate applications needed for the next generation of IMT require large contiguous blocks of spectrum available at these frequencies.
- Spectrum propagation characteristics in these bands are well suited to enable advanced antenna technologies such as MIMO and beamforming.

10.6.2   Frequency Planning Inputs for 5G

As discussed before, the frequency range in 5G will wary from sub-GHz bands to 100 GHz.

Table 10.4 National spectrum requirements of some countries.

	US	Australia	Russia	China	GSMA	India	UK
Estimation year	Until 2014	Until 2020	2020	2015,2020	2020	2017, 2020	2020
Spectrum requirements	Additional 275 MHz by 2014	1081 MHz in total	1065 MHz in total	570–690 MHz in total by 2015 1490–1890 MHz in total by 2020	1600–1800 MHz in total for some countries	Additional 300 MHz by 2017. Another 200 MHz by 2020	775–1080 MHz in total for low 2230–2770 MHz in total for high
Methodology	Original	Original	Original		Complementary to Rec. ITU-R M.1768–1	Original	Rec. ITU-R M.1768–1

Table 10.5 WRC-15 action on low-range spectrum.

Band	Action	Future activity
470–698 MHz	14 countries allocated for mobile and identified for IMT	WRC-2023 will study this band for IMT in Region 1
698–790 MHz	Identified for IMT in Region 1. Creating near global identification	
1427–1518 MHZ	Some or all of the band identified for IMT	Study of IMT compatibility for 1452–1492 MHz in Regions 1 and 3
3300–3400	Allocated for mobile and identified for IMT use in almost 50 countries	
3400–3600	Globally allocated for mobile and identified for IMT	
3600–3800 MHz	Four countries identified 3600–3700 for IMT	
5150–5925 MHz	Tabled as an agenda item for WRC-19	Whether to extend or revise mobile allocation to portions of the band

Source: Reproduced with permission of ITU.

Table 10.6 WRC-15 action on high-range spectrum (for consideration at WRC-19).

Band (GHz)	Action	Future activity
24.25–27.5	Existing allocation to mobile on primary basis	Sharing and compatibility studies for IMT identification
31.8–33.4		Consider a mobile allocation and IMT identification
37–40.5	Existing allocation to mobile on primary basis	Sharing and compatibility studies for IMT identification
40.5–42.5		Consider a mobile allocation and IMT identification
42.5–43.5	Existing allocation to mobile on primary basis	Sharing and compatibility studies for IMT identification
45.5–47	Existing allocation to mobile on primary basis	Sharing and compatibility studies for IMT identification
47–47.2		Consider a mobile allocation and IMT identification
47.2–50.2	Existing allocation to mobile on primary basis	Sharing and compatibility studies for IMT identification
50.4–52.6	Existing allocation to mobile on primary basis	Sharing and compatibility studies for IMT identification
66–76	Existing allocation to mobile on primary basis	Sharing and compatibility studies for IMT identification
81–86	Existing allocation to mobile on primary basis	Sharing and compatibility studies for IMT identification

Source: Reproduced with permission of ITU.

10.6.2.1 Spectrum Refarming

A lot of spectrum is being currently used for 2G and 3G networks (Figure 10.3). Frequencies (generally) below 3 GHz will let be 'relieved' from current technologies 2G/3G and used for 5G networks. This is a way to put the spectrum to better use without increasing the total spectrum.

10.6.2.2 New Frequency Bands

This is the result of WRC-15 where at least 19 candidates of frequency bands proposed for IMT-2020 (5G) are under 6 GHz. The final decision on the proposals will be in WRC-19. For frequencies beyond 6 GHz, very interesting work is going on in some frequency bands. This will include 14, 18, 28, 60 GHz, etc.

10.7 Network Planning Consideration in 5G

The network planning process for 5G radio and core networks is similar to that in LTE networks. However, the massively complex 5G networks in terms of multi-RATs, the inclusion of Cloud and IoT, along with virtualised network elements will be key differences. As 5G networks are yet to be deployed, let us consider the various aspects of planning the networks (RAN and core).

The 5G networks will be planned to fulfil the expectations for which they are being developed. The challenges to be considered are data rate, latency, capacity, massive device connectivity, cost/energy consumption, and user experience. Some other important elements to consider are densification of the network, big E2E spectrum and massive MIMO.

QoS given to the users will define the success of 5G networks. Network planning will be done with QoS in the primary frame focusing on VoIP traffic and real-time traffic. Data rates and latency will be a measure of QoS. Latency levels will be less than ~1 ms. The requirements to delay in control and user planes will be more rigid in the 5G network as compared with LTE.

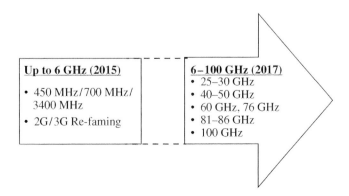

Figure 10.3 Frequency overview in 5G.

For radio network planning, let us consider the changes from the perspective of planning in LTE networks and how to plan networks to handle the challenges. For each challenge, the network requirements and impact will be different.

10.7.1 High Data Rate Networks

As compared with LTE, data rates will be increasing a thousand times more. This will impact network assumptions for cell types (macro/micro cells), spectrum requirements, air-interface, etc. Planning engineers should also understand that ~70% of the sites will be small cells located indoors – and in huge numbers. This will lead to massive connections (both devices and interfaces) leading to plans that will include mitigating challenges related to energy consumption, impact on core networks, etc. Capacity planning and frequency planning will automatically play an important role in handling the requirements of data rate, capacity, quality and latency. To get a network with all parameters playing their part right at lower costs will be a planning challenge. In 4G-LTE, the frequency selection and planning was relatively easy for under 6 GHz but now the range of frequencies increases to 6–100 GHz, thereby increasing the frequency planning challenge. As the propagation conditions will impact more at higher frequencies, small cell planning will play an important role in delivering quality services to the users.

10.7.1.1 Small Cells and HetNet

Mentioned in Release 9 for 3GPP, these mini base stations that cover small areas are usually known as pico, micro, and femto cells. The basic differentiation of these cells is given in Table 10.7. Femto cells are also known as home eNodeB and are small cells that are used in LTE networks. They consume low power, are self-optimising and have the ability to use IP back haul. These can be installed both in indoor or outdoor locations but are usually for indoor. Small cell densification in 5G will lead to ultra dense networks. Small cells will play an important role in 5G networks as they will be able to provide an increase network data capacity, improve device performance, and will be available to operate at a frequency range beyond 6 GHz. Heterogeneous networks (HetNet) will also play a significant role in 5G networks. The mobile networks will have a mix of macro cells, small cells, Cloud/virtual RAN, DAS, Wi-Fi, etc. to handle the capacity, coverage and quality of the network. HetNet will be an integral part of the 5G network evolution. In CRAN, processing of some segments (baseband and channel) takes place at centralised data centres. This helps in dynamic resource handling and efficient resource utilisation – leading to higher data capacity than traditional RAN. Challenges such as security, interference, complexity large scale deployment, etc. need to be worked on.

Table 10.7 Cell differentiation.

Category	Femto cells	Pico cells	Micro cells	Macro cells
Output power	1–250 mW	250 mW–1 W	1–10 W	10–50+ W
Cell radius	10–100 m	100–200 m	0.2–2 km	8–30 km
Users	1–30	30–100	100–2000	2000+
DAS integration	No	Yes	Yes	Yes

Due to the massive number of small cells used in 5G, the handover scheme will play an important role in balancing the load on the backhaul network. The tradeoff between hard and soft handovers need to be considered. Unlike in LTE, where handovers were between two mobiles, in 5G, it will be between two devices and elements like D2D communication will come into play. Also, usually macro cells operate in licensed bands while small cells operate in unlicensed bands. Co-channel interference will exist in overlapping cells of the same type. Network planning should cater for such problems. Also, in ultra dense networks, another crucial issue will be of handover. This can be countered by well-defined handover planning. Here, a network planned using software defined radio (SDR) technology can help.

10.7.1.2   SDR and Cognitive Radio

An SDR can be defined (IEEE P1900.1) as a radio in which some or all of the physical layer functions are software defined. Here software defined refers to use of software processing within the radio system or device to implement operating (but not control) functions. (Software for this definition refers to modifiable instructions executed by a programmable processing device.) This radio has some basic hardware components and is built on programmable chips. These chips perform functions such as modem, multiplexer, framing, scheduling, etc. An ideal SDR is based on open architecture and can operate in multiple bands and modes. Some key benefits of SDR are:

- Possible to add new features and capabilities through remote/over-the-air software upgrades without working on hardware.
- Low OPEX due to common radio platform.
- More revenues – quick capability and capacity upgrades.

Cognitive radio or CR is an extension of SDR. These radios can improve spectral efficiency and spectrum utilisation. What is additional to SDR in CR? CR can sense the environment and can change operational behaviour by using technologies such as SDR and adaptive radio.

Another aspect is the control and user plane split. Macro cells work in a licensed spectrum and give coverage to big areas while small cells work in a licensed/unlicensed spectrum giving coverage in small areas. Small cells can be dynamically switched off or can be put to sleep when not needed. This can make the network more energy efficient. Coverage and capacity can be dealt with separately enabling independent mobility of the control plane and the user plane in areas where coverage overlaps.

Some challenges can be identified on the core side – NFV/SDN. With millmetre wave technologies becoming involved in the 5G network, high precision devices will be required. Another challenge to SDN is posed by the heterogeneity of the network. It can pose problems related to mobility and resource management affecting the QoS of delivery. The interaction between virtual RAN and SDN will pose problems as the requirements of RAN in terms of performance are far superior to that of SDN.

At this juncture, when 5G standardisation work is only a couple of years away from completion, we can only look at certain challenges in network design. Network planning in such complicated multi-RAT, multi-core technologies and their integration at low cost will be a challenging task. However, development of new technologies, processes and tools will surely make 5G network design an interesting field to work in.

Appendix A

IoT (Internet of Things)

Jeevan Talegaonkar

IOT Business Consultant, Ericsson, India

A.1 Introduction

The Internet of Things (IoT) is a term which is analogous in a sense to M2M (Machine-to-Machine) communication. However, when the difference is referred to in a reference frame of entity-relationship diagrams, then, it becomes clear that M2M symbolises a one-to-one relationship between two entities or one-to-many relationships or many-to-one relationship, but, IoT symbolises many-to-many relationships among many entities. M2M usually does not have a Cloud-based middle layer that crunches data coming in from devices and sensors over cellular and noncellular networks, unlike IoT. M2M addresses isolated instances of device-to-device communication instead. This reveals that the complexity in the real world is addressed through interoperability of mutually exclusive elements in IoT to bring in elemental interdependence, interrelatedness and interactivity together for a purpose, which often is to solve a critical technology, process or a business problem. In other words, IoT embraces diversities. In Figure A.1, M2M and IoT are represented using entity relationship diagrams.

A.1.1 Constituents of IoT

IoT is an ecosystem consisting of Sensors–Devices–Gateways–Networks–Tools–Cloud–Platforms–Data Analytics Engines–Partner Onboarding and Transaction Settlement Software, Vertical Applications and Business Services Creating a Virtual Market Place.

IoT is a digital representation of the physical world; it is more horizontal in nature as there will be a free flow of data from devices and sensors of varied types used for an array of applications or specialised verticals, through specialised gateways, over various IP protocols through various cellular/noncellular network connectivity layers towards a Cloud-based middle layer, where this amount of big data will be analysed. Because of its horizontal nature and coexistence of various technologies, a collaborative Cloud-based

Fundamentals of Network Planning and Optimisation 2G/3G/4G: Evolution to 5G, Second Edition. Ajay R. Mishra.
© 2018 John Wiley & Sons Ltd. Published 2018 by John Wiley & Sons Ltd.

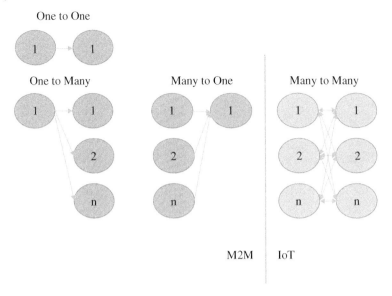

One to One

One to Many   Many to One   Many to Many

M2M   IoT

Figure A.1 M2M and IoT representation.

IoT platform will become a key that will have to be unifying in nature. This approach will generate loads of data, hence data-sorting, tabulation, analytics, retrieval, correlation and its representation will become key to informed decision making as there will not be any 'Cold Data' left in the system or organisations.

Traditional business will also digitalise in part or as a whole, drawing upon internally derived efficiencies in order to shift a part or the whole of the business models towards establishing high touch with the end customer-consumer. Collaboration will evolve among competitors as well since traditional value chains will collapse owing to a degree of disruption in every industry in pockets or as a whole. Business efficiencies will shift from isolation to collaboration, sharing economy will rise, and consortiums will be formed. This will strengthen the user experience; service; application and software differentiations thus would take the game away from physicalising to virtualisation and make future innovations in the following areas more relevant: zero touch device onboarding, low latent high speed sliced connectivity through virtualised core, machine learning, data cloning, blockchain-based storage, hypervisor-based multitenant Cloud platforms, artificial intelligence-based analytics, immersive applications, real time settlement engines and virtual market places with 'customer centricity' at their soul. Businesses working in silos will soon find the world around them changed so much so that their sheer existence might become challenging if digitalisation is not embraced quickly enough; they may find their value chains collapsing quickly with the advent of IoT.

A.1.2 Force Behind IoT

IoT as a technology is the need of the hour as many things have started getting connected now going beyond places and people. With the advent of wireless communication technology, we had started connecting places and people but, as they got connected, they started using the technology beyond voice and data for communication requirements of the machines. For example, in a manufacturing set-up, complex system links were failing and going undetected for daysr, then GSM modems were designed to be used as interjection points on critical links to send alarms upon link failures to help businesses know the points of failures, leading to intervention and repair. This reduced the time and material cost which was otherwise required to investigate such faults. With intervention of GSM modems, the cost of fault finding has come down, and the mean time to repair (MTTR) has improved significantly thus bringing in more efficiencies in the manufacturing value chain. With the latency of wireless technology going down and speeds improving many fold from GSM to 3G to LTE to 5G, more complex industrial and manufacturing problems can potentially be resolved going beyond M2M to IoT adoption.

As it is said, necessity is the mother of invention, and so it is true about the IoT solutions; they find way through the problems faced by the manufacturing industry. In other words, the potential for IoT lies in BPR (Business Process Reengineering) opportunities specific to the manufacturing industries.When they stumble across a connectivity void, IoT steps in and the IoT solution becomes a manufacturing environment specific invention over time.

Another trend that has also emerged is that of high speed computing; as the processors became faster, the need to store the data increased. Local or edge-storage became expensive, so common storage or pooled storage started which gave rise to Cloud Computing and Data Storage. Thus, communication, computation, and the Cloud formed the pillars of transformational change as far as IoT was concerned.

With the increased centralised availability of data, there emerged the need to analyse it which gave birth to Big Data over multiple IP connections. Immersive user applications on the other hand called for great user experience necessitating the use of artificial intelligence-based data analytics for consumer and industrial set-ups. Thus, IoT has become a combination of sensors, devices, Ethernet, connectivity, data, analytics, Cloud, applications, etc. This changed the customer and consumer behaviour, and the speed at which business is being done is challenged. This very change is now impacting various industries and thus putting pressure on businesses to invent new models of doing business or digitalise more and more to get closer to the customer/consumer and to get there faster. This pull has come from the behavioural change of customers and consumers which is now giving rise to IoT globally.

A.2 Types of IoT

IoT can broadly be classified as cellular and noncellular. Cellular IoT is where the device or sensor would use a mobile modem/SIM (physical or eSIM) to connect to the Internet over wireless technology (2G/3G/4G/5G). Noncellular IoT is where the connection to the Internet is made using a LPWAN (Low-Power Wide-Area Network) enabled device/ wearable to connect to Wi-Fi, Bluetooth, Zigbee, etc. A sensor or a wearable or a car or a home appliance would have its own IP to enable Internet connectivity as they would traverse over an unlicensed spectrum to tap into the Wi-Fi, Bluetooth, Zigbee network which in turn will connect to a PAN (Private Area Network) or LAN (Local Area Network) and finally aggregate over a gateway to fibre or cellular network.

The choice of a cellular or a noncellular technology will be driven by constraints of IoT application such as cost, battery life, coverage, and criticality. Cellular technologies are under licensed spectrum and noncellular technologies are under unlicensed spectrum, thus the choice will be governed by the reliability requirement of the application or use case as well.

A.2.1   Cellular IoT Technologies

A.2.1.1   LTE-M/LTE-CAT-M1
This is suitable for Low Power, Medium Speed, Wide Area (LPWA) applications, majorly mobility related, such as Connected Car. This is LTE technology using a 1.4 MHz spectrum.

A.2.1.2   NB-IoT: Narrow-Band IoT
This is suitable for low power, low speed, long battery life applications, majorly stationary, such as Water Quality Sensing or Soil Quality Sensing. This is LTE technology using a 200 kHz spectrum.

A.2.1.3   EC-GSM: Extended Coverage for GSM
This is suitable for stationary applications using GSM modems but requires more extended indoor coverage than normal.

A.2.2   Noncellular IoT Technologies

A.2.2.1   Short Range

- BLE (Bluetooth Low Energy)
- Wi-Fi (IEEE 802.11ah)
- Zigbee

A.2.2.2   Long Range

- LoRaWAN
- 6LoWPAN

Cellular IoT technologies will be used where QoS (Quality of Service), reliability, throughput, and security would matter from the application perspective. Noncellular IoT technologies will have a cost advantage. LoRaWAN and 6LoWPAN are gaining importance as outdoor, long range noncellular technologies for street lighting IoT solutions and BLE and Wi-Fi are gaining importance as indoor, short range noncellular technologies for connected home IoT applications.

A.3   IoT Use Cases

IoT will have primarily two types of use cases based on the criticality of operations involved: Critical MTC use cases and Massive MTC use cases. A few examples follow. A fully-fledged Critical MTC use case will come into reality with the low latent, high speed, high reliability 5G network roll-out.

A.3.1 Critical MTC Use Cases

A.3.1.1 Autonomous Vehicles

Driverless vehicles are controlled over a low latent 5G network through tactile feedback. They can be used in mines or sea ports for heavy lifting, etc. Future driverless cars will fall into this category.

A.3.1.2 Connected Traffic Management Including V-V and V-I Interactions

When cars become driverless, they will have machine learning-based reflexes, and any communication V-V (Vehicle-to-Vehicle) between such cars will have to happen over a low latent network. A response of such a vehicle to a message coming in from street infrastructure (Vehicle-to-Infrastructure) will be similar.

A.3.1.3 Remote Surgery

This is an example of a very Critical IoT application and must use a highly reliable, low latent, licensed wireless network preferably 5G with built-in redundancy provisions. The surgeon will have to use a haptic glove to get kinesthetic feedback from surgery.

A.3.1.4 Industrial Automation Based on Artificial Intelligence and Wireless Control of Industrial Robots and Drones

Robots in manufacturing industries run today on fibre backhaul. In the future, they may run on 5G enabled wireless backhaul, thus the network of sensors and devices and robots will be homogenous in a factory environment. There will be high degree of information interchange and interoperability possible using a cohesive IoT platform opening up more artificial intelligence-based automation possibilities.

A.3.1.5 Virtual Reality Based Immersive Event Experience

Virtual reality will be able to get, say, a film closer to the customer and provide the customer with a kinesthetic experience. The same would be the case for any sports event, thus helping raise the target rating point and hits-related IoT revenue.

A.3.2 Massive MTC Use Cases

- connected car
- variable messaging display and environment sensor integration
- smart street lighting
- smart parking
- fleet management
- smart building
- smart metering

Some examples of these IoT use cases are covered in detail below.

A.3.2.1 Connected Car

A car can be connected over a telematics module which will be an interface with an OBD 2.0 port of the car. An OBD 2.0 device with a wireless circuit housed inside (e.g. LTE/Bluetooth/GPS) can transmit the exposed OBD data from the car to the IoT Cloud platform. The data is analysed in the Cloud platform, say, for the odometer reading and a service call is placed on the owner based on the distance travelled rather than the time lapsed from the previous service date. Similarly, a service call enablement for the car could be done based on the alarms received from its critical components such as tyres, batteries, engine, etc. thus creating a new customer service experience. Even car data can be diagnosed proactively to avoid any foreseeable damage.

Once a car gets connected on the IoT platform, its data starts getting analysed and compared with the OEM (Original Equipment Manufacturer) ERP (Enterprise Resource Planning) and Dealer Management System (DMS) data. Then a business logic could be run within the Cloud platform to make many more use cases for the OEM and several ecosystem players to create new revenue streams on a virtual market place by onboarding such partners, e.g. genuine part suppliers, insurance providers, advertising partners, etc. Suppose a critical component of the car has two suppliers, say 'A' and 'B', and the part supplied by 'A' is wearing out faster than that from 'B', then, supplier ranking could be achieved by rating them as 'Tier 2' and 'Tier 1' suppliers, respectively. Similarly, 'after-market sale' of car parts could be augmented by onboarding the suppliers on the same platform. The worn out genuine-part price could be sent immediately with the location of service centre details to the owner's mobile app. thus creating a differentiated service experience for the owner, and at the same time avoiding the possibility of a spurious part getting into the original product. Further, all such sales transactions happening on the Cloud platform could be captured in a real-time settlement engine by the OEM to generate a parallel revenue stream for offering an indirect-channel of sales to its genuine-part suppliers over the virtual marketplace of the IoT platform. Through diagnostics of car data, say, powertrain, any extraneous performance augmentation of an engine part could be noticed and evidenced as a breach of warranty by the OEM. Thus, there could be many use cases of a connected car with an IoT platform. However, the security of such platform remains of paramount importance as its breach could lead the control of the car to be taken for the wrong reasons. Hence, 'Connected Car' is treated as an application on the boundary of Massive and Critical MTC and must run on a cellular IoT platform as the reliability of wireless backhaul remains of paramount importance alongside data security (Figure A.2).

A.3.2.2 Variable Messaging Display (VMD) and Environment Sensor Integration

VMD is used in cities to communicate traffic updates or environmental parameters and emergency messages to citizens and to generate advertising revenue. A digital advertising display (LED) can be erected on an outdoor cantilever. The form factor of the display may vary depending on city requirement (e.g. $1\,m \times 1\,m$ or $2\,m \times 1\,m$) and viewing distance, traffic obtrusion, etc. The display content could be a photo, a text or a video file. Display is connected through a media player which will store and stage the content as per schedule. The media player will connect through a PoE switch to fibre and through a L2 switch to the command and control room where VMD content push software server would reside; traffic data of the city could be analysed and relayed to the outdoor display unit based on Cloud sourcing of GPS content from a public mapping platform. There might as well be a requirement to relay individual messages to a select number of displays in a city or to mass release the messages to all displays in one go. There could be a requirement to integrate output of an environment sensor to the VMD content push software and showcase the results of environmental parameters on the

VMD screen as a ticker. IoT will play its part when such complex integration and correlation of data coming in from different application software will be called for. However, the absence of such data may not pose a direct life threat to the users of such an application, hence such an IoT application is categorised as a Massive MTC type and can use noncellular IoT technology for its functioning; it could use LoRaWAN backhaul instead of fibre or a cellular backhaul (Figure A.3).

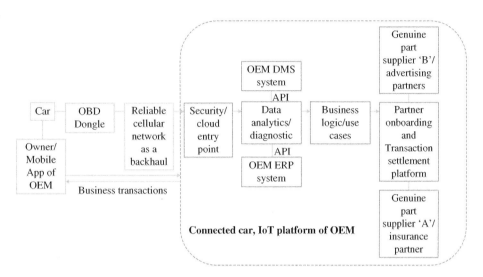

Figure A.2 Connected car, IoT platform of OEM.

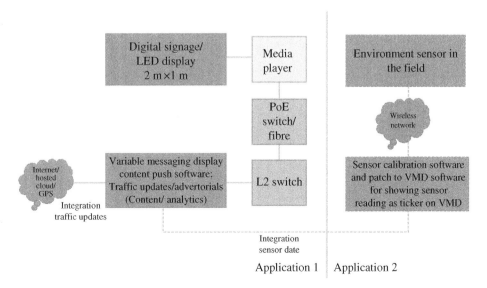

Figure A.3 Architecture of a VMD and environment sensor application integration.

A.3.2.3 Smart Street Lighting

Another example of Massive and noncellular MTC is 'smart street lighting' operated through LED lights and LED controllers over Zigbee or 6LoWPAN gateway. It provides mesh networking through Zigbee controllers, which is based on the 802.15.4 standard. It has a lower bit rate of 250 kbps for low power implementations and allows LED fixtures to connect to each other for distributed control. The mesh network is controlled from a Zigbee gateway or hub that acts as the control centre for the pool of street lights. Using this IoT technology, efficient energy saving is obtained as the lights can be remotely controlled using software features and different pre-settings.

A.4 IoT Value Chain

Devices or sensors form the starting element or bottom of the IoT value chain, then comes the connectivity layer of cellular or noncellular technologies which is followed by an IoT enablement platform consisting of analytics and transaction settlement engines on the Cloud followed by an application layer serving different industrial applications or specialised verticals. The top overarching layer is that of a virtual market place which forms a business service layer handling different business and revenue models.

A higher value is delivered to the customer or business as we move up from the device layer to the business services layer as the bottom of the value chain would be a one-time investment with incremental operational expense requirement. However, applications and business services would touch the end consumers and industries and hence would have higher value to be delivered based on evolving experiential requirement of the customer and consumer. ADM (Application Development Maintenance), LCM (Life Cycle Management), and SLAs (Service Level Agreements) hence become bedrock of such long-tail (multi-year) contracts as innovations and functionalities will reside in business services, software and applications rather than hardware as the consumer touch-points will change significantly with the IoT value chain and ICT led digital transformation. Interestingly, the IoT Volume Pyramid and Value Pyramid are inverse to each other (Figure A.4).

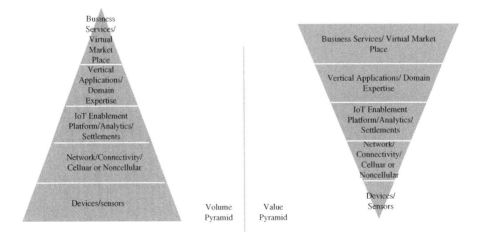

Figure A.4 IoT value chain.

A.5 Phases of IoT Maturity

The following can be considered as the three phases of IoT maturity:

Phase 1: Infrastructure (Communication/Computation/Cloud)
Phase 2: Platforms (Analytics/Functional Virtualization/Applications)
Phase 3: Business Models (Collaborative/Sharing Business Models)

For now, we are in Phase 1 as far as IoT maturity is concerned as it is evolving and embracing changes to succeed in the future. The realisation has started to occur to business leaders and industries as the technologies are maturing and their adoption is taking place in a few mature pockets of industries and geographies. The successes are encouraging others to follow suit. Transformation is happening at all levels but the pace is uneven; this is because of the push–pull dynamics that exist in different industry verticals having different IoT drivers. There is an interesting example from the utilities (power) sector. It was witnessing power theft so that the need to deploy smart meters was felt urgently to arrest revenue loss, namely deployment of smart grids to detect substation link failures, as the losses incurred on account of distribution were much higher than generation or transmission in the power value chain. So, where the business need is felt more pressingly to change the way of working by using appropriate intervention of sensors, low latent connectivity, efficient enablement platforms, suitable applications, settlement engines, or new responsive business service models, there IoT is embraced faster than in other pockets. The same is true for other industrial verticals such as automotive, public transportation, public safety, oil and gas, healthcare, logistics, manufacturing, financial services, retail, media, agriculture, etc.

Another example of the above trend could be in robotics. It is adopted as technology by the automotive industry faster than in other industries, primarily because the pressing need to change was felt by the automotive industry first for lifting heavy parts, painting metal sheets in a hazardous environment, etc. and it had the capital to invest as well. Hence, the development of advanced robotics happened there.

Industrial automation will grow and sharpen further with the advent of low latent, high speed, high bandwidth, more reliable technologies like 5G. Hence, 5G and IoT will together bring in another era of industrial revolution where we will witness Phases 2 and 3 getting traction.

Appendix B

Introduction to MIMO and Massive MIMO

Swapnaja Deshpande

Research Fellow, NorthCap University, Haryana, India

B.1   Introduction

This topic briefly reviews the technology of multiple-input multiple-output (MIMO) in wireless communication. It also discusses the rationale behind why MIMO evolves into massive MIMO with the evolution of wireless technology.

MIMO is a multiple antenna technique used in wireless communication. Multiple antenna techniques are categorised as 'diversity techniques' and 'spatial multiplexing techniques'. In the former technique, the same signal is received through multiple antennas at the receiver or transmitted through multiple antennas at the transmitter thereby increasing transmission reliability. Thus, multiple antennas can be used to provide diversity gain and increase the reliability of wireless links if both transmit and receive diversity are considered. In spatial multiplexing, multiple independent streams can be transmitted simultaneously thereby increasing transmission speed. In the spatial multiplexing technique, maximum achievable transmission speed can be same as the capacity of the MIMO channel, but in the case of the diversity technique the achievable transmission speed is much lower than the capacity of the MIMO channel. There are several different uses of multiple antennas in wireless communication. With channel state information at the transmitter, multiple transmit antennas provide channel gain with transmit beamforming. Herein, a new concept of MIMO, i.e. multiple antennas at transmitter and receiver, is introduced. The wireless communication industry is currently experiencing a prolonged exponential growth in the demand for network traffic with no sign of it slowing down; the same is expected for the number of connected devices. Current networks are reaching their capacity limits with respect to. the physical layer (because of the so-called 'spectrum deficit' or 'data tsunami'), especially in highly populated urban areas with a high density of connected devices.

So, the big question in the wireless communications community is how to increase the network capacity to match the exponentially increasing traffic demand. The total capacity of a wireless network is directly related to the area throughput (in bps per unit area) of the network, which is a combination of three multiplicative factors:

$$\underbrace{Area\ Throughput}_{\text{bps area}^{-1}} = \underbrace{Available\ Spectrum}_{\text{Hz}}.\underbrace{Cell\ Density}_{\text{Cells area}^{-1}}.\underbrace{Spectral\ Efficiency}_{\text{bps Hz}^{-1}\text{cell}^{-1}}$$

Thus, higher area throughput can be, and traditionally has been, achieved by allocating more frequency spectrum (Hz) for wireless communications, increasing cell density (more cells per area), and improving the spectral efficiency (bps Hz^{-1}cell^{-1}). So, with MIMO technology we improve spectral efficiency by increasing the capacity of the wireless communication system without any increase in available bandwidth. In wireless communication, MIMO has become an essential element of different communication standards including IEEE 802.11n (Wi-Fi), IEEE 802.11ac (Wi-Fi), HSPA+ (3G), Wi-MAX (4G), and LTE (4G). Recently, it has been used for Power Line Communication for three-wire installations as a part of ITU G.hn standard and Home Plug AV2 specification. Figure B.1 shows a cellular base station and a Wi-Fi access point, both equipped with multiple antennas which is an example of MIMO technology used in our everyday life.

B.2 Multi-user MIMO

Let us briefly review MIMO systems. Generally, MIMO systems are divided into two categories: single-user MIMO (SU-MIMO), as depicted in Figure B.2a, and multi-user MIMO (MU-MIMO), as depicted in Figure B.2b.

(a)   (b)

Figure B.1 (a) A cellular base station tower with multiple antennas; (b) a Wi-Fi access point with multiple antennas. *Source*: www.linksys.com.

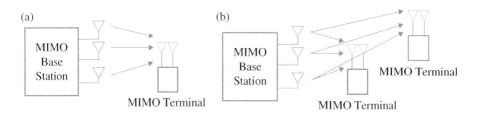

Figure B.2 (a) Single-user MIMO; (b) multi-user MIMO.

In SU-MIMO, the transmitter and receiver are equipped with multiple antennas. Performance gain in terms of coverage, link reliability and data rate can be achieved through techniques such as beamforming, diversity-oriented space–time coding, and spatial multiplexing of several data streams. These techniques cannot be fully used at the same time, thus we typically find a trade-off between them. For example, adaptive switching between spatial diversity and multiplexing schemes is adopted in LTE. The situation with MU-MIMO is radically different. The wireless channel is now spatially shared by different users, and the users transmit and receive without joint encoding and detection among them. By exploiting differences in spatial signatures at the base station antenna array induced by spatially dispersed users, the base station communicates simultaneously to the users. As a result, performance gains in terms of sum rates of all users can be impressive. A major challenge is, however, the interference among the co-channel users. Signal processing in MU-MIMO often aims at suppressing inter-user interference, so spatial channel knowledge becomes more crucial compared with SU-MIMO. In general, by exploiting the spatial domain of wireless channels, MIMO has many key advantages compared with single-antenna systems. Better coverage can be obtained through beamforming, improved link reliability through diversity, higher capacity through spatial multiplexing, and decreased delay dispersion are some of the major advantages of MU-MIMO as compared with SU-MIMO. Now, let us consider different MIMO configurations, i.e. SISO, SIMO, MISO, and MIMO.

B.2.1 Single-Input Single-Output (SISO)

Consider an antenna transmitting a signal x at a frequency f. The signal is faded as it propagates through an environment, which is modelled as a multiplicative coefficient h. The received signal y will be hx. The capacity of a SISO channel is given by:

$$C = \log_2\left(1 + \rho|h|^2\right) \text{ bps Hz}^{-1}$$

The SISO configuration is shown in Figure B.3.

B.2.2 Single-Input Multiple-Output (SIMO)

Consider two receiving antennas with received signals y_1 and y_2 with different fading coefficients h_1 and h_2. The SIMO configuration is shown in Figure B.4. The effect upon the signal x for a given path is called a channel. Herein multiple receive antennas will provide a stronger signal through diversity without any increase in channel capacity. The capacity of a SIMO channel is given by:

$$C = \log_2\left(1 + \rho\sum_{i=1}^{M_r}|h_i|^2\right) \text{ bps Hz}^{-1}$$

Figure B.3 Single-input single-output (SISO).

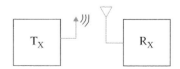

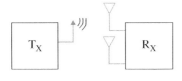

Figure B.4 Single-input multiple-output (SIMO).

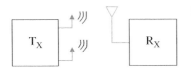

Figure B.5 Multiple-input single-output (MISO).

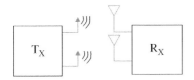

Figure B.6 Multiple-input multiple-output (MIMO).

B.2.3 Multiple-Input Single-Output (MISO)

Consider two transmitting antennas with signals x_1 and x_2 and one receiving antenna with received signal y_1. Here channel capacity is increased but to a smaller extent. The capacity of a MISO channel is given by:

$$C = \log_2\left(1 + \frac{\rho}{M_t}\sum_{i=1}^{M_t}|h_i|^2\right) \text{ bps Hz}^{-1}$$

The MISO configuration is shown in Figure B.5.

B.2.4 Multiple-Input Multiple-Output (MIMO)

Consider two transmitting antennas and two receiving antennas, which will add a degree of freedom. Here, channel coefficients are assumed to be independent and uncorrelated so that signal recovery, i.e. x_1 and x_2, will be easy at the receiver. The capacity of a MIMO channel will increase linearly due to the spatial multiplexing gain. The MIMO configuration is shown in Figure B.6.

In radio, MIMO is the use of multiple antennas at Tx and Rx to improve communication performance. MIMO offers significant increase in data throughput and link range without additional bandwidth or increased transmit power. MIMO achieves this goal by spreading the same total transmit power over the antennas to achieve an array gain that improves spectral efficiency, diversity gain that improves the link reliability. MIMO is different from beamforming and diversity, it improves the SNR (signal to noise ratio) of the wireless communication system. MIMO refers to a particular technology, for sending and receiving more than one data signal, with the same radio channel simultaneously via multipath propagation. Natural multipath propagation can be exploited to transmit multiple independent data streams using co-located antennas and multidimensional processing known as spatial multiplexing. Spatial multiplexing will give multiplexing gain which leads to capacity enhancement of the system as per $C = N. B. \log(1 + SNR)$, where N is the number of transmit antennas, B is the available bandwidth, and C is the capacity of the system.

B.3 Spatial Multiplexing of Deterministic MIMO Channels

Consider, $X \in \mathbb{C}^{n_t}$ as transmitted signal, $Y \in \mathbb{C}^{n_r}$ as received signal, $W \sim \mathcal{CN}\left(0, N_0 I_{n_r}\right)$ as white Gaussian noise, then, the time-invariant channel is defined as: $Y = HX + W$,

Consider, the following MIMO channel model as shown in Figure B.7.

The channel matrix $H \in \mathbb{C}^{n_t \times n_r}$ is given by

$$H = \begin{bmatrix} h_{11} & \cdots & h_{1r} \\ h_{21} & \cdots & \vdots \\ h_{t1} & \cdots & h_{rt} \end{bmatrix}$$

where h_{ij} is the channel coefficient between the ith receiver antenna and the jth transmitter antenna.

The channel matrix is deterministic and known to both transmitter and receiver. It is assumed to be constant at all times. This is a vector Gaussian channel. The capacity is calculated by decomposing the vector channel into a set of parallel independent scalar Gaussian subchannels. Capacity is computed at the receiver through multidimensional signal processing, which is very difficult and rather practically impossible. For enabling easy calculations, a one-to-one channel matrix is realised without interference, i.e. $h_{11}, h_{22}, \ldots$

To satisfy Figure B.8, consider the following equation:

$$\begin{bmatrix} y_1 \\ \vdots \\ y_r \end{bmatrix} = \begin{bmatrix} h_{11} & 0 & 0 \\ 0 & h_{22} & 0 \\ 0 & 0 & h_{33} \end{bmatrix} \begin{bmatrix} x_1 \\ \vdots \\ x_t \end{bmatrix} + \begin{bmatrix} n_1 \\ \vdots \\ n_r \end{bmatrix}$$

where H is a diagonal matrix.

This diagonal matrix can be obtained by using singular value decomposition (SVD). We can represent H using SVD as:

$$H = U \wedge V^*$$

Figure B.7 MIMO channel matrix.

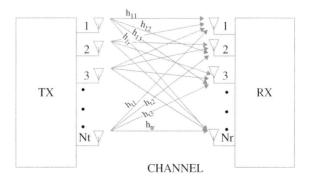

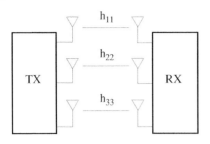

Figure B.8 One-to-one channel matrix.

Figure B.9 Pre-processing at the TX
and post-processing at the RX.

where $U \in \mathbb{C}^{n_r \times n_r}$ and $V \in \mathbb{C}^{n_t \times n_t}$ are unitary matrices and $\wedge \in \mathbb{C}^{n_r \times n_t}$ is a diagonal matrix whose off-diagonal elements are zero and diagonal elements represented by $\lambda_1 \geq \lambda_2 \geq \ldots \geq \lambda_{nmin}$ are singular values of channel matrix H, where, $n_{min} = \min(n_t, n_r)$. Now, the one-to- one channel is realised if we cancel U and V^* from the above equation, i.e. $U\,U^* = I$, $VV^* = I$. Thus, pre-processing at the transmitter and post-processing at the receiver is required, i.e. multiply X by U^* and Y by V. This is summarised in Figure B.9.

In SVD, the input and output relationship is very simple; input is expressed in terms of a coordinate system defined by the columns of V and the output is expressed in terms of a coordinate system defined by the columns of U. The capacity of the time invariant MIMO channel is thus given by:

$$C = \sum_{i=1}^{n_{min}} \log\left(1 + \frac{P_i^* \lambda_i^2}{N_0}\right) \text{bps Hz}^{-1}$$

where P_i^* are water filling power allocations at the transmitter given by:

$$P_i^* = \left(\mu - \frac{N_o}{\lambda_i^2}\right)^+$$

Thus, power is allocated at the transmitter using the water filling algorithm. According to the water filling algorithm, power allocated to the channel is more if channel noise is less and vice versa. This is explained in Figure B.10.

As shown in Figure B.10, power is distributed below μ and μ is selected to satisfy the total power constraint $\sum_i P_i^* = P$. Each singular value, i.e. λ_i represent single channel and each non-zero channel can support a data stream. In this way, the MIMO channel supports the spatial multiplexing of multiple streams. Figure B.11 shows the SVD-based architecture for MIMO communication.

Figure B.10 Power allocation through the water filling algorithm.

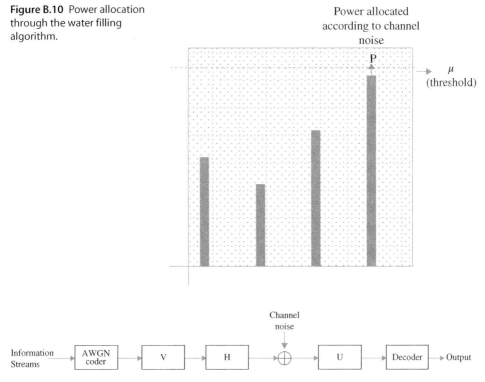

Figure B.11 SVD architecture for MIMO.

B.4 MIMO performance improvements

MIMO results in four major performance improvements, i.e. array gain, diversity gain, spatial multiplexing gain, and interference reduction. Array gain depends on transmit and receive antennas and requires channel state information of the transmitter and the receiver. It is obtained due to increase in average SNR due to coherent combining. Diversity gain can be achieved without knowledge of channel state information. Fading effects in wireless links can be easily mitigated with diversity.

In spatial multiplexing independent data streams are transmitted from individual antennas. Data streams can be easily extracted at the receiver under uncorrelated fading channel conditions. With spatial multiplexing, capacity increases linearly without any additional power and bandwidth. The spatial multiplexing technique makes the receivers very complex, and therefore it is typically combined with OFDM, where the problems created by the multi-path channel are handled efficiently. Interference is reduced by exploiting the difference between spatial signatures of the desired channel and co-channel signal, but for that channel state information of the desired signal is required.

B.4.1 Combined Advantages of MIMO

Throughput, signal level and coverage of a wireless communication system increases with MIMO technology, as shown in Figure B.12.

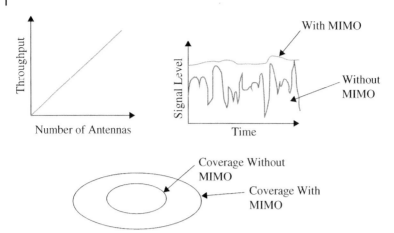

Figure B.12 Combined advantages of MIMO.

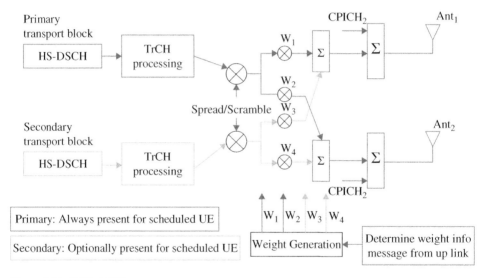

Figure B.13 HSPA+MIMO.

B.4.2 Applications

B.4.2.1 HSPA + MIMO
MIMO was introduced in the form of a double transmit antenna array (D-TxAA) for the high-speed downlink shared channel (HS-DSCH) (Figure B.13).

B.4.2.2 LTE Downlink
The basic concept for LTE in downlink is OFDMA (uplink: SC-FDMA), while MIMO technologies are an integral part of LTE. The modulation modes are QPSK, 16QAM, and 64QAM. Peak data rates of up to 300 Mbps (4×4 MIMO) and up to 150 Mbps (2×2 MIMO) in the downlink and up to 75 Mbps in the uplink are specified (Figure B.14).

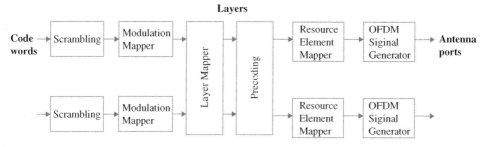

Figure B.14 LTE – downlink.

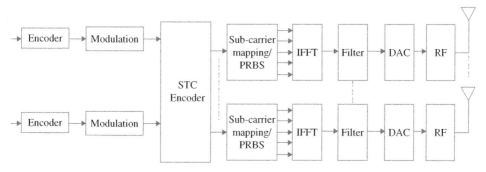

Figure B.15 WiMAX downlink.

B.4.2.3 WiMAX Downlink

WiMAX promises a peak data rate of 74 Mbps at a bandwidth of up to 20 MHz. The modulation modes are QPSK, 16QAM, and 64QAM (Figure B.15).

B.5 Massive MIMO

A further promising way to improve spectral efficiency is now commonly referred to as massive MIMO or large-scale MIMO. The idea is to dramatically increase the number of antennas at the base station (in the order of hundreds to thousands). This technology is based on invoking large scale statistical effects that (in optimal conditions) eliminate small scale fading, interference, and noise from the communication system, as well as focus the transmitted energy only at the intended target. This allows many more UTs to be scheduled than is possible today, hence immensely increasing overall spectral efficiency. Interestingly, the assumption of hundreds of antennas might not be so far-fetched, as a 4G/LTE-A base station at maximal size already has 240 antenna elements. Massive MIMO has attracted much attention from the research community and its potential is being investigated by many. Crucially, many approaches to making the basic premise more practical have been discovered. Advances have been made on the problem of CSI estimation of the hundreds of channels (each BS antenna to each user antenna). The problem of the computational cost for the precoding schemes is being

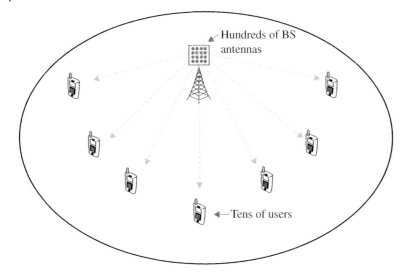

Figure B.16 Basic concept of massive MIMO.

treated and energy efficiency aspects are being looked at. Even hardware impairments, that are certainly important as the cost per transceiver needs to be reduced when using hundreds of them, are being investigated.

B.5.1 Evolution of MIMO Towards Massive MIMO

In both SU-MIMO and MU-MIMO, theoretically, the more antennas the transmitter and/or receiver are equipped with, the larger is the scale on which the spatial domain can be exploited. This leads to better performance in terms of the above-mentioned MIMO advantages. By letting the number of base station antennas grow without limit in MU-MIMO scenarios, the first important phenomenon is that the effects of additive receive noise and small-scale fading disappear, as does intracellular interference among users. The only remaining impediment is intercellular interference from transmissions that are associated with the same pilot sequence used in channel estimation. It is concluded that the throughput per cell and the number of terminals per cell are independent of the cell size, the spectral efficiency is independent of the system bandwidth, and required transmit energy per bit vanishes. So, this is an important direction in which cellular systems may evolve.

Scaling up MIMO provides many more degrees of freedom in the spatial domain than any of today's systems. This rescues us from the situation where the wireless spectrum has become congested and expensive, especially in frequency bands below 6 GHz. In contrast to conventional MU-MIMO with up to eight antennas, we call MIMO with a large number of antennas 'massive MIMO', 'very-large MIMO', or 'large-scale MIMO'. The basic concept of massive MIMO is explained in Figure B.16.

In massive MIMO operation, we consider a MU-MIMO scenario, where a base station equipped with a large number of antennas serves many terminals in the same time-frequency resource. Processing efforts can be mostly made at the base station side, and terminals have simple and cheap hardware. Until now, many theoretical and experimental studies have been done in the massive MIMO context. These studies have shown that massive MIMO can greatly improve spectral efficiency while decreasing radiated output power. Massive MIMO has the potential to meet these future requirements. In frequency bands below 6 GHz, massive MIMO is a candidate for smooth evolution from LTE to pre-5G or so-called 4.5G. In high frequency bands, e.g. in millimetre-wave transmission, using many antennas is a potential solution to overcome high propagation losses. Thanks to the large array gain, massive MIMO is also considered a technique to improve wireless network coverage. From another point of view, radiated power from both base stations and terminals can be scaled down, making massive MIMO also a candidate for 'green communications'.

Appendix C

Blockchain Technology

Priyanka Ray

Principal Consultant, Ericsson, India

C.1 Overview

Driven by digital trend explosions, today businesses are under tremendous pressure from new technologies and competition and are constantly looking for new ways to increase their revenue. The existing value chain is in the verge of collapsing in favour of effective ways to store, exchange and track assets in a secure, efficient and trusted manner in the digital multi-vendor ecosystem. This is the reason why *blockchain* has gained momentum in the enterprise world.

With cloud computing and the Internet, people are able to have global access to computer applications from a variety of devices. Even though cloud computing architecture is decentralised in terms of hardware, it has given an application level centralisation. Currently we are seeing a transition from centralised computing, storage and processing to decentralised architecture and *distributed ledger technology* is one of the key innovations making this shift possible.

Blockchain has the power to drive digital transformation across industries.

C.2 Distributed Ledger Technology (DLT)

A distributed ledger is a type of secure data structure which resides across multiple computer devices, spread across locations or regions with no centralised controller. This is a new way of keeping track of who owns financial, physical or electronic assets.

DLT includes a number of innovations such as database entries that cannot be reversed, automated data synchronisation, privacy and security and transparency. While distributed ledgers existed prior to Bitcoin, the Bitcoin blockchain marks the convergence of a host of technologies, including timestamping of transactions, Peer-to-Peer (P2P) networks, cryptography, and shared computational power, along with a new consensus algorithm.

Fundamentals of Network Planning and Optimisation 2G/3G/4G: Evolution to 5G,
Second Edition. Ajay R. Mishra.
© 2018 John Wiley & Sons Ltd. Published 2018 by John Wiley & Sons Ltd.

C.3 What Is Blockchain?

Blockchain technology is a specific form or subset of DLT which constructs a chronological chain of blocks. Blockchain technology is behind the popular application called Bitcoin. It was first implemented in 2009 by Satoshi Nakamoto as the core component in Bitcoin. Blockchain is a P2P distributed ledger supported by consensus, smart contract, cryptography and other assistive technologies. It is a continuously growing list of records or blocks that are linked and secured by cryptography.

The *Harvard Business Review* defines it as 'an open, distributed ledger that can record transactions between two parties efficiently and in a verifiable and permanent way'.

A block refers to a set of transactions that are bundled together and added to the chain and has a given number of transactions.

The different types of blockchain are permissionless or public (such as Bitcoin or Ethereum) and permissioned or private (such as different Hyperledger frameworks). Any trustless participant node can join a permissionless blockchain network while a permissioned blockchain requires pre verification of the participating nodes within the network.

A permissioned blockchain is chosen for trusted participants such as in supply chain while a permissionless blockchain is chosen when a network can commoditise trust where participants can transact without having to verify each other's identity such as in Bitcoin.

Each block in a blockchain contains a hash pointer that links to the previous block, a timestamp, a nonce and transaction data. So, by design, it is difficult to modify the data of a block in a blockchain and is an example of distributed computing system with *Byzantine fault tolerance*. By storing data across the network, blockchain eliminates the risk of *node failure vulnerability*. It has no central point of failure and data are more transparent to all the parties in the network (Figure C.1).

Blockchain as a Data Structure

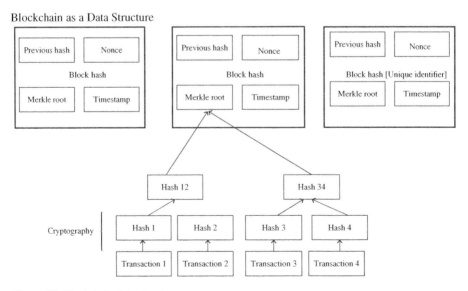

Figure C.1 Blockchain data structure.

The blockchain database consists of two kinds of records: transactions and blocks. Blocks hold valid transactions that are hashed and encoded in a *Merkel tree*. A transaction is a record of an event, cryptographically secured with a digital signature that is verified, ordered, and bundled together into blocks in the blockchain. Each block includes the hash of the previous block, thus linking the two. The linked block forms the chain. This linking iteration confirms the integrity of the previous block all the way to the first block or the genesis block. A smart contract within a blockchain may be programmed to automatically execute a transaction upon meeting predefined criteria.

The Merkle tree, also known as a binary hash tree, is a data structure that is used to store hashes of the individual data in large datasets in a way to make the verification of the dataset efficient. It is an anti-tamper mechanism to ensure that the large dataset has not been changed.

The blockchain security method is public key cryptography. The public key is the address on the blockchain while the private key is like a password that gives the owner access to their digital data. Thus, data in a blockchain is considered immutable and incorruptible.

Blockchain technology removes the need of intermediaries since data is distributed among all the participant nodes. Every participant or node in the blockchain can keep a copy of the data and they are updated automatically every time a new transaction occurs.

The most important key features of a blockchain are P2P network, shared ledger, consensus algorithm, privacy, and smart contract.

P2P consists of computer systems which are connected to each other via the Internet without a central server. Peers contribute to the computing power and storage that is required for the upkeep of the network. The P2P network has no single point of attack or failure and hence is considered to be more secure than a centralised network. A permissioned P2P network has to guarantee uptime and requires a high level of quality of service while the permissionless P2P network has some nodes offline.

Immutability or unchanging over time is the most powerful reason for deploying blockchain-based solutions across a variety of industries such as banking and financial services, insurance, communications, manufacturing and technology, energy and utilities, retail and consumer goods, healthcare and life sciences, and government. Once the data is written on a block it is difficult to change or if changed then it will be visible to all. It is extremely difficult to change a transaction in a blockchain, since each block is linked to the previous block by including the hash of the previous block. The hash includes the Merkle root hash of the all the transactions in the previous block. If a single transaction is changed, the Merkle root hash and the hash contained in the changed block, and so all the subsequent blocks, need to be updated to reflect the change. This discrepancy would be noticed if the hash is recalculated. This feature of blockchain makes it suitable for accounting, financial transactions, identity management, asset ownership, management and transfer, etc.

Smart contracts are computer programs/codes stored in a blockchain network (on each participant's database) that executes predefined actions if the conditions are met or based on a trigger of events. Smart contracts provide the language of the transaction and felicitate the exchange and transfer of anything valuable such as money, content, assets, etc. They are self-verifying and self-executing agreements that work autonomously.

dApp are blockchain-enabled websites and smart contracts allow them to connect to the blockchain network. The traditional web applications use HTML, Javascript, etc. to render a page and uses API to connect to the backend database:

Frontend → API → Database

dApp are similar to the traditional web applications; the front end uses the same technology to render a page. The main difference is the calling of the back end blockchain platform; it uses a smart contract to connect to the database:

Frontend → Smart Contract → Blockchain

A blockchain dApp should have the following criteria:

- Application must be open source.
- Application data and records of operation must be cryptographically stored.
- Application must use cryptographic token.
- Application must generate token.

Consensus in a network means the agreement between all the network participants as to the correct state of the data. Consensus leads to all the participants sharing the same data. Thus, the consensus algorithm ensures that the data on the ledger is the same for all the participant nodes and prevents malicious nodes from manipulating the data.

Different types of consensus algorithm are: Practical Byzantine Fault Tolerance (PBFT), Proof of Work (PoW), Proof of Stake (PoS), Delegated Proof of Stake (DPoS), Proof of Burn (PoB), Proof of Capacity (PoC), Proof of Elapsed Time (PoET), and Proof of Authority (PoA). For example, Bitcoin uses PoW as a consensus algorithm while Hyperledger uses PoS.

The *PoW* consensus algorithm involves solving a computational puzzle to create a new block in the blockchain network. The process is known as mining and the nodes in the network involved in mining are called miners. Here the miners compete to add the next block to the chain by racing to complete a very difficult cryptographic puzzle. The first who solves the puzzle gets rewarded with tokens. PoW requires a huge amount of energy and the latency of transaction validation is high. POW is 51% susceptible to network attacks, i.e. the blockchain can be attacked by miners who have more than 50% of the network's computing power.

In the *PoS* consensus algorithm, the nodes are called validators and instead of mining the blockchain, the nodes validate the transactions to earn a transaction fee. All the coin exists from day 1, so no mining is required. The nodes are randomly selected to validate a transaction and the probability of selection of the node depends on the amount of stake held. So, if node X owns two coins and node Y owns one coin, node X is twice as likely to be called upon to validate a block of transactions. The specific implementation of PoS can vary, depending on the use case, or as a matter of software design. Instances include Proof of Deposit and PoB. The PoS algorithm saves expensive computational resources that are spent in mining under a PoW consensus regime.

PoET is developed by Intel. Hyperledger Sawtooth implementation uses the PoET consensus algorithm. It is a hybrid of a random lottery and first-come first-served basis. Each validator is given a random time to wait and the validator with the shortest wait time is made the leader and the leader gets to create the next block in the chain.

PoA is a consensus algorithm which can be used for permissioned ledgers. It uses a set of authorities, which are designated nodes that are allowed to create new blocks and secure the ledger. Ledgers using PoA require sign-off by a majority of authorities in order for a block to be created.

C.4 Blockchain Versus Database

A database uses client server architecture. A client has permission on their account to change entries that are stored on a centralised server. Whenever a client accesses a database, they will always get the updated version of the database entry. In a database, data can easily be deleted or modified. There is database administrator who has permission to change any part of the data or the structure. They are designed for centralised controlled architecture.

In the blockchain database, each participant/node maintains and calculates new entries into the database. All nodes work together ensuring that they a consensus on the data, providing in-built security for the network. A blockchain is a write only data structure and the new entry gets appended to the end of the ledger. There are no administrative functions or permissions in a blockchain that allow the editing or deletion of a block. Blockchain is designed for decentralised architecture (Figure C.2).

Blockchains are well-suited as a system of records for certain functions, while a centralised database is entirely suitable for different kinds of functions and use cases (Table C.1).

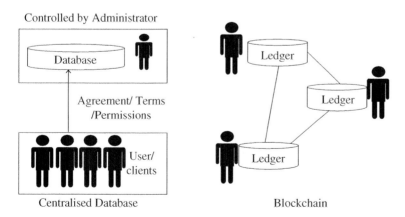

Figure C.2 Difference between centralised database and blockchain.

Table C.1 Difference between centralised database and blockchain.

Feature	Blockchain	Database
Decentralised control	• Allows different parties who do not trust each other to share information without central administrator • Eliminates risk of centralised control	• Trust is put on an administrator who has centralised control • Costly option
History	• Keeps information from the beginning of the chain till present (history of themselves) • Data is immutable	• Centralised databases keep information for a particular period of time (snapshot of a moment) • Very costly option
Performance	• Slow	• Fast
Confidentiality	• Permissionless blockchain is read and write uncontrolled • Permissioned blockchain is read and write controlled	• The same as blockchain

C.5 Telcos in Blockchain

Telcos worldwide are slowly realising the significance of blockchain technology and now are increasingly investing in blockchain development and implementation. They are entering into partnerships or collaboration agreements with technology companies and investing in funds dedicated to blockchain technology research and development.

Telcos across the globe have invested in blockchain technology projects. For example, Sprint partnered SoftBank and a software company TBCASoft to develop a blockchain solution for telco carriers; NTT DOCOMO Ventures Inc. along with ORIX Corporation, ORIX Bank Corporation, The Shizuoka Bank, Ltd., and NTT Data Corporation joined together to fund research for the development of new financial services using blockchai. Verizon is working on a blockchain application to store passcodes for digital content. Orange Silicon Valley's initiative, ChainForce, is to facilitate collaboration between companies and start-ups to develop new blockchain technologies and use cases. Orange Digital Ventures invested in a blockchain start-up, Chain, which is developing enterprise-grade blockchain solutions for the financial industry and other transactional services. Du announced a pilot programme which uses blockchain technology to facilitate the secure transmission of electronic health records. Telstra is researching blockchain companies who can assist them in profiting from the mobile banking boom.

Potential use cases are shown in Table C.2.

C.6 Blockchain in Other Industries

Industries such as finance, healthcare, retail, manufacturing, etc. are all investing in blockchain technology to reap its benefit and compete in the growing market and with digital trends. Blockchain technology has energised these industries globally. The

Table C.2 Telco use case examples.

Identity management services

New revenue streams could be generated by offering blockchain identity management services to subscribers and business partners

Data management services

A Telco could also use blockchain technology to provide data storage and verification services to subscribers

Fraud prevention services

'Roaming fraud' and 'identity fraud' could be mitigated through the use of blockchain technology

Mobile wallets

Telcos could partner with other service providers to offer their subscribers 'international mobile wallets'

Internet of Things

This allows for highly secure peer-to-peer self-managed mesh networks using a sufficiently large number of nodes

Smart cities

Public blockchain services

Digital asset transactions

Implement micropayment-based business models for digital assets

5G

Blockchain technology to provide subscribers with a swift and reliable 5G service

Roaming

This enables Telcos to automatically enter into 'micro contracts' for roaming services

Connectivity provisioning

Telcos could use blockchain technology to make public Wi-Fi more readily accessible to subscribers

---

concept has brought a disruption in them. So, the companies are standardising the technology, talent, and platforms that will drive future initiative and then coordinate and integrate multiple blockchains working together across the value chain.

> Blockchain technology continues to redefine not only how the exchange sector operates, but the global financial economy as a whole. (Bob Greifeld, Chief Executive of NASDAQ)

Potential use cases are shown in Table C.3.

C.7 Bitcoin

Bitcoin is digital currency created and held electronically. It is not controlled by anyone. It is created digitally by people from all around the world running software programs that can solve complex mathematical problems. Bitcoins are mined, using computing

Table C.3 Potential use case examples.

Energy, Communications, and Services
- Cargo assurance
- Inventory reconciliation
- Market data subscription
- Wholesale energy supply

Financial services
- Asset leasing
- Cross border payments
- Identity management
- Invoice financing
- KYC
- P2P payments
- Global remittance
- Mortgage record management
- Mortgage fraud management
- Title registration
- Corporate bonds

Healthcare and Lifescience
- Clearing house
- Health records
- Healthcare fraud
- Real-time eligibility check
- Patient registration
- Pharma supply chain transparency

Insurance
- Commercial insurance
- P2P insurance
- Smart contract for flight travels
- Smart contract for term insurance

Automobile
- Contract documents
- Mileage documents

Manufacturing and High Tech
- Digital asset tracking
- Partner management
- Service order shipment
- Supply chain logistics

Retail and Consumer packaged goods
- Distributed market place
- Food safety in supply chain
- Inventory tracking
- Store credit and voucher

power in a distributed network. This distributed network processes transactions made with the virtual currency, making bitcoin its own payment network.

Satoshi Nakamoto, a software developer, proposed Bitcoin. The idea for this was to produce a currency independent of any central authority, transferable electronically and with very low transaction fees

The basis of Bitcoin is blockchain technology. It usesP2P technology with no central authority or banks managing the transactions or issuing of Bitcoin. It is open source and design for the public; nobody owns or controls Bitcoin and everyone can take part.

Bitcoin cryptocurrencies are kept using public and private keys. The public key (comparable with a bank account number) serves as the address which is published to the public. The private key (comparable with an ATM PIN) is not shared and is kept with the owner, and used to authorise Bitcoin transmissions.

The individuals who are part of the Bitcoin network which has the governing computing power, also known as miners, are rewarded by Bitcoins based on their work done to validate a transaction. New Bitcoins are released to the miners at a fixed and periodically declining rate, so that the total supply of Bitcoins approaches 21 million.

Bitcoin mining is a process which involves solving a difficult mathematical puzzle to find the next block in a blockchain and receiving a reward in the process. With increase in Bitcoins there is increase in the difficulty of the mining process, i.e. the computing power is increased. The mining difficulty began at 1.0 with Bitcoin's debut back in 2009; at the end of the year, it was only 1.18. By April 2017, the mining difficulty reached 4.24 *billion*. Initially, an ordinary desktop computer was enough for the mining process; now, to combat the difficulty level, miners use faster hardware such as Application-Specific Integrated Circuits (ASIC) or Graphic Processing Units (GPUs), etc.

As with everything else, Bitcoin is not risk averse. There are number of risks that should be considered:

- *Regulatory risk:* Since Bitcoin is a strong rival to government currency, a lot of risks are possible, such as money laundering, black market, illegal activities, or tax evasion.
- *Security risk:* Bitcoin is a high security risk since it is a digital exchange of digital currency so Bitcoin are at high risks from hacker, malware and operational glitches.
- *Fraud:* Bitcoin are also prone to fraud and Ponzi schemes. As for any other investments, Bitcoin value fluctuates depending on the market buying and selling.

C.8   Benefits of Blockchain

Blockchain can bring many business benefits. It breaks down barriers that lead to greater efficiency, greater accountability, greater auditability, traceability and reliability, security, and lower cost. This brings about boundary-less collaboration with greater visibility. It creates an autonomous network and enables automatic workflow.

Blockchain gives user empowerment since the users are in total control of their data.

The quality of data is maintained as the data from the blockchain is consistent, timely, and accurate.

Blockchain is highly available and fault tolerant technology makes it perfect for mission critical applications.

Blockchain can be trusted and intermediaries can be removed. Public blockchain is accessible to everyone which ensures transparency. The transactions are immutable.

C.9   Challenges of Blockchain

As for all new technologies, blockchain has its own share of challenges. First, since blockchain is very nascent, it is yet to have widespread applications.

Regulatory compliances are a big challenge in blockchain implementation as many governments have still not made a concrete policy surrounding blockchain-based systems.

Blockchain implementation on a large-scale basis is very expensive and hence is a deterrent for small businesses.

Appendix D

3GPP Releases

Guninder Preet Singh

Product Management, Ericsson, India

D.1 Introduction

The ever-increasing demand for data services and the advent of new applications on devices compelled telecom service providers to look for a technology which can keep pace with such phenomenal evolution and at the same time provides for a common technology platform for a complete ecosystem. LTE technology evolved over the years in 3GPP to address these widely different technical requirements from mobile broadband to IoT.

The following sections describe the journey of LTE RAN in 3GPP specifications from Release 8 to Release 14.

D.2 Release 8

Release 8 introduced LTE for the first time, with a completely new radio interface enabling substantially improved data performance compared with previous 2G-, 3G- and CDMA-based systems. LTE radio systems met user demand for higher data rates and quality of service, were optimised for the packet switch system and avoided fragmentation of technologies for paired (FDD) and unpaired (TDD) band operation. The main requirements for the new access network were high spectral efficiency, the capability to support high peak data rates, short round trip time, and flexibility in frequency and bandwidth. The salient features introduced in Release 8 include:

- OFDMA in downlink, SC-FDMA in uplink. OFDM is a multicarrier technology subdividing the available bandwidth into a multitude of mutual orthogonal narrowband subcarriers. In uplink, the SC-FDMA solution is deployed, which generates a signal with single carrier characteristics, solving the issue of high peak to average power ratio (PAPR).

Fundamentals of Network Planning and Optimisation 2G/3G/4G: Evolution to 5G, Second Edition. Ajay R. Mishra.
© 2018 John Wiley & Sons Ltd. Published 2018 by John Wiley & Sons Ltd.

- Peak data rates up to 300 Mbps downlink and 75 Mbps uplink. Support for larger bandwidth of up to 20 MHz to support higher peak rates.
- Flexible bandwidth implementation in bandwidths of 1.4, 3, 5, 10, 15 or 20 MHz, to allow for different deployment scenarios.
- Support for different frequency bands. Release 8 defined multiple frequency bands including existing deployed frequency bands in 2G and 3G technologies. In Release 8 there are 15 bands specified for FDD and 8 bands for TDD.
- Support for spatial multiplexing and different MIMO antennas techniques – 2X2 MIMO, 4X2 MIMO and 4X4 MIMO. Spatial Multiplexing, 4X4 MIMO, 64 QAM modulation, and 20 MHz bandwidth allowed support of up to 300 Mbps in downlink for FDD mode.
- The LTE network becomes a network of base stations, without any centralised intelligent controller such as RNC or BSC and the eNBs are interconnected via the X2 interface and towards the core network by the S1 interface making the architecture flat (Figure D.1).
- The base stations or eNBs become more intelligent in LTE leading to speed up in the connection set-up time and reduction in time required for a handover. For certain end-user applications, the connection set-up time could be critical, e.g. on-line gaming. The time for a handover is essential for real-time services where end-users tend to end calls if the handover takes too long. The MAC scheduler resides in eNB, in line with HSDPA in UMTS where the MAC scheduler resides in NodeB, leading to fast turn-around and decision making between the network and the UE.
- The key component for efficient utilisation of radio resources, the MAC scheduler has the transmission time interval (TTI) set to only 1 ms which is much lower than the previous radio systems.

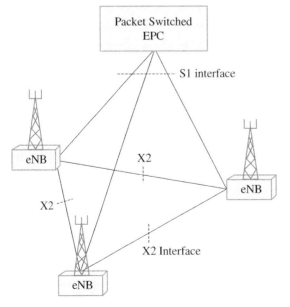

Figure D.1 LTE network interface overview.

- LTE defined reduction in latency down to 10 ms and supported different QoS requirements amongst the UEs. LTE supports both delay sensitive real-time services as well as mobile broadband services requiring high data peak rates.

D.3 Release 9

Release 9 brought new developments to the network architecture and a number of refinements to Release 8 features. Prominent changes are:

- Multimedia Broadcast/Multicast Services (eMBMS), also known as LTE Broadcast were introduced in Release 9. eMBMS allows for the efficient delivery of the same multimedia content to multiple devices within a cell. It is a point-to-multipoint interface specification with target applications such as mobile TV broadcasting, live streaming, and the emergency alerts.
- Home eNB (low power eNB) or the Femtocell Concept got integrated in 3GPP in Release 9. Release 8 specified the basic functionalities for the support of Home eNB (HeNB).
  Release 9 added further functionalities that enabled mobile operators to provide more advanced services as well as improving the user experience.
- Release 8 supports single layer beamforming. It is extended to multi-layer beamforming in Release 9. LTE Release 8 supports beamforming based on a user-specific reference which is further evolved in Release 9 to support multi-layer beamforming.
- SON introduced in Release 8 was extended with self-optimisation functionalities such as mobility robustness optimisation and mobility load balancing optimisation.
- LTE in Release 8 did not define support for positioning methods which was added in Release 9 such as assisted GPS (A-GPS), observed time difference of arrival (OTDOA) and enhanced cell ID (E-CID).

D.4 Release 10

Release 10 is known as the first release with LTE-A specifications which included significant features and improvements to fulfil ITU IMT-A requirements:

- Carrier aggregation (CA) first got introduced in Release 10. This feature subsequently underwent various enhancements to adapt to varied industry requirements of different bands and bandwidth combinations. CA allowed the combination of up to five separate carriers to enable bandwidths up to 100 MHz. Both contiguous and noncontiguous carriers could be aggregated.
- Enhanced downlink multiple antenna transmission allowed for supporting downlink SU-MIMO with up to eight-layer spatial multiplexing. Along with enhanced downlink MU-MIMO the need for UE-specific reference signal was identified in Release 10.

- Enhanced uplink multiple antenna transmission allowed for supporting uplink SU-MIMO with up to four-layer spatial multiplexing. Additionally, Release 10 allowed frequency selective scheduling (FSS) in uplink.
- Support of relay nodes to to decrease coverage holes. Relaying provides extended LTE coverage in targeted areas at low cost by extending the coverage of main eNB in low coverage. The relay nodes are connected to donor eNB (DeNB) or the main macro site.
- Support of enhanced inter-cell interference coordination (eICIC) for heterogeneous network deployments.
- SON enhancements. Coverage and capacity optimisation (CCO) functionality was included and enhancements in mobility robustness optimisation and mobility load balancing were part of Release 10.
- Additional LTE bands such as 2600 MHz for TDD, 2 GHz band and L-band for LTE.

D.5   Release 11

Release 11 focused on the features introduced in Release 10 and made further enhancements in them. Some key items included are:

- CA underwent further enhancements in Release 11 with introduction of intra-band CA for TDD mode. Further, multiple timing advances for uplink carrier aggregation were introduced. Other enhancements included definition of new band combinations for CA.
- Release 11 defined the uplink time difference of arrival (UTDOA) as a network-based positioning technique in LTE (i.e. transparent to the UE).
- Additional sub-GHz bands in the 700 and 850 MHz frequency bands were defined.
- LTE RAN enhancements for diverse data application to address RAN-level improvements within the existing RAN architecture.

D.6   Release 12

LTE continued to evolve in Release 12 with the introduction of many key feature and enhancements:

- Release 12 introduced CA between FDD and TDD carriers for the first time. With FDD carriers available with operators generally in lower bands compared with TDD carriers, it helped to extend the coverage of TDD carriers through FDD and TDD CA. There were further enhancements in CA with the definition of new band combinations for CA in downlink and uplink.
- The dual connectivity concept got introduced for the first time. Dual connectivity allowed UE radio resources to be provided from at least two different network points that are connected through non-ideal backhaul.
- To achieve higher network capacity under co-channel interference, network-assisted interference cancellation and suppression (NAICS) was defined to enable more effective and robust UE interference cancellation and/or suppression.
- Introduction of device-to-device communication for proximity services.
- Additional bands defined in FDD in 2300 and 450 MHz frequency bands.

D.7   Release 13

The following are the key enhancements:

- LTE got defined in in unlicensed spectrum to meet the growing traffic demand. Through licensed-assisted access (LAA), unlicensed spectrum as the secondary cell can be aggregated with the primary cell on licensed spectrum to provide potential operational cost saving, improved spectral efficiency and a better user experience.
- Devices defined to optimise cost and enhance coverage to support machine-type communication (MTC) via the LTE network to provide the opportunity for new revenue generation for mobile operators. Specific enhancements include 1 Rx antenna, downlink and uplink maximum TBS size of 1000 bits, reduced downlink channel bandwidth of 1.4 MHz, ultra-long battery life via power consumption reduction techniques and enhancements to support coverage improvement up to 15 dB for FDD.
- Earlier 3GPP evaluations for beamforming and MIMO had been antenna arrays that exploit the azimuth dimension. In Release 13, 3GPP initiated studies on how two-dimensional antenna arrays can improve the spectral efficiency by exploiting the vertical dimension for beamforming and MIMO operations.
- Release 13 studies the performance of existing positioning methods in indoor environments. It also evaluated the improvements to these existing methods or specifying the new positioning methods to achieve better indoor positioning accuracy.
- Introduction of NB-IoT specifications providing the air-interface for ultra-low complexity UEs which can connect to the network in massive numbers, in extremely challenging coverage conditions, and have long battery life.

D.8   Release 14

Release 14 is considered as the beginning of 5G network evolution. LTE evolved further in Release 14 with the following key features and enhancements:

- Release 14 provided specification support for large active antenna systems (AAS) comprising up to 64–128 antenna elements and higher order MU spatial multiplexing by enhancing reference signals (RSs), CSI (Channel State Information) reporting mechanism, and transmission schemes.
- Support for UL 256 QAM was introduced in Release 14. This support led to the introduction of new UL UE categories that would utilise the newly added modulation order in UL. Further, support for TTI bundling for TDD UL/DL configurations #2 and #3 were introduced.
- NB-IoT introduction for new bands. The purpose was that when a new band got specified or an operator wished to operate NB-IoT in an existing band, there was no need to specify a new WI for each band.
- LTE Release 13 introduced improved support for machine-type communications in the form of a low-complexity UE category, coverage enhancement modes and extended discontinuous reception. It got broadened in Release 14 with the range of use cases that can be addressed by LTE MTC by providing higher data rate support, multicast support, improved positioning, VoLTE enhancements, and mobility enhancements.

- Release 14 improved the mobility and throughput performance under high speed up to and above $350\,\mathrm{km\,h}^{-1}$ by enhancing the requirements for UE RRM, UE demodulation and base station demodulation.
- Release 14 specified UL support for LAA SCell operation in unlicensed spectrum by specifying UL CA for LAA SCell(s) including channel access mechanisms, core and RF requirements for base stations and UEs, and RRM.

D.9 Release 15 and Future Releases

At this stage, let us have a look at the roadmap of standardisation towards 5G. There are two phases of the work towards 5G standardisation: Phase 1 is expected to be completed by mid-2018 while Phase 2 is expected to be released in late 2019. While the former is focusing on the immediate industry requirements, the latter will cover all use cases and requirements (as seen in Chapter 10) (Figure D.2).

These releases will focus on NR (New Radio of 5G). NR shall eventually address all requirements and use cases that have been identified (see Chapter 10) including aspects such as NR forward compatibility.

Jun-17   Jun-18   Dec-19

Rel-14   Rel-15   Rel-16

Figure D.2 5G standards roadmap.

Appendix E

A Synopsis on Radio Spectrum Management

Ramy Ahmed Fathy[1] and Asit Kadayan[2]

[1] *Director, Telecom Regulatory, Egypt*
[2] *Director, Telecom Regulatory, India*

E.1   Introduction to Spectrum

When we use the word spectrum in telecom space it refers to the radio frequency (RF) spectrum. Radio is a general term applied to the use of radio waves. The RF spectrum refers to the continuum of frequencies that characterises radio signals. Radio waves mean electromagnetic waves of frequencies lower than 3000 GHz and propagating in space without artificial guide.

Over the past few years, there has been a lot of debate about spectrum shortages, spectrum allocations, interference, etc. Methods of allocation of spectrum in a transparent manner which maximises government revenue was also equally debated. To get more spectrum to expand the capacity of radio networks, Telecom Service Providers (TSPs) are looking at various ways and are even proposing mergers just to achieve it.

E.2   Spectrum Management

Considered as a scarce resource by many administrations worldwide, radio spectrum management and its efficient use remain a top priority for the mobile and wireless communications industries. Since the early 1990s, and with the rapid increase in mobile radiotelephones, there has been an urgent need for new frequency allocations in the, back then, unchartered parts of the radio spectrum (from 1 to 3 GHz). It was evident that the usefulness of various parts of the spectrum varies. The choice of the frequency spectrum bands was dependent on the application and use case.

Economics has been a major factor in the band selection, with factors such as antenna size and configuration of antenna, radiating power, and height of antenna tower influencing the selection. The estimated number of sites required for meeting the targeted coverage levels and the planned capacity affected the overall economics. As a result, it

Fundamentals of Network Planning and Optimisation 2G/3G/4G: Evolution to 5G,
Second Edition. Ajay R. Mishra.

built pressure to use limited spectrum with more and more densification of sites which meant lower and lower inter site distances leading to higher and higher interference between systems resulting in reduction in the overall capacity of the information conveyed. To manage spectrum across multiple players effciently and optimally, the deployment of national radio spectrum management and regulations were a necessity.

Radio spectrum management falls within the mandate of national administrations, with Telecom Regulatory Authorities (TRAs) or equivalent bodies used to accomplish such a task. It is the role of national administrations through spectrum management and regulation to ensure that no wireless service, used by any incumbent, causes unplanned interference over others. In addition, specific requirements are stipulated by the state to ensure that specific targets related to service provisioning are met (e.g. coverage obligations, emergency services accessibility, and quality of service). Spectrum management also needs to address requirements for the new use cases being discovered every year and the use of radio for many purpose. Proper radio regulations could stimulate innovation and remove entry barriers to new comers, having a direct impact on competition.

While assigning spectrum, national administrations may permit planned interference, which may be required for a variety of reasons such as reuse of spectrum, practical design considerations, etc. This is required to serve the high levels of demand with limited availability of the spectrum. Exclusive use of the spectrum is granted by means of a licensing regime, where spectrum is assigned with obligations to meet a certain set of regulations. These obligations may cover legal, technical, or financial aspects. Examples for legal aspects include the right to access the spectrum; protection from interference caused by other spectrum users operating in the same or adjacent bands or bands which may interfere due to harmonics in the band spectrum assignee. Other spectrum users may be licensee or operating in unlicensed bands or secondary users of the spectrum band. Legal regulations may also include prohibitions or restrictions or require prior authorisation as in the case of trading or leasing of spectrum or reselling of the spectrum for uses or services other than those specified to the incumbent.

On the other hand, technical regulatory obligations may include aspects such as maximum effective isotropic radiative power (EIRP), guard bands, requirements of out of band (OOB) emissions, spurious emissions and other metrics usually meant to minimise any potential interference effects on neighbouring bands assigned to other users or allocated to other services.

Finally, financial aspects of regulatory obligations usually include spectrum fees which in principle cover costs associated with the regulatory burden to ensure that the spectrum assignees are meeting their obligations. In some situations, the spectrum fees are used by the administration as a revenue generating tool. Other cases include fees needed to evacuate existing incumbents operating in a particular band to reuse it for other purposes (e.g. a different service).

The protection of incumbents from interference, and the stipulation of regulations to ensure that service obligations are met, are considered fundamental requirements of national administrations in managing the use of spectrum made within their geographical territory. The same or adjacent spectrum bands are used in other countries and regional and international aspects of radio spectrum management become important for many aspects such as interference across international borders, end user devices roaming internationally, and development of a harmonised ecosystem. To ensure that a government undertakes to ensure that radio stations within its jurisdiction do not cause harmful interference to other radio stations situated in other countries, end user devices may easily work during international roaming and avoid fragmentation of market, there was a need for international coordination on spectrum issues.

Cross border mobility and interoperability may be required while mobile users are roaming internationally, e.g. radio equipment installed in ships, aircraft, etc. moving from one country to another, and they need to communicate with radio stations of other countries. Radio links used to interconnect between different stations, if used in a cross-border setting to connect between countries, need harmonisation. Furthermore, stations placed on the borders need to operate in frequency bands, or under specific operating conditions, that would minimise interference. This calls for international and/or regional coordination. Radio Regulations are considered to be one of the most important tools in international spectrum management to date.

E.3 Role of National Administrations in Spectrum Management

Spectrum is a sovereign asset and radio spectrum management falls within the mandate of national administrations. Use of the spectrum in each country is overseen by the government or the designated national regulatory authority, which manages the radio spectrum and issues spectrum licenses. The spectrum management functions could fall under the responsibility of more than a single administrative unit due to the variety of spectrum users within national boundaries and the legal and regulatory regime for these users. For example, military services using spectrum could be managed by an authority or an agency different to that used for civil systems. On the other hand, specialised services of radio, such as broadcasting, may be assigned to a third organisation.

With the emergence of new technologies and with the astounding growth of telecommunication services all over the world, the spectrum management process has become extremely complex and intricate. With increase in usage of telecom services in terms of increase in number of subscriptions, and more and more types of services, there is always demand of additional spectrum by telecom service providers for meeting the requirements of both coverage and capacity to connect more people and offer faster speeds. It is evident during the last few decades from the use of spectrum for telecom services that it has enormous potential to generate economic value and social benefit. This requires identifying more spectrum bands which can be suitable for particular applications and are available to be released to the concerned spectrum users. National regulators or administrators are required to release sufficient, affordable spectrum in a timely manner for the spectrum users.

Before release of spectrum to users, there are several factors that are involved in spectrum planning. For example, at the international level, the International Telecommunication Union (ITU) and regional bodies are deeply involved in agreeing and assigning future spectrum bands for mobile, bound by international treaty. National regulatory authorities are concerned with interference that could arise from incompatible spectrum use along borders, which must be managed or negotiated with neighbouring countries. At the national level, even after reallocating a particular spectrum band for mobile services, there is the work of migrating incumbent spectrum users, such as broadcasters or defence programmes, out of the band in a practical, managed way. Finally, equipment manufacturers need to develop affordable devices that work seamlessly within new frequency bands. Each of these steps can take years to achieve before new spectrum can be licensed and used for mobile services.

Governments are giving great importance to making available spectrum to users and make available more and more spectrum to meet their requirements but in view of the scarcity of available and useful spectrum, it is necessary that all the spectrum and orbit users, whether government or private, work in the spirit of mutual understanding and cooperation and utilise these resources in the most optimal manner with self-discipline.

E.4   The International Frequency Allocation Table

Allocations of frequencies globally have been divided into three Regions with a capital 'R'. Within the Radio Regulations of the ITU, those three Regions are defined for purposes of frequency allocation. Figure E.1 illustrates the ITU-defined Regions.

Uniform frequency allocations worldwide are essential to reduce costs in the industry (for realising interoperability of equipment in different countries), simplify coordination, especially in mitigating cross-border interference.

The radio spectrum is subdivided into nine frequency bands, which are designated by progressive whole numbers in accordance with Table E.1. As the unit of frequency is the hertz (Hz), frequencies shall be expressed:

- in kilohertz (kHz), up to and including 3000 kHz;
- in megahertz (MHz), above 3 MHz, up to and including 3000 MHz;
- in gigahertz (GHz), above 3 GHz, up to and including 3000 GHz.

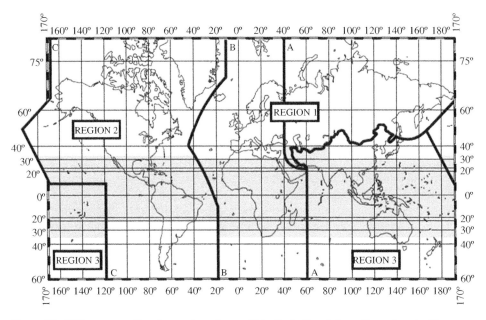

Figure E.1 ITU frequency allocation regions. *Source*: ITU – Radio Regulations. Reproduced with permission of ITU.

Table E.1 Frequency band symbols and ranges.

Band number	Symbols	Frequency range (lower limit exclusive, upper limit inclusive)	Corresponding metric subdivision
4	VLF	3–30 kHz	Myriametric waves
5	LF	30–300 kHz	Kilometric waves
6	MF	300–3000 kHz	Hectometric waves
7	HF	3–30 MHz	Decametric waves
8	VHF	30–300 MHz	Metric waves
9	UHF	300–3000 MHz	Decimetric waves
10	SHF	3–30 GHz	Centimetric waves
11	EHF	30–300 GHz	Millimetric waves
12	THF	300–3000 GHz	Decimillimetric waves

The idea behind frequency allocation lies in the division of the radio spectrum into bands, where each band is then being allocated to one or more class of radio systems, called a radio service. Table E.2 illustrates a sample of frequency allocation in the 2194–3230 kHz band.

Some clarifications which may be needed while examining Table E.2 are as follows:

- In Region 2, bands from 2300 to 2495 kHz are allocated to more than one service. In this case, the bands are said to be 'shared'.
- We note some differences in the allocations in different Regions. This indicates that different ITU Regions dictate different requirements and needs, as reflected by different services allocated for the same bands. Equipment manufacturers targeting compatability of different Regions should incorporate technical elements complying with regulations of different Regions.
- In some allocation tables, entries are qualified by footnotes. Footnotes are important to specify cases when one or more countries have allocations in a band that differ from those of the rest of the world.

There are about 40 radio service defined by the ITU used to group systems together for setting common allocations, and regulations. Services specify rules common to some specific telecommunication purposes. The following is a brief description of some of the radio services commonly found in a telecommunication network operator:

1) Fixed service: A radiocommunication service between specified fixed points.
2) Fixed-satellite service: A radiocommunication service between earth stations at given positions, when one or more satellites are used; the given position may be a specified fixed point or any fixed point within specified areas; in some cases this service includes satellite-to-satellite links, which may also be operated in the inter-satellite service; the fixed-satellite service may also include feeder links for other space radiocommunication services.
3) Mobile service: A radiocommunication service between mobile and land stations, or between mobile stations.

Table E.2 Sample of frequency allocation in 2194–3230 kHz.

Allocation to services		
Region 1	Region 2	Region 3
2194–2300 FIXED MOBILE except aeronautical Mobile (R) 5.92 5.103 5.112	**2194–2300** FIXED MOBILE 5.112	
2300–2498 FIXED MOBILE except aeronautical mobile (R) BROADCASTING 5.113 5.103	**2300–2495** FIXED MOBILE BROADCASTING 5.113	
2498–2501 STANDARD FREQUENCY AND TIME SIGNAL (2500 kHz)	**2495–2501** STANDARD FREQUENCY AND TIME SIGNAL (2500 kHz)	
2501–2502	STANDARD FREQUENCY AND TIME SIGNAL Space Research	
2502–2625 FIXED MOBILE except aeronautical mobile (R) 5.92 5.103 5.114	**2502–2505** STANDARD FREQUENCY AND TIME SIGNAL **2505–2850** FIXED MOBILE	
2625–2650 MARITIME MOBILE MARITIME RADIO NAVIGATION 5.92		
2650–2850 FIXED MOBILE except aeronautical mobile (R) 5.92 5.103		
2850–3025	AERONAUTICAL MOBILE (R) 5.111 5.115	
3025–3155	AERONAUTICAL MOBILE (OR)	
3155–3200	FIXED MOBILE except aeronautical mobile (R) 5.116 5.117	
3200–3230	FIXED MOBILE except aeronautical mobile (R) BROADCASTING 5.113 5.116	

Source: ITU – Radio Regulations. Reproduced with permission of ITU.

4) Mobile-satellite service: A radiocommunication service:

- between mobile earth stations and one or more space stations, or between space stations used by this service; or
- between mobile earth stations by means of one or more space stations.

This service may also include feeder links necessary for its operation.

5) Land mobile service: A mobile service between base stations and land mobile stations, or between land mobile stations.
6) Land mobile-satellite service: A mobile-satellite service in which mobile earth stations are located on land.
7) Broadcasting service: A radiocommunication service in which the transmissions are intended for direct reception by the general public. This service may include sound transmissions, television transmissions, or other types of transmission.
8) Broadcasting-satellite service: A radiocommunication service in which signals transmitted or retransmitted by space stations are intended for direct reception by the general public.

In the broadcasting-satellite service, the term 'direct reception' shall encompass both individual reception and community reception.

The mobile service consists of mobile radio stations, stations at fixed positions on land that communicate directly with them, in addition to the radio links used between them. These mobile stations can be portable or on land vehicles. Direct communication between one mobile station and another is also part of the mobile service.

Links between mobile stations and satellites or between mobile stations with routes transited via satellites are considered part of the mobile-satellite service. However, if routes are terminated on a fixed location, which communicates as well with a mobile radio station, that particular link (satellite to fixed radio station) is considered part of another service, the fixed-satellite service.

Table E.2 also indicates the presence of more than one service in some bands. The joint use of bands is denoted by band 'sharing'. Sharing is usually governed by some constraints related to the system parameters used by different services to minimise any potential interference between them. Alternatively, strict frequency coordination procedures can be mandated to measure the estimated levels of interference before a particular station is licensed to operate. Two types of allocations are stipulated by the Radio Regulations. The first type is the primary service, which gains protection from any interference caused by other primary and/or secondary services. The second is the secondary service which, on the other hand, must not cause any harmful interference at a station in another country that has a primary service allocation. Secondary services gain no protection status. (Primary service allocations are written in all capital letters, while secondary services are written with only the first letter capitalised in Table E.2.)

E.5 Spectrum Bands and Their Impact on the User

Spectrum bands have different characteristics, and this makes them suitable for different purposes. In general, low-frequency transmissions can travel greater distances before losing their integrity, and they can pass through dense objects more easily. Less data can be transmitted over these radio waves, however. Higher frequency transmissions carry more data but are poorer at penetrating obstacles. National regulatory authorities have a difficult job, therefore, to allocate and license appropriate spectrum to the services and sectors that need it, maximising the value generated by this finite resource.

After allocation of frequency bands, allotment of a RF or RF channel is done for use by one or more administrations for a terrestrial or space radiocommunication service in one or more identified countries or geographical areas and under specified conditions. Allotment means entry of a designated frequency channel in an agreed plan, adopted by a competent conference.

E.6 NFAP (National Frequency Allocation Plan): India

Allotted frequency bands or frequency channels for a region may be authorised by national administration to users or applications for use under specified conditions. This is known as assignment of a RF or RF channel. In India, the Department of Telecommunications (DoT) publishes the National Frequency Allocation Plan (NFAP).

Although spectrum allocation is available from 9 kHz to 1000 GHz in the radio regulations the usable part of the spectrum is much less due to availability of equipment and economy of scale. Therefore, spectrum review is basically for the frequency band below 100 GHz.

The RF spectrum is shared by various radiocommunication services for a variety of applications including public telecom services, aeronautical/maritime safety communications, radars, seismic surveys, rocket and satellite launching, earth exploration, and natural calamities forecasting.

The NFAP has made certain provisions for new technologies/applications such as ultra-wide-band devices, short range, low power devices, intelligent transport systems, E band, etc.

As per NTP-99, the NFAP is to be revised generally every two years in line with the decisions of the World Radiocommunication Conference (WRC). In order to meet spectrum requirements of fast emerging new wireless technologies, the NFAP is revised. NFAP is also developed with special emphasis to encourage or promote indigenous manufacturing and technologies by provisioning of small chunks of the spectrum in certain frequency bands, sub-bands in limited geographical area.

Spectrum management is the combination of administrative and technical procedures necessary to ensure the efficient operation of radiocommunication services. Therefore, efficient spectrum management needs to be the art and science of carefully planning spectrum allocation in a coordinated manner without compromising national interests and efficiently assigning frequencies for the benefit of users and with minimum scope of harmful interference. It is equally important that users should plan, establish and operate their radiocommunication networks optimally using spectrum efficient technologies with optimal technical parameters and should take all necessary measures for coexistence and for optimal sharing of the resources. Radiocommunication networks are like a global society necessitating appropriate discipline. Therefore, spectrum management is carried out in four levels of regulatory framework: international allocation of frequency bands for 41 different types of services defined in Radio Regulations; Regional allocation of frequency bands; NFAP; and Licensing.

RF spectrum and satellite orbits including geostationary satellite orbits are a scarce natural resource, susceptible to harmful interference and are international in character since radio waves cannot be confined to national boundaries. As for any other natural resource, they cannot be owned but are used or shared amongst countries, services, users, technologies, etc. without any element of exclusiveness. No ownership of any frequency band is conferred on any entity. It is essential that these scarce resources are used rationally, optimally, efficiently, and economically so that equitable access could be available to large radiocommunication networks in an interference-free radio environment.

E.6.1   Allocation of Spectrum

Mobile communication systems today have access to about 400–1000 MHz total bandwidth of dedicated spectrum. There are large variations between countries. Essentially all spectrum that could be feasible for mobile radiocommunication today is already allocated to other radiocommunication services. The airwaves are indeed very crowded.

The allocated spectrum is distributed over frequencies from around 450 MHz up to 3.6 GHz. Additionally, there is about 500 MHz of shared spectrum available for unlicensed or licensed exempt short-range use, e.g. remote controls, home video streaming. and Wi-Fi.

In today's society, radio spectrum is becoming increasingly important for all walks of life and therefore there is a need for its efficient and effective management. Whilst spectrum management has always been important, in recent years the complexity of the task has been compounded by the proliferation of both traditional and entirely new radio spectrum frequency-using services. The growing demand for information-rich content, faster access speeds and mobility by both business and private users is increasingly being met by broadband wireless applications.

The economic importance of radio spectrum has been vividly demonstrated in recent years by the outcome of market-based licensing processes and the proliferation of radio-based applications. There is a clear relationship between the range of radio applications, the number of users and the value of spectrum to society. This has been further evidenced by the government revenue earned in 3G and BWA auctions.

The long lead time needed for the introduction of major new services necessitates the requirement of long term planning. This is often conducted in the absence of certainty with respect to whether the envisaged new services will actually come to market. For example, the first global allocation of spectrum for 3G mobile services, then known as FPLMTS (Future Public Land Mobile Telecommunications System), was agreed at the ITU World Administrative Radiocommunication Conference in 1992, about 18 years before the entry into commercial operation of the service. The preparatory work in various international bodies began considerably earlier than that.

Technology convergence between fixed and mobile telecommunications services as well as broadcasting is increasingly causing the traditional boundaries between these services to become blurred and challenging the allocation categories for radio spectrum. Near term developments such as short range, low power ultra-wide-band technology may also have profound implications for spectrum management.

E.7   Efficient Use of Spectrum

There is the need to have enough spectrum secured to meet the consumer driven growth of mobile broadband usage for 4G and the next generation 5G networks, alongside the needs of other spectrum users such as the broadcast industry. However, there will be large regional variations. Efficient use of spectrum, licensed and unlicensed as well as multimedia distribution spectrum, is needed.

Mobile broadband has proven to be unmatched in terms of consumer demand and penetration and it is empowering people and enriching their lives, stimulating progress for citizens in villages and cities in all countries of the world. There is a need to satisfy the coverage requirements in both under-served and metropolitan areas, as well as bridging the digital divides between regions and people. There are many benefits to be gained from finding solutions to the issues faced with regard to radio spectrum – for both individuals and society at large.

Access to sufficient spectrum is of paramount importance in terms of providing affordable mobile broadband and meeting the tremendous growth in mobile data traffic.

E.8   Future Needs of Spectrum

Mobility is key for people, digital enterprises and also for connected machines. A new mobile broadband subscription is activated every second and spectrum is an essential resource in meeting this tremendous growth in mobile traffic. By 2020, more than 10 billion connected devices will be Machine-to-Machine (M2M) and consumer electronic devices.

From 2020, additional spectrum will also be required to support mobile media distribution as well as a wide range of 5G use cases, such as smart infrastructure with connected transport, live TV at scale and anywhere, remote control of heavy machines or drones, remote surgery or human interaction with IoT-like surveillance or tactile Internet.

The entire telecommunications industry is increasingly demanding more and more spectrum from finite resources. In the ITU's next WRC in November 2019, referred to as WRC-19, mobile network operators are keen to get new spectrum bands for mobile network deployments and also keen to get large chunks of C-band spectrum to be earmarked for future 5G networks. Higher frequencies, including some of the K_a-band spectrum favoured by numerous high-throughput satellite projects today, have also caught the attention of mobile operators for 5G.

A huge amount of spectrum is on the agenda of WRC-19 and the industry is looking for a total of more than 30 GHz of spectrum bandwidth, which is about 15 times what is currently used by 2G, 3G, and 4G.

The industry has greatest interest in harmonised spectrum, i.e. bands already prescribed for the same purpose either globally or across a large geographic areas. Bands which are already harmonised, such as satellite bands or broadcasting bands, are very good candidates for this purpose.

For the 3.4–4.2 GHz of spectrum that comprises the C-band, mobile network operators gained limited access to the lower portion (3.4–3.6 GHz) during WRC-07. At

WRC-15, the mobile industry obtained near-global access to that spectrum, as well as access to 3.6–3.7 GHz in a small number of countries.

Europe, in particular, has a clear desire to purpose 3.4–3.8 GHz, half of C-band, for 5G. Europe also wants to designate 26 GHz, the lower portion of the K_a-band (which stretches from 26 to 40 GHz), for 5G. However, the USA, Japan, and South Korea want to use 28 GHz for 5G.

It should be noted that 26 GHz offers more spectrum for 5G than 28 GHz (over 2 GHz compared with a few hundred megahertz). Regional regulatory organisations, including the Arab Spectrum Management Group and the Eastern Europe and Central Asia-focused Regional Commonwealth in the Field of Communications, are targeting 26 GHz bands as one of the options.

5G will be a cooperative undertaking by the mobile broadband community, the satellite community and HAPS (High Altitude Platform Stations) to provide a ubiquitous service for broadband. It is an opportunity for the satellite industry to be part of this major undertaking to provide 5G everywhere to everything, not only everyone.

Appendix F

Artificial Intelligence

Pieter Geldenhuys

Managing Director, ITSI, South Africa

F.1   Machine Learning

Machine Learning (ML)will prove to be one of the foundational technologies that will not only redefine the processes used in organisations but will be responsible for the overhaul of entire industry-based business models. As entire books are written on the subject, this appendix will focus onsome basic elementary concepts in ML, followed by a few future business scenarios in which ML will play a foundational impact.

One advantage that ML has above human pattern recognition is that it looks at the patterns in the data without the filter of human subjectivity. The massive amount of processing power ML has at its disposal also allows it to use a process of evolution to obtain the fittest algorithm for the task at hand. Where humans satisfy, ML optimises. It is this ability to parse a tremendous amount of data that enables it to unlock untold opportunities. A sample of these new opportunities will be presented at the end of this appendix.

F.2   The History of Artificial Intelligence and Machine Learning

Throughout human history, people have used technology to model themselves. There is evidence of this from ancient China, Egypt, and Greece that bears witness to the universality of this activity. Each new technology has, in its turn, been exploited to build intelligent agents or models of mind. Clockwork, hydraulics, telephone switching systems, holograms, analogue computers, and digital computers have all been proposed both as technological metaphors for intelligence and as mechanisms for modelling mind.

Fundamentals of Network Planning and Optimisation 2G/3G/4G: Evolution to 5G,
Second Edition. Ajay R. Mishra.
© 2018 John Wiley & Sons Ltd. Published 2018 by John Wiley & Sons Ltd.

The terms artificial intelligence (AI) and ML have become much more widespread than ever before. They are often used interchangeably and promise all sorts of applications from smarter home appliances to robots doing our jobs.

But while AI and ML are very much related, they are not quite the same thing. AI is a branch of computer science *attempting to build machines capable of intelligent behaviour*, while Stanford University defines ML as 'the science of getting computers to act without being explicitly programmed.'

F.2.1   The Difference Between Human Learning and ML

The most valuable resource we have in the universe is intelligence, which is simply information and computation; however, in order to be effective, technological intelligence has to be communicated in a way that helps humans take advantage of the knowledge gained. The optimal way to solve this problem is a combination of human and machine intelligence working together to solve the problems that matter most.

Human (or any other animal for that matter) brain computational power is defined (and limited) by basic evolution imperatives: survival and procreation. Our physiology and psychology only had to evolve to allow us to perform a set of basic actions. Human Learning is the act of acquiring new, or modifying and reinforcing existing, knowledge, behaviours, skills, values, or preferences which may lead to a potential change in synthesising information, depth of the knowledge, attitude or behaviour relative to the type and range of experience.

In contrast, ML at its foundation is the practice of using algorithms to analyse data, learn from it, and then make a prediction about something in the world. So, rather than creating software routines with a specific set of instructions to accomplish a specific task, the machine is 'trained' using large amounts of data and algorithms that give it the ability to learn how to perform the task.

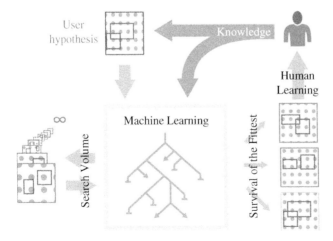

One of the most important departure points in the AI and ML journey is to develop a deeper understanding of the difference between Human Learning and ML. This understanding is one of the most important foundations as it highlights the differences of the outcomes of each approach. It also allows us to understand the limitations of each approach, and allow us to optimise our endeavours in order to obtain the optimal value of each approach. In the end, it is the synergy that will be created in the joint understanding of the outcomes of Human Learning and ML that will lead to the overall success of AI in the workplace.

F.2.2 The Difference between Supervised, Unsupervised and Reinforcement ML Algorithms

There are three main AI and ML categories under which algorithms are classified. The various ML algorithms categories are as follows:

- *Supervised ML algorithms* make predictions on a given set of data samples. They search for patterns within the value labels assigned to data points.
- *Unsupervised ML algorithms* are normally used when there are no labels associated with data points. They organise the data into a group of clusters to describe its structure. This allows complex data to appear simple and is organised for further analysis.
- *Reinforcement ML algorithms* are similar to the evolutionary characteristics of an organism that undergoes mutation. If a mutation performs better than the original, it replaces it, until the most optimal mutation remains for the environment it competes in. The abundance of processing power allows multiple mutations of an algorithm to occur. This process leads to the emergence of the most effective algorithm in a specific environment. In data analysis terms, these algorithms choose an action, based on each data point and later learn how good the decision was. Over time, the algorithm changes its strategy to learn better and achieve the best reward.

F.3 Artificial Intelligence is about Statistical Relationships, not Cause and Effect Relationships

Newtonian Physics, or Linear Predictive physics holds the view that the outcome of an event can be predicted if sufficient insight into the action that precedes the event is known. The Cause and Effect relationship is therefore predictable and repeatable.

The functional relationship between force, mass, and acceleration is due to gravity. With this relationship ($F=ma$), many things can be predicted in a deterministic fashion from that equation.

AI does not work on the same basis. In a Big Data scenario, with AI, thousands or millions of different types of things would be observed falling from different types of things and from different heights. Based on sensing all that happens (being trained) the AI system could then infer about how other things will fall. The AI system is based in a statistical model. Specifically, it will correctly and incorrectly make inferences about the data. The question is how 'accurate' is the AI system in making that prediction or inference?

In business systems, the more important question is: how accurate does the AI system need to be? In the current state of AI, it seems, that with more and more data and computing power, training an AI system can become more accurate. However, more and more data and computing power, and more specialised skills come at greater cost.

The accuracy required for a specific business system directly relates to data, effort, training time and computing power needed, and so is a critical part of business cases.

F.4 Artificial Intelligence Created User Behaviour Mapping Versus Traditional Market Research

Market research is traditionally conducted to obtain deeper insight into the true state of the market, and the emotions and perspectives the agents within the market have of a company and its products. A serious limitation of traditional research is the categorisation of market segments that are done before the data is gathered. In a number of instances, research is done to see how effective mapping of data is done onto pre-existing categories. Another limitation is that people are not rational decision makers, although rational decisions are often communicated via market questionnaires.

ML can be used to obtain far better insight into markets, as it looks at actions, rather than opinions. ML can be fed raw data without any pre-constructed categorisations. The ML engine will be able to construct far more applicable categorisations from the data it has been fed. The key to success will however be the sources of data, rather than simply the algorithms used. ML will provide far richer insight into market dynamics as it is not restricted by outdated categorisations that do not effectively map an individual's needs.

The use of ML to define human behaviour is often referred to as computational social science. Computational social science refers to the academic subdisciplines concerned with computational approaches to the social sciences. This means that computers are used to model, simulate, and analyse social phenomena.

F.5 Artificial Intelligence Created User Behaviour Mapping Versus Narrative Inquiry

Narrative inquiry is a way of understanding and inquiring into experience through 'collaboration between researcher and participants, over time, in a place or series of places, and in social interaction with milieus'.

Narrative analysis therefore can be used to acquire a deeper understanding of the ways in which a few individuals organise and derive meaning from events. It can be particularly useful for studying the impact of social structures on an individual and how that relates to identity, intimate relationships, and family. Narrative inquiry can also be used to allow new behaviour patterns to emerge as new coherences between artifacts and archetypes within markets are mapped.

F.6 The New Digital Ecosystem and Future of Artificial Intelligence

A digital ecosystem is an interdependent group of enterprises, people and/or things that share standardised digital platforms for a mutually beneficial purpose (such as commercial gain, innovation or common interest). Digital ecosystems enable you to interact with customers, partners, adjacent industries – even your competition.

Automata theory is a theoretical branch of computer science. At an elemental level, automata theory deals with the logic of computation with respect to simple machines, referred to as automata. Through automata, computer scientists are able to understand how machines compute functions and solve problems and, more importantly, what it means for a function to be defined as computable or for a question to be described as decidable.

Automatons are abstract models of machines that perform computations on an input by moving through a series of states. At each state of the computation, a transition function determines the next configuration on the basis of a finite portion of the present configuration. As a result, once the computation reaches an accepting configuration, it accepts that input.

F.7 The Role of Machine Learning: Headlines from the Future

A number of highly possible scenarios are given in the following. At the time of writing (January 2018), these scenarios are envisaged in the not to distant future. None of them would be possible without ML.

F.7.1 21 June 2019: Photo Identikit Created from DNA Sample Found at Crime Scene – Suspect Arrested

In an astounding development, the San Francisco Police Department (SFPD) arrested a murder suspect based on a photo identikit created from a DNA sample found at the crime scene. Once the photo of the subject was created from the DNA sample, it was distributed to inhabitants of the Richmond District via Social Media. An anonymous caller identified the suspect close to the corner of 22nd Ave and Clement Street. The police was notified last Friday and the suspect was apprehended. A comparison between the photo identikit and the suspect's picture had a 93% match rating of dominant facial features.

Police Superintendent Lykes Donutt was predictably upbeat in his statement to local and international news media. 'The photo identikit was developed by our new Computational Social Science department', he boasted. 'The latest Machine Learning techniques were used to make an accurate prediction of the suspect's facial features, hair, eye-colour and age.' Superintendent Donutt used the media exposure to release five new photo identikits of suspects in other crimes, also based on DNA samples collected at various crime scenes. He believes that if this trial is successful, a number of other Police Departments will follow suit.

Anonymous sources in the Police Department said that this initiative has been months in the making. Ever since personal DNA analysis became a must-have for individuals to manage their health and diet, companies like 23andMe and Ancestry.com processed millions of DNA samples in the preceding years. The sites also offered to use your DNA to link your profile with unknown relatives, and the opportunity to link a photo to your DNA profile. In the process, hundreds of thousands of members gave permission for the anonymised data to be shared. It was this initiative that opened the door for what transpired today. A variety of Machine Learning tools were used to work through the multitude of pictures and DNA data points to try and determine patterns between the DNA and associated facial features. Within a short period of time, the DNA strains that link to specific facial features have been identified. An algorithm for facial feature prediction was launched by the University of Stanford in October 2018, which led to the interest of the SFPD shortly thereafter.

The lawyer for the suspect, Hylie Suspisus, is adamant that there were no grounds for his client to be arrested. Photo identikits have traditionally been created by eyewitnesses to the crime. In this case, there was no eyewitness account, and the photo identikit could therefore not be used to arrest his client. He is adamant that the case will be dismissed in his favour. Asked if the suspect is prepared to be interrogated by JAMES3000, the ML bot that looks at micro-expressions and brain-scans to determine if a suspect is lying, Mr Suspisus declined. Although JAMES3000 has a 99.999% record in correctly measuring brain activity, Mr Suspisus said that guilt should not be established via the search for subjective truth but rather through the correct application of legal precedent.

F.7.2 Stockholm, 12 August 2018: Voice Assistant in Car Leads to Driver's Arrest

A driver was arrested in Stockholm when he exceeded the speed limit by $90\,\mathrm{km\,h^{-1}}$ in Storskogen, a residential suburb of Stockholm. The driver, Räcklösa Rumphål, became argumentative with the traffic official, Föreskrifter Ochre. Mr Rumphål would not adhere to the requests of the traffic official and started questioning the accuracy of the speed camera used to determine his speed. He wanted to see the quality assurance certificate used for the equipment, the maintenance certificate and the competence certificate of the official using the equipment. His protestations were however short-circuited when the official directly asked the car what its highest speed was in the past hour. The car responded with a figure that was exactly the same as the speed camera reading. The driver was arrested on the spot, partly based on the independent confirmation from the car that the reading was correct. The report did not indicate if it was Amazon Alexa, Google Assistant or another UPA (Ubiquitous Personal Assistant) that spilled the beans on its owner.

The lawyer for the driver, Ms Otillförlitlig Shyster expressed her displeasure with both the traffic official and the company behind the UPA. 'Firstly, the traffic official had no right to enter into a conversation with the vehicle without the express prior approval of the owner', said Ms Shyster. 'It is a clear case of impersonation, as the UPA in the car was unaware that it was divulging private information to a third party. We believe that the conversation that a user has with his UPA is a private one, be it in connection with the UPA linked to his house, phone, car, door, air conditioner or fridge'. She further stated that there is legal protection that agents of the government cannot listen in on private conversations without a court order, even if it is between an individual and his UPA.

> We believe that this applies to agents of the government impersonating an individual and allowing their UPA to divulge information that may lead to their owner being compromised in any way.

UPAs became extremely popular in homes from 2015 onwards, and exploded into various other forms including cars, fridges, cameras, drones and even suitcases. As voice replaced touch as the primary interface into the Internet and into the Internet of Things, a variety of strange and unforeseen incidents have challenged the wisdom behind some of the applications of the UPA. Amazon originally did not require a password if someone wanted to purchase anything via Alexa. This however changed quickly as the technology was abused. It seems that in this case, some changes to the UPA might be forthcoming.

F.7.3   21 July 2020: Online Dating Service Now Offers DNA Matchmaking

'Matching individuals whose offspring will have a high likelihood of getting a serious heredity disease is *nothing less than criminal*', said Pierre Château-de-Maison, founder of DNA-Match, and online dating service that matches the DNA of those looking for a romantic life partner. His words refer to the practice of online dating services who simply match people based on their looks and interests. He further stated: 'Ignorance of the importance of DNA is no longer an excuse. If anyone on the planet can map their DNA for less than a hundred dollars at services like 23andMe or Ancestry.com, it is inexcusable to use matching patterns that belong to the Dark Ages.'

Pierre also believes that DNA matching provides this service with a competitive advantage in introducing individuals to their perfect romantic partner. DNA matching also allows people to find their perfect match based on their olfactory senses.

A famous experiment was conducted by Claus Wedekind, a researcher at the University of Bern, in Switzerland (http://www.economist.com/node/10493120). In this experiment, the DNA of both the male and female participates in the experiment was analysed. The female participants wore a plain white T-shirt for a few hours, after which it was sealed in an air-tight container. Without ever meeting any of the female participants, males were asked to smell the various T-shirts and indicate which one they were attracted to the most. The result of the research was astounding. There was an unmistaken correlation between the DNA matching of the individual based on the olfactory attraction based on the T-shirt choice. Our noses traditionally told us if the person we met was a good DNA match. This practice has however been disrupted in the last few decades due to the use of perfume, which leaves our traditional senses dumbfounded in choosing the optimal partner.

The DNA-Match website looks very much like its competitors, and allows you to shortlist prospective partners based on looks and interests. The DNA-Match website then asks you to link your DNA profile at 23andMe to their service. DNA-Match then uses a variety of ML tools and apply it to the worldwide DNA research repository at DNSpedia (which constantly updates with the latest DNA research from around the world) to determine who on your shortlist provides you with the optimal DNA match.

Pierre believes that science will provide a shortcut to meeting your Perfect Match, and give you the peace of mind that your offspring will have a very low chance of getting a hereditary disease. It seems that our destiny may not be written in our stars, but has been hidden in ourselves all along.

F.8   Conclusion

In all the possible future scenarios (headlines-from-the-future.com), ML played a foundational role in redefining the environment and the business models used within an industry segment. In years to come, the role of ML on our future will be nothing less than astounding. It does, however, have limitations.

Too often, whether it is a lack of understanding, or not believing that AI is applicable to their business, most executives are not actively involved in AI initiatives within their business. They leave important decisions to technical teams. Business Executives responsibilities are specifically around the success of their business. AI is not about technical analysis, it is about leveraging data and ML to drive business success. Without business leadership, AI success in business will only be random and limited. Active and indeed proactive involvement of business leadership is critical.

Bibliography

https://www.datasciencecentral.com/profiles/blogs/
    human-brain-vs-machine-learning-a-lost-battle
https://blogs.nvidia.com/blog/2016/07/29/
    whats-difference-artificial-intelligence-machine-learning-deep-learning-ai

Appendix G

Erlang B Tables

Fundamentals of Network Planning and Optimisation 2G/3G/4G: Evolution to 5G,
Second Edition. Ajay R. Mishra.
© 2018 John Wiley & Sons Ltd. Published 2018 by John Wiley & Sons Ltd.

N/B (%)	0.1	0.2	0.3	0.4	0.5	1	1.5	2	2.5	3	3.5	4	4.5	5
1	0	0	0	0	0.01	0.01	0.01	0.02	0.03	0.03	0.04	0.04	0.05	0.05
2	0.05	0.06	0.08	0.09	0.1	0.15	0.19	0.23	0.25	0.28	0.31	0.33	0.36	0.38
3	0.2	0.25	0.29	0.32	0.35	0.46	0.54	0.6	0.66	0.71	0.77	0.81	0.86	0.9
4	0.44	0.53	0.6	0.65	0.7	0.87	0.99	1.09	1.18	1.26	1.33	1.4	1.46	1.53
5	0.76	0.9	0.99	1.07	1.13	1.36	1.53	1.66	1.77	1.87	1.97	2.06	2.14	2.22
6	1.15	1.33	1.45	1.54	1.62	1.91	2.11	2.28	2.42	2.54	2.66	2.76	2.86	2.96
7	1.58	1.8	1.95	2.06	2.16	2.5	2.74	2.93	3.1	3.25	3.39	3.51	3.63	3.74
8	2.05	2.31	2.48	2.62	2.73	3.13	3.4	3.63	3.82	3.99	4.14	4.28	4.42	4.54
9	2.56	2.85	3.05	3.21	3.33	3.78	4.09	4.34	4.56	4.75	4.92	5.08	5.23	5.37
10	3.09	3.43	3.65	3.82	3.96	4.46	4.81	5.09	5.32	5.53	5.72	5.9	6.06	6.21
11	3.65	4.02	4.27	4.45	4.61	5.16	5.54	5.84	6.1	6.33	6.54	6.73	6.91	7.08
12	4.23	4.64	4.9	5.11	5.28	5.88	6.29	6.61	6.9	7.14	7.36	7.57	7.77	7.95
13	4.83	5.27	5.56	5.78	5.96	6.61	7.05	7.4	7.7	7.97	8.21	8.43	8.64	8.83
14	5.45	5.92	6.23	6.47	6.66	7.35	7.83	8.2	8.52	8.8	9.06	9.3	9.52	9.73
15	6.08	6.58	6.91	7.17	7.37	8.11	8.61	9.01	9.35	9.65	9.92	10.18	10.41	10.63
16	6.72	7.26	7.61	7.88	8.1	8.87	9.4	9.83	10.19	10.51	10.8	11.06	11.31	11.55
17	7.38	7.95	8.32	8.6	8.83	9.65	10.21	10.66	11.03	11.37	11.67	11.95	12.21	12.46
18	8.04	8.64	9.03	9.33	9.58	10.44	11.02	11.49	11.89	12.24	12.56	12.85	13.12	13.38
19	8.72	9.35	9.76	10.07	10.33	11.23	11.85	12.33	12.75	13.12	13.45	13.76	14.04	14.31
20	9.41	10.07	10.5	10.82	11.09	12.03	12.67	13.18	13.62	14	14.34	14.67	14.96	15.25
21	10.11	10.79	11.24	11.58	11.86	12.84	13.51	14.04	14.49	14.88	15.25	15.58	15.89	16.19
22	10.81	11.53	11.99	12.34	12.63	13.65	14.35	14.9	15.36	15.78	16.15	16.5	16.83	17.13
23	11.52	12.27	12.75	13.11	13.42	14.47	15.19	15.76	16.25	16.68	17.07	17.43	17.76	18.08

24	12.24	13.01	13.51	13.89	14.2	15.29	16.04	16.63	17.13	17.58	17.98	18.35	18.7	19.03
25	12.97	13.76	14.28	14.67	15	16.12	16.89	17.5	18.02	18.48	18.9	19.29	19.65	19.98
26	13.7	14.52	15.06	15.46	15.79	16.96	17.75	18.38	18.92	19.39	19.82	20.22	20.59	20.94
27	14.44	15.28	15.84	16.25	16.6	17.8	18.62	19.27	19.82	20.3	20.75	21.16	21.54	21.9
28	15.18	16.05	16.62	17.05	17.41	18.64	19.48	20.15	20.72	21.22	21.68	22.1	22.49	22.87
29	15.93	16.83	17.41	17.85	18.22	19.49	20.35	21.04	21.62	22.14	22.61	23.04	23.45	23.83
30	16.68	17.61	18.2	18.66	19.03	20.34	21.23	21.93	22.53	23.06	23.55	23.99	24.41	24.8
31	17.44	18.39	19	19.47	19.85	21.19	22.1	22.83	23.44	23.99	24.48	24.94	25.37	25.77
32	18.21	19.17	19.8	20.28	20.68	22.05	22.98	23.72	24.36	24.92	25.42	25.89	26.33	26.74
33	18.97	19.97	20.61	21.1	21.51	22.91	23.87	24.63	25.27	25.84	26.36	26.84	27.29	27.72
34	19.74	20.76	21.42	21.92	22.34	23.77	24.75	25.53	26.19	26.78	27.31	27.8	28.26	28.7
35	20.52	21.56	22.23	22.75	23.17	24.64	25.64	26.44	27.11	27.71	28.26	28.76	29.23	29.68
36	21.29	22.36	23.05	23.57	24.01	25.51	26.53	27.35	28.04	28.65	29.2	29.72	30.2	30.66
37	22.08	23.16	23.87	24.41	24.85	26.38	27.42	28.25	28.96	29.59	30.16	30.68	31.17	31.64
38	22.86	23.98	24.69	25.24	25.69	27.25	28.32	29.17	29.89	30.52	31.1	31.64	32.15	32.62
39	23.65	24.79	25.52	26.08	26.53	28.13	29.21	30.08	30.81	31.47	32.06	32.61	33.12	33.61
40	24.45	25.6	26.35	26.92	27.38	29.01	30.11	31	31.75	32.41	33.02	33.58	34.1	34.6
41	25.24	26.41	27.18	27.76	28.23	29.89	31.02	31.91	32.68	33.36	33.97	34.54	35.08	35.58
42	26.04	27.24	28.01	28.6	29.08	30.77	31.92	32.84	33.61	34.31	34.93	35.51	36.06	36.57
43	26.84	28.06	28.84	29.45	29.94	31.66	32.83	33.76	34.55	35.25	35.89	36.48	37.04	37.57
44	27.64	28.88	29.68	30.29	30.8	32.54	33.73	34.68	35.49	36.2	36.85	37.46	38.02	38.56
45	28.45	29.71	30.52	31.15	31.66	33.43	34.64	35.61	36.43	37.15	37.82	38.43	39.01	39.55
46	29.25	30.54	31.37	32	32.52	34.32	35.55	36.53	37.37	38.11	38.78	39.4	39.99	40.54
47	30.07	31.37	32.21	32.85	33.38	35.21	36.47	37.46	38.31	39.06	39.75	40.38	40.97	41.54
48	30.88	32.2	33.06	33.71	34.25	36.11	37.38	38.39	39.25	40.02	40.71	41.36	41.96	42.54

(Continued)

(Continued)

N/B (%)	0.1	0.2	0.3	0.4	0.5	1	1.5	2	2.5	3	3.5	4	4.5	5
49	31.69	33.04	33.91	34.57	35.11	37	38.29	39.32	40.2	40.97	41.68	42.34	42.95	43.53
50	32.51	33.88	34.76	35.43	35.98	37.9	39.21	40.25	41.15	41.93	42.65	43.32	43.94	44.53
51	33.33	34.72	35.61	36.29	36.85	38.8	40.13	41.19	42.09	42.89	43.62	44.3	44.93	45.53
52	34.15	35.56	36.47	37.16	37.73	39.7	41.05	42.12	43.04	43.85	44.59	45.28	45.92	46.53
53	34.98	36.4	37.32	38.02	38.6	40.6	41.97	43.06	43.99	44.81	45.56	46.26	46.91	47.53
54	35.8	37.25	38.18	38.89	39.47	41.5	42.89	44	44.94	45.78	46.54	47.24	47.91	48.54
55	36.63	38.09	39.04	39.76	40.35	42.41	43.82	44.93	45.89	46.74	47.51	48.23	48.9	49.54
56	37.46	38.94	39.9	40.63	41.23	43.31	44.74	45.87	46.85	47.7	48.49	49.21	49.89	50.54
57	38.29	39.79	40.76	41.5	42.11	44.22	45.67	46.82	47.8	48.67	49.46	50.2	50.89	51.55
58	39.12	40.65	41.63	42.38	42.99	45.13	46.59	47.76	48.75	49.63	50.44	51.18	51.89	52.55
59	39.96	41.5	42.49	43.25	43.87	46.04	47.52	48.7	49.71	50.6	51.42	52.17	52.88	53.56
60	40.79	42.35	43.36	44.13	44.76	46.95	48.45	49.65	50.66	51.57	52.4	53.16	53.88	54.57
61	41.63	43.21	44.23	45	45.64	47.86	49.38	50.59	51.62	52.54	53.37	54.15	54.88	55.57
62	42.47	44.07	45.1	45.89	46.53	48.77	50.31	51.53	52.58	53.51	54.35	55.14	55.88	56.58
63	43.31	44.93	45.97	46.76	47.42	49.69	51.24	52.48	53.54	54.48	55.33	56.13	56.88	57.59
64	44.15	45.79	46.84	47.65	48.31	50.6	52.17	53.43	54.5	55.45	56.32	57.12	57.88	58.6
65	45	46.65	47.72	48.53	49.2	51.52	53.11	54.38	55.46	56.42	57.3	58.11	58.88	59.61
66	45.84	47.51	48.59	49.41	50.09	52.43	54.04	55.33	56.42	57.39	58.28	59.11	59.88	60.62
67	46.69	48.38	49.47	50.3	50.98	53.35	54.98	56.28	57.38	58.37	59.27	60.1	60.88	61.63
68	47.54	49.24	50.35	51.18	51.87	54.27	55.92	57.22	58.34	59.34	60.25	61.09	61.89	62.64
69	48.39	50.11	51.22	52.07	52.77	55.19	56.85	58.18	59.31	60.31	61.24	62.09	62.89	63.65
70	49.24	50.98	52.1	52.96	53.66	56.11	57.79	59.13	60.27	61.29	62.22	63.08	63.9	64.67

71	50.09	51.85	52.98	53.85	54.56	57.04	58.73	60.08	61.24	62.27	63.21	64.08	64.9	65.68
72	50.94	52.72	53.87	54.74	55.46	57.96	59.67	61.03	62.2	63.24	64.19	65.08	65.9	66.69
73	51.8	53.59	54.75	55.63	56.35	58.88	60.61	61.99	63.17	64.22	65.18	66.07	66.91	67.71
74	52.65	54.46	55.63	56.52	57.25	59.8	61.55	62.95	64.14	65.2	66.17	67.07	67.91	68.72
75	53.51	55.34	56.52	57.42	58.15	60.73	62.49	63.9	65.11	66.18	67.16	68.07	68.92	69.74
76	54.37	56.21	57.4	58.31	59.05	61.66	63.44	64.86	66.08	67.16	68.14	69.06	69.93	70.75
77	55.23	57.09	58.29	59.2	59.96	62.58	64.38	65.81	67.04	68.13	69.14	70.06	70.94	71.77
78	56.09	57.96	59.18	60.1	60.86	63.51	65.32	66.77	68.01	69.11	70.12	71.06	71.95	72.78
79	56.95	58.84	60.07	61	61.76	64.44	66.27	67.73	68.98	70.1	71.11	72.06	72.95	73.8
80	57.81	59.72	60.95	61.9	62.67	65.36	67.21	68.69	69.95	71.08	72.11	73.06	73.96	74.82
81	58.67	60.6	61.85	62.79	63.57	66.29	68.16	69.65	70.92	72.06	73.1	74.06	74.97	75.84
82	59.54	61.48	62.74	63.69	64.48	67.22	69.1	70.61	71.89	73.04	74.09	75.06	75.98	76.86
83	60.4	62.36	63.63	64.59	65.39	68.15	70.05	71.57	72.87	74.02	75.08	76.06	76.99	77.87
84	61.27	63.24	64.52	65.5	66.29	69.08	71	72.53	73.84	75.01	76.08	77.07	78	78.89
85	62.14	64.13	65.41	66.4	67.2	70.02	71.95	73.49	74.81	75.99	77.07	78.07	79.01	79.91
86	63	65.01	66.31	67.3	68.11	70.95	72.89	74.45	75.79	76.98	78.06	79.07	80.03	80.93
87	63.87	65.9	67.21	68.2	69.02	71.88	73.85	75.41	76.76	77.96	79.05	80.07	81.04	81.95
88	64.74	66.78	68.1	69.11	69.93	72.81	74.79	76.38	77.73	78.94	80.05	81.08	82.05	82.97
89	65.61	67.67	69	70.01	70.84	73.75	75.74	77.34	78.71	79.93	81.04	82.08	83.06	83.99
90	66.48	68.56	69.9	70.92	71.76	74.68	76.7	78.31	79.69	80.92	82.04	83.09	84.07	85.01
91	67.36	69.44	70.8	71.82	72.67	75.62	77.65	79.27	80.66	81.9	83.03	84.09	85.09	86.04
92	68.23	70.33	71.69	72.73	73.58	76.56	78.6	80.23	81.64	82.89	84.03	85.09	86.1	87.06
93	69.1	71.22	72.59	73.64	74.49	77.49	79.55	81.2	82.61	83.87	85.03	86.1	87.11	88.08
94	69.98	72.11	73.49	74.55	75.41	78.43	80.5	82.17	83.59	84.86	86.02	87.11	88.13	89.1
95	70.85	73	74.39	75.46	76.33	79.37	81.46	83.13	84.57	85.85	87.02	88.11	89.14	90.12

(Continued)

(Continued)

N/B (%)	0.1	0.2	0.3	0.4	0.5	1	1.5	2	2.5	3	3.5	4	4.5	5
96	71.73	73.89	75.3	76.36	77.24	80.31	82.41	84.1	85.55	86.84	88.02	89.12	90.16	91.15
97	72.61	74.79	76.2	77.27	78.16	81.24	83.37	85.07	86.53	87.82	89.02	90.12	91.17	92.17
98	73.48	75.68	77.1	78.19	79.07	82.18	84.32	86.04	87.5	88.81	90.01	91.13	92.15	93.19
99	74.36	76.57	78.01	79.09	79.99	83.12	85.28	87	88.48	89.8	91.01	92.14	93.2	94.22
100	75.24	77.47	78.91	80.01	80.91	84.07	86.24	87.97	89.46	90.79	92.01	93.15	94.22	95.24
101	76.12	78.37	79.81	80.92	81.83	85	87.19	88.94	90.44	91.78	93.01	94.16	95.23	96.26
102	77	79.26	80.72	81.83	82.75	85.95	88.15	89.91	91.42	92.78	94.01	95.16	96.25	97.29
103	77.88	80.16	81.63	82.75	83.67	86.89	89.11	90.88	92.41	93.76	95.01	96.17	97.27	98.31
104	78.77	81.05	82.53	83.66	84.59	87.83	90.06	91.85	93.39	94.76	96.01	97.18	98.29	99.34
105	79.65	81.95	83.44	84.58	85.51	88.77	91.02	92.82	94.37	95.75	97.01	98.19	99.3	100.36
106	80.53	82.85	84.35	85.49	86.43	89.72	91.98	93.79	95.35	96.74	98.01	99.2	100.32	101.39
107	81.42	83.75	85.26	86.41	87.35	90.66	92.94	94.76	96.33	97.73	99.01	100.21	101.34	102.42
108	82.3	84.65	86.17	87.32	88.28	91.6	93.9	95.73	97.31	98.72	100.01	101.22	102.36	103.44
109	83.19	85.55	87.08	88.24	89.2	92.55	94.85	96.71	98.29	99.71	101.01	102.23	103.37	104.47
110	84.07	86.45	87.99	89.16	90.12	93.49	95.82	97.68	99.28	100.71	102.02	103.24	104.39	105.49
111	84.96	87.35	88.9	90.08	91.05	94.44	96.77	98.65	100.26	101.7	103.02	104.25	105.41	106.52
112	85.85	88.25	89.81	90.99	91.97	95.38	97.74	99.63	101.25	102.69	104.02	105.26	106.43	107.55
113	86.73	89.15	90.72	91.91	92.89	96.33	98.7	100.6	102.23	103.69	105.03	106.27	107.45	108.57
114	87.62	90.06	91.63	92.83	93.82	97.28	99.66	101.57	103.21	104.68	106.03	107.28	108.47	109.6
115	88.51	90.96	92.54	93.75	94.74	98.22	100.62	102.54	104.2	105.68	107.03	108.3	109.49	110.63
116	89.4	91.86	93.46	94.67	95.67	99.17	101.58	103.52	105.18	106.67	108.03	109.31	110.51	111.66
117	90.29	92.77	94.37	95.59	96.6	100.12	102.54	104.49	106.17	107.66	109.04	110.32	111.53	112.69
118	91.18	93.67	95.29	96.51	97.53	101.07	103.51	105.47	107.15	108.66	110.04	111.33	112.55	113.71

119	92.07	94.58	96.2	97.44	98.45	102.01	104.47	106.44	108.14	109.65	111.05	112.34	113.57	114.74
120	92.96	95.48	97.11	98.36	99.38	102.96	105.43	107.42	109.13	110.65	112.05	113.36	114.59	115.77
121	93.86	96.39	98.03	99.28	100.31	103.91	106.4	108.4	110.11	111.64	113.06	114.37	115.61	116.8
122	94.75	97.3	98.95	100.2	101.24	104.86	107.36	109.37	111.1	112.64	114.06	115.38	116.63	117.83
123	95.64	98.21	99.86	101.13	102.17	105.81	108.33	110.35	112.08	113.64	115.06	116.4	117.65	118.86
124	96.54	99.11	100.78	102.05	103.1	106.76	109.29	111.32	113.07	114.64	116.07	117.41	118.68	119.89
125	97.43	100.02	101.7	102.98	104.03	107.71	110.26	112.3	114.06	115.63	117.07	118.42	119.7	120.92
126	98.33	100.93	102.62	103.9	104.96	108.66	111.22	113.28	115.05	116.63	118.08	119.44	120.72	121.95
127	99.22	101.84	103.53	104.83	105.89	109.62	112.19	114.25	116.03	117.63	119.09	120.45	121.74	122.98
128	100.12	102.75	104.45	105.75	106.82	110.56	113.15	115.23	117.02	118.62	120.09	121.47	122.76	124.01
129	101.02	103.66	105.37	106.68	107.75	111.52	114.12	116.21	118.01	119.62	121.1	122.48	123.79	125.04
130	101.91	104.57	106.29	107.6	108.69	112.47	115.09	117.19	119	120.62	122.11	123.5	124.81	126.07
131	102.81	105.48	107.21	108.53	109.62	113.42	116.05	118.17	119.99	121.62	123.11	124.51	125.33	127.1
132	103.71	106.39	108.13	109.46	110.55	114.37	117.02	119.15	120.98	122.62	124.12	125.52	126.86	128.13
133	104.6	107.3	109.05	110.39	111.48	115.33	117.99	120.12	121.97	123.61	125.13	126.54	127.88	129.16
134	105.5	108.21	109.97	111.31	112.41	116.28	118.95	121.1	122.96	124.61	126.13	127.56	128.9	130.19
135	106.4	109.13	110.89	112.24	113.35	117.24	119.92	122.08	123.94	125.61	127.14	128.57	129.93	131.22
136	107.3	110.04	111.81	113.17	114.28	118.19	120.89	123.07	124.93	126.61	128.15	129.59	13C.95	132.25
137	108.2	110.96	112.74	114.1	115.22	119.15	121.86	124.04	125.92	127.61	129.16	130.61	131.97	133.29
138	109.1	111.87	113.66	115.03	116.15	120.1	122.83	125.02	126.91	128.61	130.16	131.62	133	134.32
139	110	112.78	114.58	115.96	117.09	121.05	123.79	126	127.91	129.61	131.18	132.63	134.02	135.35
140	110.9	113.7	115.51	116.89	118.02	122.01	124.77	126.98	128.9	130.61	132.18	133.65	135.05	136.38
141	111.81	114.61	116.43	117.82	118.96	122.96	125.74	127.96	129.89	131.61	133.19	134.67	136.07	137.41
142	112.71	115.53	117.35	118.75	119.9	123.92	126.7	128.95	130.88	132.61	134.2	135.69	137.1	138.44
143	113.61	116.44	118.28	119.68	120.83	124.87	127.68	129.93	131.87	133.61	135.21	136.7	138.12	139.48

(Continued)

(Continued)

N/B (%)	0.1	0.2	0.3	0.4	0.5	1	1.5	2	2.5	3	3.5	4	4.5	5
144	114.51	117.36	119.2	120.61	121.77	125.83	128.64	130.91	132.86	134.61	136.22	137.72	139.15	140.51
145	115.42	118.28	120.13	121.54	122.7	126.79	129.62	131.89	133.85	135.61	137.23	138.74	140.17	141.54
146	116.32	119.19	121.06	122.47	123.64	127.75	130.59	132.87	134.85	136.61	138.24	139.76	141.19	142.57
147	117.22	120.11	121.98	123.41	124.58	128.7	131.56	133.85	135.84	137.61	139.25	140.78	142.22	143.61
148	118.13	121.03	122.9	124.34	125.52	129.66	132.53	134.84	136.83	138.62	140.26	141.79	143.25	144.64
149	119.04	121.95	123.83	125.27	126.46	130.62	133.5	135.82	137.82	139.62	141.27	142.81	144.27	145.67
150	119.94	122.87	124.76	126.2	127.39	131.58	134.47	136.81	138.81	140.62	142.28	143.83	145.3	146.71
151	120.85	123.78	125.69	127.14	128.33	132.53	135.44	137.79	139.81	141.62	143.29	144.85	146.33	147.74
152	121.75	124.7	126.61	128.07	129.27	133.49	136.42	138.77	140.8	142.62	144.3	145.87	147.35	148.77
153	122.66	125.62	127.54	129.01	130.21	134.45	137.39	139.75	141.79	143.62	145.31	146.89	148.38	149.81
154	123.57	126.54	128.47	129.94	131.15	135.41	138.36	140.74	142.79	144.63	146.32	147.9	149.4	150.84
155	124.47	127.46	129.4	130.87	132.09	136.37	139.33	141.72	143.78	145.63	147.33	148.92	150.43	151.88
156	125.38	128.38	130.33	131.81	133.03	137.33	140.31	142.71	144.77	146.63	148.34	149.94	151.46	152.91
157	126.29	129.3	131.25	132.74	133.98	138.29	141.28	143.69	145.77	147.64	149.35	150.96	152.48	153.94
158	127.2	130.22	132.18	133.68	134.92	139.25	142.25	144.67	146.76	148.64	150.36	151.98	153.51	154.98
159	128.1	131.15	133.12	134.62	135.86	140.2	143.22	145.66	147.75	149.64	151.37	153	154.54	156.01
160	129.02	132.06	134.05	135.55	136.8	141.17	144.2	146.64	148.75	150.64	152.39	154.02	155.56	157.04
161	129.93	132.99	134.97	136.49	137.74	142.13	145.17	147.63	149.75	151.65	153.4	155.04	156.59	158.08
162	130.83	133.91	135.9	137.43	138.68	143.09	146.15	148.61	150.74	152.65	154.41	156.06	157.62	159.12
163	131.74	134.83	136.83	138.36	139.62	144.05	147.12	149.6	151.74	153.66	155.42	157.08	158.64	160.15
164	132.65	135.75	137.77	139.3	140.57	145.01	148.09	150.58	152.73	154.66	156.43	158.1	159.67	161.19
165	133.57	136.68	138.7	140.24	141.51	145.97	149.07	151.57	153.72	155.66	157.45	159.12	160.7	162.22
166	134.48	137.6	139.63	141.18	142.45	146.93	150.04	152.55	154.72	156.67	158.46	160.14	161.73	163.26

167	135.39	138.53	140.56	142.12	143.39	147.89	151.02	153.54	155.72	157.67	159.47	161.16	162.76	164.29
168	136.3	139.45	141.49	143.05	144.34	148.86	151.99	154.53	156.71	158.68	160.49	162.18	163.79	165.33
169	137.21	140.37	142.42	143.99	145.28	149.82	152.97	155.51	157.71	159.68	161.5	163.2	164.81	166.36
170	138.12	141.3	143.36	144.93	146.22	150.78	153.94	156.5	158.7	160.69	162.51	164.22	165.84	167.4
171	139.04	142.22	144.29	145.87	147.17	151.74	154.92	157.49	159.7	161.69	163.52	165.24	166.87	168.43
172	139.95	143.15	145.22	146.81	148.11	152.71	155.89	158.47	160.7	162.69	164.54	166.26	167.9	169.47
173	140.86	144.07	146.15	147.75	149.06	153.67	156.87	159.46	161.69	163.7	165.55	167.28	168.92	170.5
174	141.77	145	147.09	148.69	150	154.63	157.85	160.44	162.69	164.71	166.56	168.3	169.96	171.54
175	142.69	145.92	148.02	149.63	150.95	155.6	158.82	161.43	163.69	165.71	167.58	169.33	170.99	172.58
176	143.6	146.85	148.96	150.57	151.89	156.56	159.8	162.42	164.68	166.72	168.59	170.35	172.01	173.61
177	144.52	147.78	149.89	151.51	152.84	157.52	160.78	163.41	165.68	167.72	169.6	171.37	173.04	174.65
178	145.43	148.7	150.83	152.45	153.78	158.49	161.75	164.39	166.68	168.73	170.62	172.39	174.07	175.68
179	146.35	149.63	151.76	153.39	154.73	159.45	162.73	165.38	167.67	169.73	171.63	173.41	175.1	176.72
180	147.26	150.56	152.7	154.33	155.68	160.42	163.71	166.37	168.67	170.74	172.65	174.43	176.13	177.76
181	148.18	151.49	153.63	155.27	156.62	161.38	164.69	167.36	169.67	171.75	173.66	175.46	177.16	178.79
182	149.09	152.41	154.57	156.21	157.57	162.35	165.66	168.35	170.67	172.75	174.68	176.48	178.19	179.83
183	150.01	153.34	155.5	157.16	158.52	163.31	166.64	169.34	171.66	173.76	175.69	177.5	179.22	180.87
184	150.92	154.27	156.44	158.1	159.47	164.28	167.62	170.32	172.66	174.77	176.7	178.52	180.25	181.9
185	151.84	155.2	157.38	159.04	160.41	165.24	168.6	171.31	173.66	175.77	177.72	179.54	181.28	182.94
186	152.76	156.13	158.31	159.98	161.36	166.21	169.57	172.3	174.66	176.78	178.73	180.57	182.31	183.98
187	153.68	157.06	159.25	160.92	162.31	167.17	170.55	173.29	175.66	177.78	179.75	181.59	183.34	185.02
188	154.59	157.99	160.19	161.87	163.25	168.14	171.53	174.28	176.66	178.79	180.76	182.61	184.37	186.05
189	155.51	158.91	161.12	162.81	164.2	169.1	172.51	175.27	177.65	179.8	181.78	183.63	185.4	187.09
190	156.43	159.84	162.06	163.76	165.15	170.07	173.49	176.26	178.65	180.81	182.79	184.66	186.43	188.13
191	157.35	160.78	163	164.7	166.1	171.03	174.47	177.25	179.65	181.82	183.81	185.68	187.46	189.17

(Continued)

(Continued)

N/B (%)	0.1	0.2	0.3	0.4	0.5	1	1.5	2	2.5	3	3.5	4	4.5	5
192	158.27	161.71	163.94	165.64	167.05	172	175.45	178.24	180.65	182.82	184.83	186.7	188.49	190.2
193	159.18	162.64	164.87	166.59	168	172.97	176.43	179.23	181.65	183.83	185.84	187.73	189.52	191.24
194	160.1	163.56	165.81	167.53	168.95	173.93	177.41	180.22	182.65	184.84	186.86	188.75	190.55	192.28
195	161.02	164.5	166.75	168.48	169.89	174.9	178.39	181.21	183.65	185.85	187.87	189.77	191.58	193.32
196	161.94	165.43	167.69	169.42	170.85	175.87	179.37	182.2	184.65	186.85	188.89	190.8	192.61	194.35
197	162.86	166.36	168.63	170.37	171.79	176.83	180.35	183.19	185.65	187.86	189.9	191.82	193.64	195.39
198	163.78	167.29	169.57	171.31	172.74	177.8	181.32	184.18	186.65	188.87	190.92	192.84	194.68	196.43
199	164.7	168.22	170.51	172.26	173.7	178.77	182.31	185.17	187.65	189.88	191.94	193.87	195.71	197.47
200	165.62	169.15	171.45	173.2	174.64	179.74	183.29	186.16	188.65	190.89	192.95	194.89	196.74	198.51

Bibliography

3GPP, Technical Specification Group 23.907, Services and System Aspects, QoS Concepts.

3GPP, Combined GSM and Mobile IP Mobility Handling in UMTS IP CN 3G TR 23.923, V3.0, 2000-05.

3GPP, Study on Rel. 2000 Services and Capabilities (3G TS 22.796), V2.0.0, 2000-06.

3GPP, Technical Specification Group Radio Access Network; Requirements for Evolved UTRA (E-UTRA) and Evolved UTRAN (E-UTRAN) (Release 9), 3GPP technical report V9, 2009-12.

3GTS, 25.213, Spreading and Modulation (FDD).

3GTS, 25.223, Spreading and Modulation (TDD).

Ahmadi, S. (2011). User-plane/control-plane latency and handover interruption time. In: *Mobile WiMAX: A Systems Approach to Understanding IEEE 802.16m Radio Access Technology*, 686.

Akhtar, S. (2009). 2G-5G networks: Evolution of technologies, standards, and deployment. In: *Encyclopedia of Multimedia Technology and Networking*, 2nde.

Allen, K.C. (1987). Observation of specific attenuation of millimetre waves by rain, IEE A&P Conference.

Amzallag, D., Bar-Yehuda, R., Raz, D. and Scalosub, G. (2008). Cell selection in 4G cellular networks, Conference on Computer Communications, September 2008.

Anas, M. et al. (2007). Performance evaluation of received signal strength based hard handover for UTRAN LTE, IEEE Vehicular Technology Conference.

Andrews, J.G. et al. (2014). What will 5G be? *IEEE J. Sel. Areas Commun.* 32: 1065–1082.

Anfossi, D., Bacci, P., and Longhetti, A. (1960). An application of Lidar technique to the study of nocturnal radiation inversion. *Atmos. Environ.* 8: 483–494.

Ansari, A.J. and Evans, B.G. (1982). Microwave propagation in sand and dust storms. *IEE Proc., Part F* 129: 315–322.

Bean et al. (1966). *ESSA Monograph*. US Government Printing Office.

Beck, R. and Panzer, H. (1989). Strategies for handover and Dynamic Channel Allocation in Micro-cellular Mobile Radio System, in IEEE Vehicular Conference, pp. 178–185.

Boccardi, F. et al. (2014). Five disruptive technology directions for 5G. *IEEE Commun. Mag.* 52: 74–80.

Castro, J.P. (2001). *The UMTS Network and Radio Access Technology*. John Wiley & Sons, Ltd.

Fundamentals of Network Planning and Optimisation 2G/3G/4G: Evolution to 5G,
Second Edition. Ajay R. Mishra.
© 2018 John Wiley & Sons Ltd. Published 2018 by John Wiley & Sons Ltd.

Chen, X. and Lin, X. (2014). Big Data Deep Learning: Challenges and perspectives. *IEEE Access* 2: 514–525.

Corici, C., Fiedler, J., Magedanz, T. and Vingarzan, D. (2011). Evolution of the resource reservation mechanisms for machine type communication over mobile broadband evolved packet core architecture, 2011 IEEE Globecom Workshops, 5 December 2011, pp. 718–722.

Craig, K.H. and Kennedy, G.R. (1987). Studies of microwave propagation on a microwave line-of-sight link. IEE A&P Conference.

Crane, R.K. (1977). Prediction of attenuation due to rainfall on satellite systems. *Proc. IEEE* 65: 456–474.

Crane, R.K. (1981). Fundamental limitations caused by RF propagation. *Proc. IEEE* 69: 196–209.

del Apio, L. et al. (2011). Energy efficiency and performance in mobile networks deployments with femtocells, 22nd IEEE International Symposium on Personal Indoor and Mobile Radio Communications, Toronto, Canada, 11–14 September 2011, pp. 107–111.

Dhillon, H.S., Ganti, R.K. and Andrews, J.G. (2011). A tractable framework for coverage and outage in heterogeneous cellular networks, Information Theory and Applications Workshop, San Diego, CA, 6–11 February 2011.

Doble, J. (1996). *Introduction to Radio Propagation for Fixed and Mobile Communication*. Artech House.

Effenberger, J.A., Strickland, R.R., and Joy, E.B. (1986). The effect of rain on a Radome's performance. *Microwave Journal* 29: 261–274.

El-Beaino, W., El-Hajj, A.M. and Dawy, Z. (2012). A proactive approach for LTE radio network planning with green considerations, IEEE International Conference on Telecommunications, 23–25 April 2012.

ETSI, Digital Cellular Telecommunication System (Phase 2+), Radio Network Planning Aspects, GSM 03.30.

ETSI Digital Cellular Telecommunication System (Phase 2+), General Packet Radio Service (GPRS); Mobile Station (MS)- Base Station System (BSS) Interface; Radio Link Control/ Medium Access Control Protocol, GSM 04.60.

ETSI, Digital Cellular Telecommunication System (Phase 2+), Radio Transmission and Reception, GSM 05.05.

Feng, S. and Seidel, E. (2008). *Self-Organizing Networks (SON) in 3GPP Long Term Evolution*. Nomor Research.

Fritze, G. (2012). *SAE: The Core Network for LTE*. Ericsson.

Gabriel, L. et al. (2011). Economic benefits of SON features in LTE networks, IEEE Samoff Symposium.

Gordejuela-Sanchez, F. and Zhang, J. (2008). Practical design of IEEE 802.16e networks: A mathematical model and algorithms, Global Communications Conference (GLOBECOM'08).

Gu, L. et al. (2014). Cost minimization for Big Data processing in Geo-distributed data venters. *IEEE Trans. Emerging Top. Comput.* 2: 314–323.

Halonen, T., Romero, J., and Melero, J. (2007). *GSM, GPRS and EDGE Performance*. John Wiley & Sons, Inc.

Hata, M. (1980). Empirical formula for propagation loss in Land Mobile Radio Services. *IEEE Trans. Vehicular Technol.* 29: 317–325.

Haykin, S. (1983). *Communication System*. John Wiley & Sons, Inc.

Heifiska, K. and Kangas, A. (1996). Microcell propagation model for network planning, in Proceedings of PIMRC'96, Taipei, Taiwan, 18 October 1996, pp. 148–152.

Holma, H. and Toksala, A. (2000). *WCDMA for UMTS*. John Wiley & Sons, Ltd.

Huang, C.Y. and Yates R.D. (1992). Call admission control in cellular radio system. *IEEE Trans. Vehicular Technol.*, **I**.

Huawei (2013). MBB Best User Experience, White Paper.

IETF (2001). TCP over 2.5G and 3G Wireless Networks, Internet-draft, October.

Ishimaru, A. (1985). Introduction to wave propagation and scattering in random media. *IEEE Antennas and Propagation Soc. Newsl.* 27: 4–6.

Islam, M. and Mitschele-Thiel, A. (2012). Reinforcement learning strategies for self-organized coverage and capacity optimization, IEEE Conference on Wireless Communications and Networking, 1–4 April 2012, pp. 2818–2823.

ITU (1995). Okumara–Hata propagation model, in Prediction Methods for the Terrestrial Land Mobile Services in the VHF and UHF Bands, ITU-R Recommendation P.529-2, pp. 5–7.

ITU-R P.530-9: Propagation data and prediction methods required for the design of terrestrial line-of sights system.

ITU-R P.837-3: Characteristics of propagation for propagation modelling.

ITU-T G.826: Error performance and objectives for international, constant bit rate digital paths at or above primary rate.

ITU-T G.827: Availability parameters and objectives for path elements of international constant bit rate digital paths at or above the primary rate.

Jakes, W.C. Jr. (ed.) (1974). *Microwave Mobile Communications*. Wiley-Interscience.

Janaswamy, R. (2000). *Radio Propagation and Smart Antennas for Wireless Communications*. Kluwer Academic.

Kaaresoja, T. and Ruutu, J (1998). Synchronization and cell loss in cellular ATM evaluation system, in Proceedings of the 5th International Workshop on Mobile Communication, Berlin, Germany, 12–14 October 1998.

Khan, F. and Pi, J. (2011). Millimeter-S Unleashing 3–300 GHz spectrum, technical report, Samsung.

Knorr, J.B. (1985). Guided EM Waves with atmospheric ducts. *Microwave and RF*, May.

Ko, H. (1985). A practical guide to anomalous propagation. *Microwave and RF*, **April**.

Koutitas, G., Karousos, A., and Tassiulas, L. (2012). Deployment strategies and energy efficiency of cellular networks. *IEEE Trans. Wireless Commun.* 11: 2552–2563.

Lee, H.-W. and Chong, S. (2010). Downlink resource allocation in multicarrier systems: Frequency-selective vs. equal power allocation. *IEEE Trans. Wireless Commun.* 7: 3738–3747.

Lee, W.C.Y. (1990). *Mobile Cellular Telecommunication Systems*. McGraw-Hill.

Lee, W.C.Y. (1993). *Mobile Communication Design Fundamentals*. John Wiley & Sons, Inc.

Lei, L. et al. (2013). Challenges on wireless heterogeneous networks for mobile cloud computing. *IEEE Wireless Commun.* 20: 34–44.

Lempiainen, J. (2001). *Radio Interface Planning for GSM/GPRS/UMTS*. Kluwer Academic.

Lomotey, R.K. and Deters, R. (2014). Towards knowledge discovery in Big Data, IEEE International Symposium on Service-Oriented System Engineering (SOSE), pp. 181–191.

Lorincz, J., Capone, A., and Begusic, D. (2011). *Optimized Network Management for Energy Savings of Wireless Access Networks*. Elsevier.

LTE, Requirements for further advancements for Evolved Universal Terrestrial Radio Access (E-UTRA) (LTE-Advanced) (3GPP TR 36.913 version 9.0.0 Release 9), ETSI technical report v.9, February 2010.

Luketic, I., Simunic, D. and Blajic, T. (2011). Optimization of coverage and capacity of Self-Organizing Network in LTE, The 34th International Convention MIPRO, Opatija, Croatia, 23–27 May 2011, pp. 612–617.

MacCartney, G.R., Zhang, J., Nie, S. and Rappaport, T.S. (2013). Path loss models for 5G millimeter wave propagation channels in urban microcells, IEEE Global Communications Conference, Exhibition and Industry Forum, December 2013.

McAllister, L.G. et al. (1969). Acoustic sounding – A new approach to the study of atmospheric structure. *Proc. IEEE* 57: 579.

Mehrotra, A. (1994). *Cellular Radio Performance Engineering*. Artech House.

Mehrotra, A. (1997). *GSM System Engineering*. Artech House.

Meng, Y.S. and Lee, Y.H. (2010). Investigations of foliage effect on modern wireless communication systems: A review. *Progr. Electromag. Res.* 105: 313–332.

Metis (2013). The 5G Mobile and Wireless Communications System, technical report.

Mishra, A.R. (1999). Observation of the fading phenomenon on the western coast of India, on a 7GHz terrestrial path, IMOC Conference, Rio de Janeiro.

Mishra, A.R. (2001). Cellular transmission network optimisation. *ADVANCE (Nokia Research Centre Journal)*, **March**.

Mishra, A.R. (2003a). EDGE network analysis & optimisation, WPMC.

Mishra, A.R. (2003b). Transmission network planning and optimisation in third generation access networks. *ADVANCE (Nokia Research Centre Journal)*, **June**.

Mishra, A.R. et al. (2011). Planning hybrid telecom networks for enhanced multimedia services, 2nd International Conference on Wireless Communications, Vehicular Technology, Information Theory and Aerospace & Electronic System Technology (WIRELESS VITAE), Chennai, India, 28 February–3 March 2011.

Mitra, A.P. et al. (1977). Tropospheric disturbance of 17–21 Dec.1974 and its effect on the microwave propagation. *Boundary Layer Meteor* 11: 103.

Mouley, M. and Pautet, M.-B. (1992). *The GSM System for Mobile Communications*. Telecom Publishing.

Musolesi, M. (2014). Big mobile data mining: Good or Evil? *IEEE Internet Comput.* 18: 78–81.

NSN (2013). Future Works, White Paper.

Oguchi, T. (1983). Electromagnetic wave propagation and scattering in rain and other hydrometers. *Proc. IEEE* 71: 1029–1078.

Ojanpera, T. and Prasad, R. (1999). *Wideband CDMA for Third Generation Mobile Communication*. Artech House.

Ojanpera, T., Prasad, R. and Harada, H. (1998). Qualitative comparision of some multi-user detector algorithms for Wideband CDMA in IEEE Vehicular Conference, pp. 46–50.

Okumara, Y., Ohmori, E., Kawano, T., and Fukuda, K. (1968). Field Strength and its variability in VHF and UHF land mobile radio service. *Rev. Electr. Commun. Lab.* 16: 825–873.

Pajukoski, K. and Savusalo, J. (1997). Wideband CDMA Test System, in Proceedings of PIMRC'97, Helsinki, Finland, 1–4 September 1997, pp. 669–672.

Parsons, D. (1992). *The Mobile Radio Propagation Channel*. Pentech Press.

Premnath, K. and Rajavelu, S. (2011). Challenges in Self Organizing Networks for wireless telecommunications, International Conference on Recent Trends in Information Technology, 3–5 June 2011, pp. 1331–1334.

Qiu, Q. et al. (2009). LTE/SAE model and its implementation in NS 2, 5th International Conference on Mobile Ad-hoc and Sensor Networks, 14–16 December 2009, pp. 299–303.

Qualcomm (2013). The 1000x Mobile Data Challenge, White Paper.

Ramaprasath, A., Srinivasan, A. and Lung, C. (2015). Performance optimization of Big Data in mobile networks, Proc. IEEE Canadian Conf. Electrical and Computer Engineering (CCECE), 3–6 May 2015, pp. 1364–1368.

Rummler, W.D. (1974). A new selective fading model: Application to propagation data. *BSTJ*.

Samulevicius, S., Pedersen, T.B. and Sorensen, T.B. (2015). MOST: Mobile broadband network optimization using planned spatio-temporal events, in Proc. IEEE Vehicular Technology Conference, Glasgow, UK, pp. 1–5.

Sarkar, S.K. (1978). Radio climatological effect on tropospheric radiowave propagation over the Indian sub-continent. PhD thesis, University of Delhi.

Siomina, I. and Di, Y. (2012). Analysis of cell load coupling for LTE network planning and optimization. *IEEE Trans. Wireless Commun.* 6: 2287–2297.

Sipila, K., Laiho-Steffens, J., Wacker, A. and Jasperg, M. (1999). modelling the impact of the fast power control on the WCDMA uplink, in Proceedings of VTC'99, pp.1266–1270.

Siwiak, K. (1998). *Radio Propagation for Antennas for Personal Communications*. Artech House.

Tran, T.-T., Shin, Y., and Shin, O.-S. (2012). Overview of enabling technologies for 3GPP LTE-advanced. *EURASIP J. Wireless Commun. Netw.* doi: 10.1186/1687-1499-2012-54.

Tsurimi, H. and Suzuki, Y. (1999). Broadband RF Stage architecture for software defined radio in handheld terminal application. *IEEE Commun. Mag.* 37: 90–95.

UMTS 22.01 Service Aspects-Service Principles.

UMTS 22.25 Quality of Service and Network Performance.

UMTS 22.05 Service Capabilities.

UMTS 23.05 Network Principles.

Vigants, A. (1974). Space diversity engineering. *BSTJ*.

Vigants, A. (1981). Microwave obstructive fading. *BSTJ* 60.

Walfish, J. and Bertoni, H.L. (1988). Theroretical model of UHF propagation in urban environments. *IEEE Trans. Antenna Propag.* AP-36: 1788–1796.

Wu, X. et al. (2014). Data mining with Big Data. *IEEE Trans. Knowl. Data Eng.* 26: 97–107.

Xia, H.H. et al. (1994). Micro-cellular propagation characteristics for personal communication in urban and sub-urban environments. *IEEE Vehicular Technol.* 43: 743–752.

Yaacoub E. and Dawy, Z. (2014). LTE radio network planning HetNets: BS placement optimization using simulated annealing, IEEE Mediterranean Electrotechnical Conference, Beirut, Lebanon, 13–16 April 2014.

Yaacoub, E., Imran, A. and Dawy, Z. (2014). A generic simulation-based dimensioning approach for planning heterogeneous LTE cellular networks, IEEE Mediterranean Electrotechnical Conference, Beirut, Lebanon, 13–16 April 2014.

Yang, S.C. (1998). *CDMA RF System Engineering*. Artech House.

Zander, J. (1996). Radio Resource Management – An overview, in Proc. IEEE Vehicular Technology Conference, Atlanta, GA, pp. 661–665.

Zhang, J., Sun, C., Yi, Y. and Zhuang, H. (2013). A hybrid framework for capacity and coverage optimization in Self-Organizing LTE Networks, IEEE 24th International Symposium on Personal, Indoor and Mobile Radio Communications: Mobile and Wireless Networks.

Zhao, H. et al. (2013). 28 GHz millimeter wave cellular communication measurements for reflection and penetration loss in and around buildings in New York City, IEEE International Conference on Communications, Budapest, Hungary, 9–13 June 2013.

Zheng, K. et al. (2015). Heterogeneous vehicular networking: a survey on architecture, challenges and solutions. *IEEE Commun. Surv. Tutor* 17: 2377–2396.

Index

Fundamentals of Network Planning and Optimisation 2G/3G/4G: Evolution to 5G,
Second Edition. Ajay R. Mishra.
© 2018 John Wiley & Sons Ltd. Published 2018 by John Wiley & Sons Ltd.